Lecture Notes in Computer Science   13076

More information about this subseries at http://www.springer.com/series/7409

Tran Khanh Dang · Josef Küng ·
Tai M. Chung · Makoto Takizawa (Eds.)

Future Data and Security Engineering

8th International Conference, FDSE 2021
Virtual Event, November 24–26, 2021
Proceedings

Springer

Editors
Tran Khanh Dang ⓘ
HCMC University of Technology
Ho Chi Minh City, Vietnam

Josef Küng
Johannes Kepler University of Linz
Linz, Austria

Tai M. Chung
Sungkyunkwan University
Suwon, Korea (Republic of)

Makoto Takizawa
Hosei University
Koganei-shi, Tokyo, Japan

ISSN 0302-9743   ISSN 1611-3349 (electronic)
Lecture Notes in Computer Science
ISBN 978-3-030-91386-1   ISBN 978-3-030-91387-8 (eBook)
https://doi.org/10.1007/978-3-030-91387-8

LNCS Sublibrary: SL3 – Information Systems and Applications, incl. Internet/Web, and HCI

This Springer imprint is published by the registered company Springer Nature Switzerland AG
The registered company address is: Gewerbestrasse 11, 6330 Cham, Switzerland

Preface

In LNCS volume 13076 and CCIS volume 1500 we present the accepted contributions for the 8th International Conference on Future Data and Security Engineering (FDSE 2021). The conference took place during November 24–26, 2021, in an entirely virtual mode (from Ho Chi Minh City, Vietnam). The proceedings of FDSE have been published in the LNCS and CCIS series by Springer. Besides DBLP and other major indexing systems, the FDSE proceedings have also been indexed by Scopus and listed in the Conference Proceeding Citation Index (CPCI) of Thomson Reuters.

The annual FDSE conference is a premier forum designed for researchers, scientists, and practitioners interested in state-of-the-art and state-of-the-practice activities in data, information, knowledge, and security engineering to explore cutting-edge ideas, to present and exchange their research results and advanced data-intensive applications, and to discuss emerging issues on data, information, knowledge, and security engineering. At FDSE, researchers and practitioners are not only able to share research solutions to problems of today's data and security engineering themes but are also able to identify new issues and directions for future related research and development work.

The two-round call for papers resulted in the submission of 168 papers. A rigorous peer-review process was applied to all of them. This resulted in 24 accepted papers (an acceptance rate of 14.3%) and two keynote speeches for LNCS volume 13076, and 36 accepted papers (including eight short papers, an acceptance rate of 21.4%) for CCIS volume 1500, which were presented online at the conference. Every paper was reviewed by at least three members of the International Program Committee, who were carefully chosen based on their knowledge and competence. This careful process resulted in the high quality of the contributions published in these two volumes. The accepted papers were grouped into the following sessions:

- Advances in Machine Learning for Big Data Analytics (LNCS)
- Big Data Analytics and Distributed Systems (LNCS and CCIS)
- Blockchain and Access Control (CCIS)
- Blockchain and IoT Applications (LNCS)
- Data Analytics and Healthcare Systems (CCIS)
- Machine Learning and Artificial Intelligence for Security and Privacy (LNCS)
- Security and Privacy Engineering (CCIS)
- Industry 4.0 and Smart City: Data Analytics and Security (LNCS and CCIS)
- Emerging Data Management Systems and Applications (LNCS)
- Short Papers: Security and Data Engineering (CCIS)

In addition to the papers selected by the Program Committee, four internationally recognized scholars delivered keynote speeches:

- Artur Andrzejak, Heidelberg University, Germany
- Johann Eder, Alpen-Adria-Universität Klagenfurt, Austria

- Tai M. Chung, Sungkyunkwan University, South Korea
- Thanh Thi Nguyen, Deakin University, Australia

The success of FDSE 2021 was the result of the efforts of many people, to whom we would like to express our gratitude. First, we would like to thank all authors who submitted papers to FDSE 2021, especially the invited speakers for the keynotes. We would also like to thank the members of the committees and additional reviewers for their timely reviewing and lively participation in the subsequent discussion in order to select such high-quality papers published in these two volumes. Last but not least, we thank the Organizing Committee members for their great support of FDSE 2021 even during the COVID-19 pandemic time.

November 2021

Tran Khanh Dang
Josef Küng
Tai M. Chung
Makoto Takizawa

Organization

Honorary Chair

Tomas Benz — Vietnamese-German University, Vietnam

Program Committee Chairs

Tran Khanh Dang	Ho Chi Minh City University of Technology, Vietnam
Josef Küng	Johannes Kepler University Linz, Austria
Tai M. Chung	Sungkyunkwan University, South Korea
Makoto Takizawa	Hosei University, Japan

Steering Committee

Dirk Draheim	Tallinn University of Technology, Estonia
Dinh Nho Hao	Institute of Mathematics, Vietnam Academy of Science and Technology, Vietnam
Fukuda Kensuke	National Institute of Informatics, Japan
Dieter Kranzlmüller	Ludwig Maximilian University of Munich, Germany
Fabio Massacci	University of Trento, Italy
Erich Neuhold	University of Vienna, Austria
Silvio Ranise	Fondazione Bruno Kessler, Italy
A Min Tjoa	Technical University of Vienna, Austria
Manuel Clavel	Vietnamese-German University, Vietnam

Publicity Committee

Nam Ngo-Chan	University of Warsaw, Poland
Tran Minh Quang	Ho Chi Minh City University of Technology, Vietnam
Le Hong Trang	Ho Chi Minh City University of Technology, Vietnam
Tran Tri Dang	RMIT University, Vietnam

Program Committee

Artur Andrzejak	Heidelberg University, Germany
Pham The Bao	Saigon University, Vietnam
Hyunseung Choo	Sungkyunkwan University, South Korea
Manuel Clavel	Vietnamese-German University, Vietnam
H. K. Dai	Oklahoma State University, USA
Vitalian Danciu	Ludwig Maximilian University of Munich, Germany
Quang-Vinh Dang	Industrial University of Ho Chi Minh City, Vietnam
Nguyen Tuan Dang	Saigon University, Vietnam

Eric Pardede	La Trobe University, Australia
Cong Duc Pham	University of Pau, France
Vinh Pham	Sungkyunkwan University, South Korea
Nhat Hai Phan	New Jersey Institute of Technology, USA
Thanh An Phan	Ho Chi Minh City University of Technology, Vietnam
Nguyen Van Sinh	International University - VNU-HCM, Vietnam
Erik Sonnleitner	Johannes Kepler University Linz, Austria
Ha Mai Tan	National Taiwan University, Taiwan
Nguyen Hoang Thuan	RMIT University, Vietnam
Michel Toulouse	Hanoi University of Science and Technology, Vietnam
Ha-Manh Tran	Ho Chi Minh City University of Foreign Languages and Information Technology, Vietnam
Truong Tuan Phat Tran	Viettel High Technology Industries Corporation, Vietnam
Thien Khai Tran	Ho Chi Minh City University of Foreign Languages and Information Technology, Vietnam
Le Hong Trang	Ho Chi Minh City University of Technology, Vietnam
Tran Minh Triet	Ho Chi Minh City University of Science, Vietnam
Hai Truong	Singapore Management University, Singapore
Takeshi Tsuchiya	Tokyo University of Science, Japan
Le Pham Tuyen	Kyunghee University, South Korea
Le Thi Kim Tuyen	Heidelberg University, Germany
Hoang Huu Viet	Vinh University, Vietnam
Edgar Weippl	SBA Research, Austria
Wolfram Woess	Johannes Kepler University Linz, Austria
Honguk Woo	Sungkyunkwan University, South Korea
Kok-Seng Wong	VinUniversity, Vietnam
Sadok Ben Yahia	Tallinn University of Technology, Estonia
Szabó Zoltán	Corvinus University of Budapest, Hungary

Local Organizing Committee

Tran Khanh Dang	Ho Chi Minh City University of Technology, Vietnam
Josef Küng	Johannes Kepler University Linz, Austria
La Hue Anh	Ho Chi Minh City University of Technology, Vietnam
Nguyen Le Hoang	Ho Chi Minh City University of Technology, Vietnam
Ta Manh Huy	Ho Chi Minh City University of Technology, Vietnam
Nguyen Dinh Thanh	Ho Chi Minh City University of Technology, Vietnam

Additional Reviewers

Phuong Hoang Ai	Trung Ha
Xuan Tinh Chu	Pham Nguyen Hoang Nam
Vipin Deval	Thi Ai Thao Nguyen
Bhavya Gera	Manh-Tuan Nguyen

Le Hoang Nguyen   Van Hau Tran
Chau D. M. Pham   Tan Dat Trinh
Huy Ta   Chibuzor Udokwu
Cong Tran

Contents

Industry 4.0 and Smart City: Data Analytics and Security

Blockchain and IoT Applications

Machine Learning and Artificial Intelligence for Security and Privacy

Emerging Data Management Systems and Applications

Invited Keynotes

Federated Learning: Issues in Medical Application

Joo Hun Yoo, Hyejun Jeong, Jaehyeok Lee, and Tai-Myoung Chung[✉]

College of Computing and Informatics, Sungkyunkwan University,
Suwon, Republic of Korea
{andrewyoo,june.jeong,esinfam99}@g.skku.edu, tmchung@skku.edu

Abstract. Since the federated learning, which makes AI learning possible without moving local data around, was introduced by google in 2017 it has been actively studied particularly in the field of medicine. In fact, the idea of machine learning in AI without collecting data from local clients is very attractive because data remain in local sites. However, federated learning techniques still have various open issues due to its own characteristics such as non identical distribution, client participation management, and vulnerable environments. In this presentation, the current issues to make federated learning flawlessly useful in the real world will be briefly overviewed. They are related to data/system heterogeneity, client management, traceability, and security. Also, we introduce the modularized federated learning framework, we currently develop, to experiment various techniques and protocols to find solutions for aforementioned issues. The framework will be open to public after development completes.

Keywords: Federated learning · Medical application · Data privacy · Heterogeneity · Incentive mechanism · Security

1 Introduction

Since the advent of Artificial Intelligence-based learning techniques, machine learning, have been extensively studied in the medical field such as radiology, pathology, neuroscience, and genetics. Instead, machine learning or deep learning techniques identify hidden multi-dimensional patterns to predict results, which come as a huge gain in disease diagnosis that requires a higher understanding of nearly human-unidentifiable correlation. However, most medical machine learning researches often ignore data privacy.

Federated learning has emerged in reaction to the strengthened data regulations on personal data such as California's Consumer Privacy Act (CCPA), EU's General Data Protection Regulation (GDPR), and China's Cyber Security Law (CSL). Those regulations not only restrict the reckless use of personal information but prevent private data collection. Thus, centralized methods, which collect and learn based on enormous amount of data, is constrained under the

© Springer Nature Switzerland AG 2021
T. K. Dang et al. (Eds.): FDSE 2021, LNCS 13076, pp. 3–22, 2021.
https://doi.org/10.1007/978-3-030-91387-8_1

regulations. Centralized learning techniques are even more difficult to be applied in the field of medicine, where data contains a lot of sensitive information. In particular, Personal Health Information (PHI) contains a number of sensitive personal information such as names, addresses, and phone numbers, so collecting and using them for the learning process is against the worldwide privacy acts.

While primary information leakage can be prevented through various security methods such as pseudonymization, secure aggregation, data reduction, data suppression, and data masking, other variables like images or biomarkers can also identify data owners. Thus, it cannot said to be a complete solution for data privacy.

Federated learning has emerged for stronger data protection than the afore-mentioned de-identification methods. It has unique characteristics compared to centralized machine learning. Specifically, traditional machine learning requires a large volume of training data; it mandates the collection of personal data to a centralized server. Decentralized federated learning, however, generates and develops deep neural network models without data collection. It consequently allows researchers to apply machine learning methods without ruining data ownership. The most widely used federated learning framework is accomplished through repeating four steps, which will be discussed in the next section.

The structural advantages of federated learning are a huge gain in medical field. Data held by each client can be used as training data without leaving its local position. It results in utilizing more data while protecting highly sensitive medical data. In addition, the application of federated learning techniques also allows the participating hospitals to connect so as to make more accurate and generalized models. However, it is difficult to deploy the current structure due to several open issues such as data vulnerability and model poisoning as mentioned in [40].

This paper is organized as follows: in the next section, we describe how federated learning is currently applied to the medical field with specific examples. In Section three, we summarised the problems arising from the application of federated learning in healthcare and provide solutions in the following order: data and system heterogeneity, client management, traceability and accountability, and security and privacy.

2   Federated Learning in Medical Applications

Federated learning, which enables the application of various machine learning methods without data collection in distributed environments, is recently widely used in healthcare. Figure 1 illustrates two representative architectures of medical federated learning. The left figure is when patients are the local clients and the hospital is the central server, and the right one is when participating clients are the hospitals with a server at the center.

A server transmits its initial model to each client, a hospital or a patient, for example. Each local client trains the received model, and then sends only

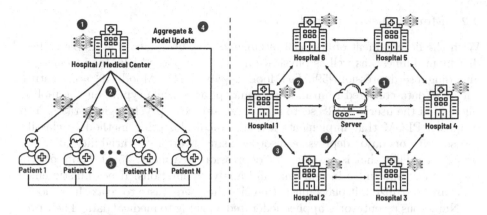

Fig. 1. Federated learning architecture in the field of medicine.

model parameters back to the server. The server then aggregates all the received model parameters to update the global model; consequently this collaborative and distributed learning can have the same effect of centralized learning. Most importantly, the global model is updated via aggregating multiple local models, so the local data privacy is isolated.

The use of medical data has become heavily available with the development of federated learning. Medical data is especially difficult to handle because it is normally in large volume, and it is restricted under data regulations. Federated learning, however, allows to utilize these data as it does not involve raw data transmission.

2.1 Physical Disorder Predictions

Deep learning techniques are used in many studies in medical field, such as finding lesions and predicting patients' disorders. With aid of federated learning, recent studies focuses not only to improve prediction accuracy but data privacy.

Liu [34] and Zhang [65] conducted experiment on diagnosing COVID-19 in both federated and centralized learning, demonstrating that both prediction accuracy are comparable. Choudhury conducted experiments to predict Adverse Drug Reactions with Electronic Health Records dataset [12] and compared the results of that used centralized learning, localized learning, and federated learning. Even Baheti's experiment showed higher accuracy in pulmonary lung nodules detection by applying federated learning [4].

Roth [44] and Jimenez [25] used mammography dataset to classify breast density and cancer, respectively. Both results showed higher classification accuracy in federated learning than centralized learning. Yi proposed SU-Net for brain tumor segmentation, which also showed a higher AUC in federated environment [60].

2.2 Mental Disorder Predictions

With the development of artificial intelligence, medical data is available to predict mental illness, as well as physical illness. Xu proposed FedMood to analyze and diagnose depression [59]. FedMood deployed DeepMood [49] architecture with the data collected from multiple smartphones while typing on the mobile phone and the user's HDRS score to analyze subjects' psychological states. Yoo proposed PFCM that implements a hierarchical clustering method to classify the severity of major depressive disorder using heart rate variability [61]. Liu proposed DTbot that is designed for depression treatment [35]. DTbot collects user's voices and video in real-time and analyzes emotions. If pessimistic emotions are recognized, it plays music that fits the users' taste to relax their mood.

Numerous recent works applied federated learning to medical data. Their prediction accuracy is higher or comparable to that of centralized machine learning. It consequently suggests the potential of federated learning, not only performance enhancement but data privacy preservation, so as to let practitioners can use it for further uses such as disease prediction.

2.3 Other Medical Applications

Along with the two aforementioned prediction tasks, federated learning is also applied for health monitoring using smart devices or life signal monitoring.

Chen [11] and Wu[55] used federated learning for activity recognition. Chen proposed FedHealth, a federated transfer learning framework for wearable devices, to build a more personalized model [11]. Wu proposed FedHome framework and GCAE method to provide a more personalized and accurate health monitoring for in-home elders [55]. Can analyzed cardiac activity data collected from smart bands for stress-level monitoring on various occasions [7]. Yuan proposed a federated learning framework for healthcare IoT devices that relaxes the computation load on them and communication overhead between the edge devices and a server [62].

All their proposed methods outperformed centralized methods. It indicates that federated learning can be used in a variety, such as disease management and prevention, addiction or mental health tracking, and real-time health monitoring [18].

3 Research Issues

3.1 Heterogeneity Issues

Although federated learning demonstrates strong performance in applying machine learning techniques while adhering to data privacy regulations, most of the works rely on an Independent and Identically Distributed (IID) assumption. However, in most real-world data environments, including the field of medicine, this assumption is not followed. We classify these non-IID situations into data heterogeneity and system heterogeneity in federated learning.

3.1.1   Data Heterogeneity

Data heterogeneity, called non-IID data distribution, means that data held by participating clients in federated learning are not uniformly distributed but are in heterogeneous distribution or characteristics. A non-IIDness can be decoupled into not identical and not independent distribution. Non-identical client distributions can be categorized into five: a feature distribution skew, label distribution skew, same label but different features, same features but different labels, and quantity skew [26]. We give practical examples of the five non-IID cases in a federated learning environment, especially in the medical field.

- *Feature distribution skew*: Even if two individuals wear the same smartwatch model and exercise for the same time duration, the features of measured values are unique because each person has different characteristics such as posture and heart rate.
- *Label distribution skew*: Frostbite is a disease that frequently occurs in cold areas because it is caused by exposure to severe cold resulting in tissue damage to body parts. Therefore, it is rare in places with relatively warm temperatures.
- *Same label but different features*: Professional medical devices are used to measure healthcare data such as neuroimages and biomarkers of patients. Medical devices, however, are made by distinct brands chosen by the different hospitals.
- *Same features but different labels*: Lung photographs damaged by the recent global pandemic COVID-19 virus are difficult to distinguish from those of pneumonia because they have common characteristics in many lesions.
- *Quantity skew*: Suppose five times more patients have visited hospital A than hospital B. The quantity of data each hospital has will also significantly differ.

Not-independent distribution is a violation of consistency of data depending on the other factors. Such violations are introduced when the data changes over time or geolocation [26].

3.1.2   System Heterogeneity

Participating devices may cause system heterogeneity depending on their hardware setting, computing power, communication cost, and network connectivity [31]. Considering the federated learning environment where multiple medical centers participate in the learning process, it can cause the differences in database or infrastructure between each hospital. For example, Samsung Medical Center has built a systematic infrastructure for efficient management of patient data through operating their own database and digital therapeutics research center. On the other hand, most clinics are more not likely to be equipped with such systems. These differences raise system heterogeneity issues when they participate in federated learning.

3.1.3 Approaches for Heterogeneity Issues

Data and system heterogeneity issues have various misleading consequences in the federated learning environment. As the distribution of each data varies, the global model does not converge to a single global model; as clients have different capabilities, the global model may be biased into a dimension. Numerous researches have been conducted to address the issue, and we group the works into the three main approaches: clustering, optimization, and model fusion.

Clustering Methods. Machine learning clustering-based techniques are the representative unsupervised learning methods for finding similarities with peripheral data for unlabeled datasets. Many researchers leveraged clustering methods to solve data heterogeneity issues because gathering data points with similar patterns from unlabeled data is analogous to grouping clients with similar weight distribution to which the server is inaccessible in federated learning.

Sattler proposed CFL that adopts a clustering method by measuring the cosine similarity of each local model [46]. They swapped the labels to fit the dataset for the non-IID environment. Their experiment results demonstrated that their proposed work achieved a reasonable performance even in extreme non-IID situations. Briggs applied a hierarchical clustering-based method to improve the performance when the clients have non-IID data [6]. They introduced a method to generate optimized clusters by comparing L1, L2, and cosine similarity distances between each cluster, demonstrating that the proposed method can reach the desired performance faster and more accurately than traditional federated learning methods. Based on [6], Yoo used Heart Rate Variability data of patients to diagnose depression severity [61]. They applied clustering-based algorithms called PFCM for new incoming participants to improve prediction accuracy and solve non-IID issues.

While there are various clustering-based techniques to deal with the performance degradation caused by data heterogeneity, there are also researches to solve the issues of model aggregation time delay caused by heterogeneous data. Chen introduced the FedCluster method to address the problem of slow convergence that occurs when the federated averaging method is applied to heterogeneous local data [9]. Each participating local client is not included in the model update at a time, but is clustered according to certain criteria, in which the cluster participates in the federated learning process for each round. Depending on the federated learning environment they want to apply, they apply the best clustering scenario of random uniform clustering, timezone-based clustering, and availability-based clustering. Experiments with MNIST and CIFAR-10 benchmark datasets demonstrated that the cyclic federated learning structure through FedCluster showed a faster convergence time than the conventional FedAvg algorithm.

As a server has no knowledge in the distribution of the participants' data, a number of works tried to address this issue from the perspective of unsupervised learning. Thus, clustering, one of the most prevalent unsupervised methods, is the most widely applied to solve the heterogeneity issues by gathering clients.

Optimization Methods. Another branch of solving the heterogeneity issue is the use of optimization algorithms. With the only use of FedAvg limits the models from converging as a single global model cannot properly represent non-IID data. This is because the data derived from different distributions diverge in various direction to represent the features of the distributions to which they belong.

Xie proposed federated SEM, a multi-center federated learning framework, which allows optimization function to find multiple local optima points [57]. The Stochastic Expectation Maximization makes it possible to find local optima point in a variety of distributions of clients in each clustered multi-center, which solves the problems that do not converge to a single global model. Reddi applied various adaptive optimizers called FEDADAGRAD, FEDYOGI, and FEDADAM, which are advantageous in hyperparameter coordination and improving the convergence rate of the federated learning model over the vanilla FedAvg method [42]. SCAFFOLD also addresses client-drifting issues in federated learning. By introducing a client control variable, Karimireddy adjust each local update in the direction of the optimal global model [27]. FedProx from [32] modifies the conventional FedAvg method in two directions; tolerating partial work and adding proximal terms. Since each participating device may differ in computing power or network bandwidth, Li solves the heterogeneity issues by assigning each device for a specific iteration. A proximal term is also used to prevent heterogeneity problems arising from excessive local update iterations.

Mixture Model of Global and Local. In addition to clustering and optimization based techniques, Hanzely [19] and Arivazhagan [1] solved heterogeneity issues by mixing global and local models generated by federated learning. Since a global model is too general to fit all clients' data and local models are too specific to be generalized, they extract representative learning layers from both global and local to create a fused model. In this way, it is possible to adopt general features from other participating clients, while adding their personalized features from own local data.

3.2 Client Management Issues

Unlike the centralized machine learning architecture, client management is another essential issue in federated learning. The server must decide which client should participate in the learning process because some free-riding clients might be looking for benefits without any contribution.

In line with this, many researchers brought incentive mechanisms to federated learning. Incentive mechanism was first introduced by system architects aiming for improved performance of repetitive loops of manufacturers by rewarding the participants [38]. Although many pieces of studies have been conducted so far, they had not utilized a proper incentive mechanism in federated learning because it is challenging to evaluate which participating client contributed how much.

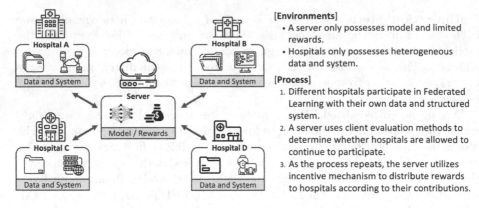

Fig. 2. Client Management in the field of medicine.

3.2.1   Client Management Based on Data Quality

Under strong data regulations, the high quality data becomes a large asset, so the local clients may want to benefit from the participation in federated learning systems. Thus, depending on the data quantity and quality, an incentive mechanism has emerged to reward participating local clients.

Chen deals with the incentive issue that occurs when multi-parties participate in federated learning to generate a better performing model collaboratively [10]. From an economic perspective, they pointed out that if other companies can grow further through the high-quality data of one good client, the client with high-quality data will not participate in federated learning because it will threaten their own profits.

3.2.2   Client Management Based on Resource Usage

In addition to the importance of client data, it is critical to identify the computational capabilities or systems they use to participate in federated learning. Zeng examined a possible multi-dimensional resource differences that occur in Mobile Edge Computing (MEC) environments [63]. Servers must recognize dedicated resources and provide appropriate incentives to encourage good clients to participate in order to ultimately improve model performance. Most federated learning researches, on the other hand, assumed that each client participating in model training has the same dedication.

3.2.3   Approaches for Client Management Issues

Generating a better performance learning model requires contributions from multiple data providers with proper quality of data and resource, though not all clients equally contribute to federated learning. Therefore, an algorithm that can identify each contribution needs to be applied. The two most widely studied approaches leveraged Shapley value and Stackelberg game theory which we will discuss in the following.

Shapley Value. Jia suggested how to perform data valuations through Shapley Value [24], which has been widely used in game theory. They listed how Shapley Value enables data evaluation in multiple machine learning analytics environments and demonstrates their approach's scalability. Similar to the Shapley Value based game-theory approach, Lim used contract theory to identify the data quality and quantity of each data owner and applied a hierarchical incentive mechanism in the federated crowd-sourcing network [33].

Stackelberg Game Theory. The Stackelberg game theory is widely used to assess each participant's contribution and construct a reward system. Sarikaya solved the problem caused by heterogeneous worker performance through the Stackelberg game-based method [45]. This measures the time it takes each participant to complete a given task for an updated gradient transfer and assigns a proper reward for each computing power based on the Stackelberg game theory. Khan also adopted a Stackelberg game-based approach in [28]. Each edge node delivers its own computation energy and latency to the model aggregator, which is in charge of incentives. The goal of the model aggregator is to minimize learning time while maximizing model performance, so it adjusts client learning level based on the clients' Stackelberg results.

Pandey proposed a two-stage of Stackelberg game by developing an optimal learning model through maximizing the utility of participating devices and MEC servers [39]. When the MEC server announces the objectives of the optimized global model it wants to create and rewards accordingly, each device participates in the federated learning by optimizing the global learning model and maximizing the yield through the local data it possesses.

Similarly to studies that deal with data values by Stackelberg game theory, auction systems were also applied to solve clients' heterogeneous resource issues. To address the problem of client management, Le used a primary-dual greedy auction mechanism [29]. When the server is assigned the task of federated learning training, each client submits a bid based on their own computation resource and transmission power. Subsequently, the server selects clients who can develop optimal models based on the bid list and provides customized rewards after completing the learning task. [63] also applied auction-based techniques on various scoring functions to allow devices with high-quality data to participate at a relatively low cost.

Giving no or the same level of incentive to all local clients will result in some participants earning rewards for providing their low-quality data and resource. Others, on the other hand, will suffer from losses while contributing high-quality data. Hence, designing federated learning without explicit incentive mechanisms may violate the purpose of federated learning, which is to collaboratively develop a high-performance learning model. Researchers must develop a more sophisticated incentive mechanism to manage local clients in a real-world federated learning environment systematically.

3.3   Traceability and Accountability Issues

The most notable advantage of federated learning is data privacy, as the global server cannot directly investigate the local clients' data. However, this presents a substantial difficulty in that it is impossible to track results or hold them accountable for learning. Rieke highlighted a method for determining who is responsible for the unexpectedly erroneous outcomes of the medical analysis caused by the federated learning algorithm [43]. Indeed, the inability to check the learning process of AI is a problem that has arisen as application of deep learning techniques has expanded. This is due to the black-box nature of the neural network, and federated learning should consider taking another step toward data privacy.

Crucial factor in the field of medicine is the explainability of what the results were based on when making a decision. That is why precision and recall are more commonly used for performance evaluation than accuracy in the medical field. Results derived from the machine learning model trained with medical data should be reviewed more carefully, though it is hardly achievable in a federated learning environment.

It would be ideal if the predictive or diagnostic model obtained by the federated learning consistently demonstrated professional-level performance, but this is not the case. Therefore, researchers should consider when it produces false-positive results. If the false-positive rate is high in the federated learning task, which participant or training round is responsible for the problem should be determined. Otherwise, federated learning cannot find out which part is causing the issue; the entire training architecture may need to be redesigned.

While medical experts can provide sufficient information during data preprocessing, such as labeling, segmentation, noise filtering, this cannot happen during the federated learning's training process [58]. As a result, federated learning can be seen as a trade-off architecture in the medical field. Although it is critical to protect PHI from arbitrary inspection, the system's most critical advantage is used as a fatal disadvantage to healthcare applications. In line with this, Explainable AI (XAI) was developed to solve the black-box problem [36], which allowed researchers to find out which parts of the deep neural networks are responsible for the performance degradation. These characteristics of XAI must also be applied when federated learning is adopted in the field of medicine to prevent medical errors caused by false-positive rates.

There have not been many studies that use XAI in the field of medical federated learning, but some researchers attempted to combine XAI and FL to increase explainability while maintaining the benefits of data privacy protections. Raza applied XAI with Federated Transfer learning to design ECG (Electrocardiography) monitoring healthcare system in [41], by adding Gradient-weighted Class Activation Mapping (Grad-CAM) module on federated learning architecture to provide ECG signal classification task. However, more extensive researches on XAI and federated learning remain an open problem [47].

3.4   Privacy and Security Issues

Although deep neural network models brought great advancement in the medical field, and federated learning prevents a model from private information leakage, various privacy and security attacks remain as unsolved problems. For instance, a medical image deep neural networks are especially susceptible to adversarial attacks due to ambiguous ground truth, highly standardized image, and many other reasons [14]. At the same time, however, the attacks can be easily detected because of the biological characteristic of the images (i.e., manipulation occurring outside the pathological region) [37]. This section will introduce various attack and defense approaches, especially those studied in federated learning environments.

Table 1. A summary of privacy and security attacks

Attack category	Attack types	Attack target	Attack methods	Attacker role
Poisoning attacks	Data Poisoning	Security (Data Integrity)	Label Flipping	Client
			Backdoor	Client
	Model poisoning	Security (Model Integrity)	Gradient Manipulation	Client
			Training Rule Manipulation	Client
Inference attacks	Membership Inference	Privacy (Information Leak)	Membership Inference	Client & Server
			Properties Inference	Client & Server
	GAN Reconstruction	Privacy (Information Leak)	Class Representative Inference	Client & Server
			Inputs and Labels Inference	Client & Server

3.4.1   Attacks

Federated learning is especially vulnerable to adversarial attacks due to the absence of raw data inspection and the collaborative training using private local data. As generally known, machine learning can be divided into the two phases: training phase and inference phase. Nevertheless, due to zero knowledge distributed nature of federated learning, the training phase attacks are more severe than those of the inference phase; as neither centralized property (i.e., server) nor the other participating clients are allowed to investigate each other's private data.

Poisoning Attacks. Poisoning attacks can be categorized into data poisoning and model poisoning attacks. The two types of poisoning attacks are different in that the former aim to compromise the integrity of the training data, while the latter aim to compromise the integrity of the model.

Data poisoning attacks include label flipping or data backdoor attacks. Label flipping attacks are one of the client-side data poisoning attacks that flip the labels of the attacker-chosen data classes to attacker-chosen labels so as to miss-classify the specific data classes. Tolpegin simulates and analyzes label flipping attacks [51], a type of data poisoning attacks. In their experiment, the class label of airplane images is flipped to bird, so the global model misclassifies airplane images to bird at inference time. Hayes introduces a contamination attack that is essentially manipulating a small set of training data [22], compromising the integrity of the data. The author suggested adversarial training as a defense, which will be discussed in Sect. 3.4.2 in detail.

The model poisoning attacks involve model backdoors and gradient and/or training rules manipulation. Although poisoning attacks can be differently categorized into two, model poisoning attacks generally include data poisoning attacks as the poisoned data ultimately leads the model to be poisoned. Therefore, we here introduce numerous previous works that are not limited to data poisoning but the hybrid approach of data and model poisoning attacks as well.

Bagdasaryan proposed model replacement to introduce a backdoor into the global model [3]. Their proposed attack kept high accuracy for both main and backdoor tasks so as to improve its persistence by evading anomaly detection. Fang manipulated the local model parameters before sending them to the global server [13]. As a result of the manipulation, the local models deviate towards the inverse direction of the global model before the attack. Bhagoji introduced a targeted model poisoning attack that poisons the model updates by explicit boosting and remains stealthy by alternating minimization [5]. Xie proposed a distributed backdoor attack that breaks down a global trigger pattern into distinct local patterns and embeds them in the training sets of several adversarial parties [56]. Their work showed that the distributed attack is more effective than the centralized backdoor attacks. Fung pointed out the vulnerabilities of federated learning, especially against the Sybil-based poisoning attacks [16]. The authors mentioned that the distributed nature increases attack effectiveness, especially when multiple malicious parties participate.

Inference Attacks. Unlike poisoning attacks, inference attacks typically hamper the privacy of private information. Inference attacks include membership inference and GAN-based reconstruction attacks that lead the system to leak information about the training data unintentionally. The recent trend of inference attacks is moving toward the GAN-based method due to its stealth and detection evasion ability. Wang achieves user-level privacy leakage through incorporating GAN with a multi-task discriminator [53]. Their proposed method discriminates category, reality, and client identity of input data samples and recovers the user-specific private data. In [64], an attacker first acts as a benign participant and stealthy trains a GAN to mimic the other participants' training samples. With the generated samples, the attacker manipulates the model update with a scaled poisoning model update so as to compromise the global model ultimately.

3.4.2 Defense Methods

The defense mechanism against the attacks in federated learning includes minimizing the influence of malicious clients and preventing the malicious clients' model parameters from incorporating into the global model. Also, with respect to privacy, the defense includes preventing private information from being leaked.

Information Leak Prevention. The famous technique to prevent the model from leaking private information, Multi-Party Computation (MPC), Homomorphic Encryption, and Differential Privacy (DP), has been widely used. PEFL solved the vulnerability by leveraging MPC [20]. Their proposed method has strength in the situation that multiple clients collude so as to prevent private data from being leaked.

The following previous works leverage a combination of those mentioned above three popular techniques. Augenstein demonstrated generative models that are trained using federated methods with differential privacy. They applied their model to both text and image, using differentially private federated GANs [2]. Ghazi also exploits differential privacy for secure aggregation via shuffled model [17]. Their proposed methods preserve privacy and relax computational complexity, which is one of the necessities for better scalability. Hao combined differential privacy and additive homomorphic encryption to obtain both performance and security [21]. Their proposed method is especially robust when not only the clients but the server is honest but curious. Truex approached similarly, combined differential privacy with multiparty computation balances the trade-off between the performance and availability [52].

Model Protection. In the perspective of security, various works have been done to prevent model from corruption. There are broadly two approaches: robust aggregation and anomaly detection. Typically, robust aggregation aims to train the global model robustly, even if malicious clients participate in the federated learning process. Anomaly detection aims to detect malicious or anomalous clients so that their model updates are not aggregated in the server and/or block them from further participation.

Fu offered a robust aggregation technique with residual-based reweighting [15]. Their reweighting strategy employed iteratively reweighted least squares to integrate repeated median regression. While, various works defended against various attacks by identifying and classifying anomalies [13,23,30,48,51]. Li proposed spectral anomaly detection mechanism based on models' low-dimensional embeddings [30]. The central server learns to detect and remove the malicious model updates by removing noisy features and retaining essential features, leading to targeted defense. A notable point of their proposed approach is that it worked in both semi-supervised or unsupervised learning and designed the protocol for encryption. FoolsGold mitigated poisoning attacks [54]. Their approach comes from the idea that the malicious clients' updates are different from those of the benign ones; thereby, the Sybils are distinguishable by measuring the contribution similarity.

The aforementioned work's objective, anomaly detection, remains the same, but several approaches leveraged clustering-based and thresholding-based approaches [8,13,23,48,50,51]. Auror dealt with targeted poisoning attacks leveraging clustering and thresholding techniques. Shen created clusters of clients and measured the pairwise distance between them, which will be used to distinguish two distinctive clusters. Within each cluster, if more than half of the clients are determined to be malicious by the predefined threshold, all the clients belonging to that cluster are then classified as malicious [48]. Fang proposed a defense mechanism, a combination of error rate-based rejection and loss function-based rejection [13]. The idea comes from that as the malicious clients tamper with the models' performance, the error rate and the loss impact are greater than those of benign clients. Therefore, if a client greatly impacts the higher error and loss rate, the clients are identified as malicious so as to aggregate only benign clients' weights. Along with the data poisoning attack, Tolpegin used PCA to visualize the spatially separable malicious clients' model updates and those of benign ones [51].

Sun defended against backdoor attacks by leveraging norm bounding and weak differential privacy [50]. They noted that the norm values of malicious clients are relatively greater than those of benign ones, so they detected malicious clients by thresholding based on the calculated norm value of each clients' weight updates. FLTrust protected the global model against byzantine attacks by making use of a ReLU-clipped cosine similarity-based trust score [8]. In their works, however, the global server had been trained on an innocent dataset called root dataset; in other words, the server did have the knowledge of benignity and malice of the client dataset, which is a breach of the no-raw-data-sharing assumption. Based on [8,50], Jeong [23] proposed ABC-FL that leverages feature dimension reduction, dynamic clustering, and cosine-similarity-based clipping to detect and classify anomalous and benign clients where the benign ones have non-IID data and IID data. Similar to the aforementioned approaches, only the benign clients' model weights are aggregated in the global server.

4   Modular Framework Under Development

Although there has been an increasing number of studies in the medical field using federated learning, a variety of open problems still remain. An easier and more convenient federated learning framework is demanding as this field of study getting more attention.

Our research team is developing a lightweight framework, *Modular FL*, that modularizes the essential component of federated learning. The Modular FL consists of the federated learning core component, serving component, and communication channel.

A federated learning core component controls the core steps of federated learning between a server and multiple clients. This component consists of four

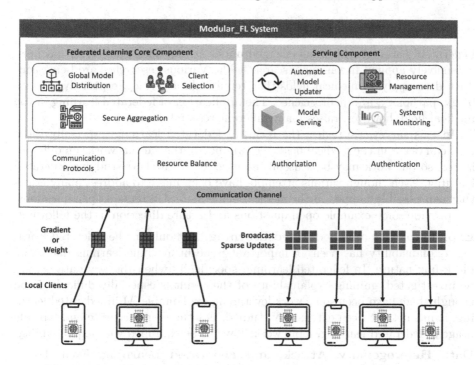

Fig. 3. The architecture of the Modular FL currently in development.

modules, namely global model distribution, client selection, and secure aggregation modules. Global model distribution module is in charge of initializing and distributing a global model to the client selected by client selection module. A secure aggregation module is responsible for aggregating the weight vectors received as a result of each local training. A serving component also consists of four modules, automatic model updater, model serving, resource management, and system monitoring, providing management related functionalities. We omit the explanation of each serving component module, as their names are self-explanatory. The server also manages communication protocols, balances shared resources, authorizes the clients, and authenticates the clients. The securely aggregated global model is advertised to local clients through communication channel and with the aid of the server's management.

Each component includes the learning and communication functions, providing researchers a structure to experiment by simply changing the functions to address the remaining issues. For example, researchers can change the client selection function to add an incentive mechanism to compensate *good quality* clients. Modular FL is also easily expandable due to its flexible structure, where model weights from learning process can be saved as Json. This characteristic leads Modular FL being not limited to a library or even a language.

5   Conclusion and Future Works

Federated learning, which has emerged under strong data protection, has been extensively applied in various research fields. Various researches are being conducted to apply to medical fields that require data privacy, especially in radiology, pathology, and neuroscience. The application of federated learning in the medical field, however, raises many different research issues.

In this survey, we review the issues of federated learning, especially those that can occur in the medical field. We summarize the various works on different issues, so our work may be a useful resource for researchers studying federated learning. Even though various attempts have been made to address many issues that can possibly happen in the medical field, there still remain open questions. We provide some example open questions and future direction in the following.

Explainable AI and Federated Learning. Not only in the federated learning, explainability has been an important issue in machine learning due to its black-box nature. In federated learning, specifically when the raw data cannot be investigated, gaining explainability of the result is especially difficult. Even though there have been some work leveraging explainable AI in federated learning, it has not yet been extensively studied. For the more accurate and fail-safe usage of federated learning, more works have to be equipped with explianability.

Data Heterogeneity, Attack, and Federated Learning. Even though numerous works on various attacks and defense mechanisms have been published, attacks with heterogeneous data has not yet been exhaustively studied. Nevertheless, the real-world data can be non-IID, attacks and defenses with those data distribution must be addressed. Since federated learning will be widely used to preserve data privacy and reduce communication complexity, researchers should study attacks and defenses even when some clients have highly non-IID data.

Acknowledgement. This research was supported by Institute of Information & communications Technology Planning & Evaluation (IITP) grant funded by the Korea government (MSIT) (No. 2020-0-00990, Platform Development and Proof of High Trust & Low Latency Processing for Heterogeneous·Atypical·Large Scaled Data in 5G-IoT Environment).

Author Contribution. J.H.Y. and H.J. are co-first authors and contributed equally to the composition and preparation of the paper. T-M.C. proposed research and theorem in the field of federated learning in medical application. J.H.Y., H.J., J.L., and T-M.C. discussed about suggested research idea. J.H.Y. and H.J. wrote the manuscript with support from J.L.. H.J. revised the entire manuscript. T-M.C. provided invaluable guidance throughout the research and the writing of the manuscript. All authors have reviewed the manuscript.

References

1. Arivazhagan, M.G., Aggarwal, V., Singh, A.K., Choudhary, S.: Federated learning with personalization layers. arXiv preprint arXiv:1912.00818 (2019)

2. Augenstein, S., et al.: Generative models for effective ml on private, decentralized datasets. arXiv preprint arXiv:1911.06679 (2019)
3. Bagdasaryan, E., Veit, A., Hua, Y., Estrin, D., Shmatikov, V.: How to backdoor federated learning. In: International Conference on Artificial Intelligence and Statistics, pp. 2938–2948. PMLR (2020)
4. Baheti, P., Sikka, M., Arya, K., Rajesh, R.: Federated learning on distributed medical records for detection of lung nodules. In: VISIGRAPP (4: VISAPP), pp. 445–451 (2020)
5. Bhagoji, A.N., Chakraborty, S., Mittal, P., Calo, S.: Analyzing federated learning through an adversarial lens. In: International Conference on Machine Learning, pp. 634–643. PMLR (2019)
6. Briggs, C., Fan, Z., Andras, P.: Federated learning with hierarchical clustering of local updates to improve training on non-IID data. In: 2020 International Joint Conference on Neural Networks (IJCNN), pp. 1–9. IEEE (2020)
7. Can, Y.S., Ersoy, C.: Privacy-preserving federated deep learning for wearable IoT-based biomedical monitoring. ACM Trans. Internet Technol. (TOIT) 21(1), 1–17 (2021)
8. Cao, X., Fang, M., Liu, J., Gong, N.Z.: FLTrust: Byzantine-robust federated learning via trust bootstrapping. arXiv preprint arXiv:2012.13995 (2020)
9. Chen, C., Chen, Z., Zhou, Y., Kailkhura, B.: FedCluster: boosting the convergence of federated learning via cluster-cycling. In: 2020 IEEE International Conference on Big Data (Big Data), pp. 5017–5026. IEEE (2020)
10. Chen, M., Liu, Y., Shen, W., Shen, Y., Tang, P., Yang, Q.: Mechanism design for multi-party machine learning. arXiv preprint arXiv:2001.08996 (2020)
11. Chen, Y., Qin, X., Wang, J., Yu, C., Gao, W.: FedHealth: a federated transfer learning framework for wearable healthcare. IEEE Intell. Syst. 35(4), 83–93 (2020)
12. Choudhury, O., Park, Y., Salonidis, T., Gkoulalas-Divanis, A., Sylla, I., et al.: Predicting adverse drug reactions on distributed health data using federated learning. In: AMIA Annual symposium proceedings, vol. 2019, p. 313. American Medical Informatics Association (2019)
13. Fang, M., Cao, X., Jia, J., Gong, N.: Local model poisoning attacks to byzantine-robust federated learning. In: 29th {USENIX} Security Symposium ({USENIX} Security 20), pp. 1605–1622 (2020)
14. Finlayson, S.G., Chung, H.W., Kohane, I.S., Beam, A.L.: Adversarial attacks against medical deep learning systems. arXiv preprint arXiv:1804.05296 (2018)
15. Fu, S., Xie, C., Li, B., Chen, Q.: Attack-resistant federated learning with residual-based reweighting. arXiv preprint arXiv:1912.11464 (2019)
16. Fung, C., Yoon, C.J., Beschastnikh, I.: Mitigating sybils in federated learning poisoning. arXiv preprint arXiv:1808.04866 (2018)
17. Ghazi, B., Pagh, R., Velingker, A.: Scalable and differentially private distributed aggregation in the shuffled model. arXiv preprint arXiv:1906.08320 (2019)
18. Hakak, S., Ray, S., Khan, W.Z., Scheme, E.: A framework for edge-assisted healthcare data analytics using federated learning. In: 2020 IEEE International Conference on Big Data (Big Data), pp. 3423–3427. IEEE (2020)
19. Hanzely, F., Richtárik, P.: Federated learning of a mixture of global and local models. arXiv preprint arXiv:2002.05516 (2020)
20. Hao, M., Li, H., Luo, X., Xu, G., Yang, H., Liu, S.: Efficient and privacy-enhanced federated learning for industrial artificial intelligence. IEEE Trans. Industr. Inf. 16(10), 6532–6542 (2019)

21. Hao, M., Li, H., Xu, G., Liu, S., Yang, H.: Towards efficient and privacy-preserving federated deep learning. In: ICC 2019–2019 IEEE International Conference on Communications (ICC), pp. 1–6. IEEE (2019)
22. Hayes, J., Ohrimenko, O.: Contamination attacks and mitigation in multi-party machine learning. arXiv preprint arXiv:1901.02402 (2019)
23. Jeong, H., Hwang, J., Chung, T.M.: ABC-FL: anomalous and benign client classification in federated learning (2021)
24. Jia, R., et al.: Towards efficient data valuation based on the shapley value. In: The 22nd International Conference on Artificial Intelligence and Statistics, pp. 1167–1176. PMLR (2019)
25. Jiménez-Sánchez, A., Tardy, M., Ballester, M.A.G., Mateus, D., Piella, G.: Memory-aware curriculum federated learning for breast cancer classification. arXiv preprint arXiv:2107.02504 (2021)
26. Kairouz, P., McMahan, H.B., et al.: Advances and open problems in federated learning. arXiv preprint arXiv:1912.04977 (2019)
27. Karimireddy, S.P., Kale, S., Mohri, M., Reddi, S., Stich, S., Suresh, A.T.: SCAFFOLD: stochastic controlled averaging for federated learning. In: International Conference on Machine Learning, pp. 5132–5143. PMLR (2020)
28. Khan, L.U., et al.: Federated learning for edge networks: resource optimization and incentive mechanism. IEEE Commun. Mag. 58(10), 88–93 (2020)
29. Le, T.H.T., et al.: An incentive mechanism for federated learning in wireless cellular network: an auction approach. IEEE Trans. Wirel. Commun. (2021)
30. Li, S., Cheng, Y., Wang, W., Liu, Y., Chen, T.: Learning to detect malicious clients for robust federated learning. arXiv preprint arXiv:2002.00211 (2020)
31. Li, T., Sahu, A.K., Talwalkar, A., Smith, V.: Federated learning: challenges, methods, and future directions. IEEE Signal Process. Mag. 37(3), 50–60 (2020)
32. Li, T., Sahu, A.K., Zaheer, M., Sanjabi, M., Talwalkar, A., Smith, V.: Federated optimization in heterogeneous networks. arXiv preprint arXiv:1812.06127 (2018)
33. Lim, W.Y.B., et al.: Hierarchical incentive mechanism design for federated machine learning in mobile networks. IEEE Internet Things J. 7(10), 9575–9588 (2020)
34. Liu, B., Yan, B., Zhou, Y., Yang, Y., Zhang, Y.: Experiments of federated learning for COVID-19 chest x-ray images. arXiv preprint arXiv:2007.05592 (2020)
35. Liu, Y., Yang, R.: Federated learning application on depression treatment robots (DTbot). In: 2021 IEEE 13th International Conference on Computer Research and Development (ICCRD), pp. 121–124. IEEE (2021)
36. Lundberg, S.M., Lee, S.I.: A unified approach to interpreting model predictions. In: Proceedings of the 31st International Conference on Neural Information Processing Systems, pp. 4768–4777 (2017)
37. Ma, X., et al.: Understanding adversarial attacks on deep learning based medical image analysis systems. Pattern Recogn. 110, 107332 (2021)
38. Mitra, S., Webster, S.: Competition in remanufacturing and the effects of government subsidies. Int. J. Prod. Econ. 111(2), 287–298 (2008)
39. Pandey, S.R., Tran, N.H., Bennis, M., Tun, Y.K., Manzoor, A., Hong, C.S.: A crowdsourcing framework for on-device federated learning. IEEE Trans. Wirel. Commun. 19(5), 3241–3256 (2020)
40. Qayyum, A., Qadir, J., Bilal, M., Al-Fuqaha, A.: Secure and robust machine learning for healthcare: a survey. IEEE Rev. Biomed. Eng. 14, 156–180 (2020)
41. Raza, A., Tran, K.P., Koehl, L., Li, S.: Designing ecg monitoring healthcare system with federated transfer learning and explainable ai. arXiv preprint arXiv:2105.12497 (2021)

42. Reddi, S., et al.: Adaptive federated optimization. arXiv preprint arXiv:2003.00295 (2020)
43. Rieke, N., et al.: The future of digital health with federated learning. NPJ Digit. Med. **3**(1), 1–7 (2020)
44. Roth, H.R., et al.: Federated learning for breast density classification: a real-world implementation. In: Albarqouni, S. (ed.) DART/DCL -2020. LNCS, vol. 12444, pp. 181–191. Springer, Cham (2020). https://doi.org/10.1007/978-3-030-60548-3_18
45. Sarikaya, Y., Ercetin, O.: Motivating workers in federated learning: a stackelberg game perspective. IEEE Netw. Lett. **2**(1), 23–27 (2019)
46. Sattler, F., Müller, K.R., Samek, W.: Clustered federated learning: model-agnostic distributed multitask optimization under privacy constraints. IEEE Trans. Neural Netw. Learn. Syst. (2020)
47. Selvaraju, R.R., Cogswell, M., Das, A., Vedantam, R., Parikh, D., Batra, D.: Grad-CAM: visual explanations from deep networks via gradient-based localization. In: Proceedings of the IEEE International Conference on Computer Vision, pp. 618–626 (2017)
48. Shen, S., Tople, S., Saxena, P.: AUROR: defending against poisoning attacks in collaborative deep learning systems. In: Proceedings of the 32nd Annual Conference on Computer Security Applications, pp. 508–519 (2016)
49. Suhara, Y., Xu, Y., Pentland, A.: DeepMood: forecasting depressed mood based on self-reported histories via recurrent neural networks. In: Proceedings of the 26th International Conference on World Wide Web, pp. 715–724 (2017)
50. Sun, Z., Kairouz, P., Suresh, A.T., McMahan, H.B.: Can you really backdoor federated learning? arXiv preprint arXiv:1911.07963 (2019)
51. Tolpegin, V., Truex, S., Gursoy, M.E., Liu, L.: Data poisoning attacks against federated learning systems. In: Chen, L., Li, N., Liang, K., Schneider, S. (eds.) ESORICS 2020. LNCS, vol. 12308, pp. 480–501. Springer, Cham (2020). https://doi.org/10.1007/978-3-030-58951-6_24
52. Truex, S., et al.: A hybrid approach to privacy-preserving federated learning. In: Proceedings of the 12th ACM Workshop on Artificial Intelligence and Security, pp. 1–11 (2019)
53. Wang, Z., Song, M., Zhang, Z., Song, Y., Wang, Q., Qi, H.: Beyond inferring class representatives: user-level privacy leakage from federated learning. In: IEEE INFOCOM 2019-IEEE Conference on Computer Communications, pp. 2512–2520. IEEE (2019)
54. Wu, C., Yang, X., Zhu, S., Mitra, P.: Mitigating backdoor attacks in federated learning. arXiv preprint arXiv:2011.01767 (2020)
55. Wu, Q., Chen, X., Zhou, Z., Zhang, J.: FedHome: cloud-edge based personalized federated learning for in-home health monitoring. IEEE Trans. Mob. Comput. (2020)
56. Xie, C., Huang, K., Chen, P.Y., Li, B.: DBA: distributed backdoor attacks against federated learning. In: International Conference on Learning Representations (2019)
57. Xie, M., Long, G., Shen, T., Zhou, T., Wang, X., Jiang, J.: Multi-center federated learning. arXiv preprint arXiv:2005.01026 (2020)
58. Xu, J., Glicksberg, B.S., Su, C., Walker, P., Bian, J., Wang, F.: Federated learning for healthcare informatics. J. Healthc. Inf. Res. **5**(1), 1–19 (2021)
59. Xu, X., Peng, H., Sun, L., Bhuiyan, M.Z.A., Liu, L., He, L.: FedMood: federated learning on mobile health data for mood detection. arXiv preprint arXiv:2102.09342 (2021)

60. Yi, L., Zhang, J., Zhang, R., Shi, J., Wang, G., Liu, X.: SU-net: an efficient encoder-decoder model of federated learning for brain tumor segmentation. In: Farkaš, I., Masulli, P., Wermter, S. (eds.) ICANN 2020. LNCS, vol. 12396, pp. 761–773. Springer, Cham (2020). https://doi.org/10.1007/978-3-030-61609-0_60
61. Yoo, J.H., et al.: Personalized federated learning with clustering: non-IID heart rate variability data application. arXiv preprint arXiv:2108.01903 (2021)
62. Yuan, B., Ge, S., Xing, W.: A federated learning framework for healthcare IoT devices. arXiv preprint arXiv:2005.05083 (2020)
63. Zeng, R., Zhang, S., Wang, J., Chu, X.: FMore: an incentive scheme of multi-dimensional auction for federated learning in MEC. In: 2020 IEEE 40th International Conference on Distributed Computing Systems (ICDCS), pp. 278–288. IEEE (2020)
64. Zhang, J., Chen, J., Wu, D., Chen, B., Yu, S.: Poisoning attack in federated learning using generative adversarial nets. In: 2019 18th IEEE International Conference on Trust, Security and Privacy in Computing and Communications/13th IEEE International Conference on Big Data Science and Engineering (TrustCom/BigDataSE), pp. 374–380. IEEE (2019)
65. Zhang, W., et al.: Dynamic fusion-based federated learning for COVID-19 detection. IEEE Internet Things J. (2021)

Time in Data Models

Johann Eder(✉)[iD], Marco Franceschetti[iD], and Josef Lubas[iD]

Department of Informatics-Systems, Universität Klagenfurt, Klagenfurt, Austria
{johann.eder,marco.franceschetti,josef.lubas}@aau.at

Abstract. Time is an essential dimension of our perception of the world and hence an important dimension for the representation of the real and social world in data models. We give an overview of the basics of representing time in data models and representing objects and processes with respect to the temporal dimension. In particular, we discuss basic concepts and novel developments in the areas of representing time, snapshot data models versus temporal versioning of data models, time-related storage of data in databases, temporal data warehouses and databases, schema evolution, and the representation and checking of temporal integrity constraints.

Keywords: Temporal data models · Temporal database · Schema evolution · Versioning · Temporal constraints · Chronon · Controllability

1 Introduction

Time is one of the fundamental dimensions of the physical space-time universe [45], it is a philosophical a-priori [32], and time is one of the most important categories for perceiving the physical and social world. Time is a very special dimension of our perception. It flows, even if we do nothing, it changes without requiring our contribution. It has a direction. And unlike space, we can never revisit a particular point in time, while we can revisit a particular point in space. Time points are in a total order.

When it comes to data modeling, We hardly observed an application domain where time is not a relevant aspect of discourse. Hence, modeling temporal aspects of a universe of discourse correctly is indispensable for data models. We are not aware of any real-world data model where no temporal information is included, where no data type *time* or *date* is used in a schema.

The Covid pandemic and the publication of pandemic related data showed us that the treatment of time in data models is frequently insufficient and leads to numerous misinterpretations as temporal aspects were not adequately modeled or communicated. So in many countries the time points of infection, of developing symptoms, of negative or positive tests, of registering symptoms, tests, etc. in databases, the time points of publishing data, etc. were not clearly distinguished and represented. This led to unnecessary and disturbing differences in statistical evaluation of available data.

© Springer Nature Switzerland AG 2021
T. K. Dang et al. (Eds.): FDSE 2021, LNCS 13076, pp. 23–35, 2021.
https://doi.org/10.1007/978-3-030-91387-8_2

In this paper, we will discuss some fundamental aspects of representing time in data models. Conceptual (data) models are essential in many areas [25] - from requirements engineering [13], software design [21,37], or database development to the meta-modeling of process enactment services [14]. These temporal models can then be implemented in various forms, e.g., in temporal databases [3,23,40, 41].

First, we show how time itself can be modeled and which data types are available for representing time, and which are the basic objects representing time: time-points and durations. Many applications have requirements relating to the concepts of time [13]. These requirements have to be checked and are mapped to data models and process models.

Most of the textbook exercises for data modeling discuss snapshot models, i.e., the representation of the state of the universe of discourse at a particular point in time, predominately, the *current* state. In real applications, nowadays, we find that it is necessary to represent the state of affairs over longer periods of time. We show the fundamental concepts of including history and temporal dimensions in data models and point out that conceptual models featuring time as fundamental modeling dimension are available (e.g., [8,31,34,36]). This has also serious consequences on data models and in particular on integrity constraints. For an example, 1-to-1 relationships in snapshot models might become n-to-m relationships in temporal data models. Hence we discuss the consequences of moving from a snapshot to a temporal representation.

In many application domains, data is typically no longer deleted or updated in situ to the current state but just declared as outdated. There is a trend to collect data over longer periods of time in order to prepare it for data mining, knowledge discovery, machine learning and other techniques of data science. On the one hand this is a tremendous source for gaining insights and to compute forecasts, but on the other hand requires an adequate modeling of the temporal dimensions to avoid erroneous inferences (see, e.g., [2,39,44]).

Data models typically can be seen as sequences of snapshots, as a series of static views on the universe of discourse. Process models, on the other hand, focus on the changes over time, the dynamics of the universe of discourse and as such naturally feature a temporal dimension (e.g., [7,12,13,17,18,22,33]). In process mining, these views are combined to derive process models and other useful information about the dynamics of organizations from the data collected and represented with a necessary temporal dimension [35,42].

Also data models, which explicitly feature time, as in most multidimensional data models which allow to stare time-series of measures frequently fall short of representing the evolution of other data in relation to the measurements. We show how temporal data warehouses [20] address this problem. In a more general view, we have to deal with the problem that not only data changes over time, but also data models evolve. Multi-version databases [26] strive to address this problem of schema evolution.

The main focus of this paper is on temporal integrity constraints which allow the representation of constraints between time-points of the data model. A particular difficulty of temporal constraints is how to check whether there are inconsistencies, i.e., whether the constraints are in conflict. As we will show, this is more difficult than checking other constraints, as satisfiability is not a sufficient criterion for specifying that it is possible to not violate some constraints. Introducing the notion of controllability we propose adequate procedures to check whether a set of constraints is in conflict.

The aim of this paper is to provide a broad overview about the issues of including time as an explicit special dimension in data models, discuss some of the problems involved and review some established and some quite recent techniques for dealing with these problems. In focal point we see in the treatment of time in data models the inherent difficulties and opportunities of integrating static and dynamic perceptions of a universe of discourse into a holistic view.

2 Representing Time

How is time measured? Do we measure time equally anywhere on Earth? How can we represent time and how can we reason about it? Let us start this discussion at the very beginning.

When we talk about time, we usually talk about *discrete time* measured in *chronons*, representing time spans of different granularity such as seconds, minutes, hours, days and so forth. In order to allow a reasonable large group of people to relate to a common time, we use calendars that are relative to some periodic event. For example the solar calendar measures time relative to earths rotation around the sun, defining a year to consist of approximately 365 days. However, there is more than one calendar system (e.g., Lunar, Lunisolar, Gregorian, ...), thus, bringing us back to one of the initial questions: "Do we measure time equally anywhere on earth?". Although there have been several proposals for an universal calendar, the simple answer to this question is no, and this fact, has to be considered when engineering time in data models.

When we reason about time, we usually refer to events that either happen at a *point in time* or within an *interval*. For example "John finished his presentation at 16:00" refers to an instantaneous event at a specific point in time, while "John was giving an hour long presentation" refers to an interval specified by a pair of time points (start and end of his presentation). Sometimes we use *temporal relations*, where we describe the time at which an event happens to be relative to some other event. For example "John started his presentation after Jane finished hers". This gives us, on one hand, more flexibility for describing points in time, and on the other hand, the tools to formulate complex temporal propositions.

From a technical perspective, we encode time as natural numbers forming a time axis starting at time point *zero*. The time at which an event happens, can be defined as time point positioned on a time axis. This position can be either absolute to *zero*, or relative to some other time point, defined by some temporal relation. As an example, let us assume that there are two time points A, B. Time point A happens at 16 (absolute) and B happens at $A+2$. We can see that there is a temporal relation between A and B stating that B happens 2 time units *after A*. Similar to the *after* relation, Allen's Interval Algebra [1]) presents in summary 13 possible temporal relations (i.e., *before, during, overlaps*, etc.).

Given the position of two arbitrary time points on a time axis, we may also compute their (temporal) distance, or in other words the *duration* or *interval*. In the above example, it is easy to see, that the duration between A and B is 2. We can use these definitions to also formulate constraints. For example we could require that an additional time point C has to occur some time in between A and B expressed by a simple inequality $A < C < B$. In Sect. 6, we give a more detailed presentation of temporal constraints in data models.

With this representation of time, the provision of abstract datatypes of *time* (or, correspondingly, of *date*), we can analyze, in which way time is included in data models. In the next section we discuss the difference between data models, which represent the universe of discourse at a particular point in time versus data models which also include past (and probably future) states.

3   Snapshots vs. Historical Data Models

Several applications require handling information that varies over time [31]. Most notable examples include medical record keeping, bank account management, and stock management. In the context of stock management in the retail industry, for instance, the availability of goods is subject to high variability over time; similarly, prices of goods may change in relation to demand. For such applications, the conceptual design of a data model is often a necessary step preceding the implementation, e.g., into a database system.

For a data model which has to adequately represent such time-varying information, it is essential to support semantics for evolving entities and relationships between entities. A fundamental distinction between data models is hence based on the existence of a notion for changes over time. The lack of such a notion characterizes *snapshot* data models; on the contrary, its awareness and the existence of modeling constructs for it enable *historical data models*.

In a snapshot data model, the represented information corresponds to a timeless state of things in the real world. There is no notion of time, hence any change to data in an instance leads to the loss of past information, which may be undesirable in several applications.

In a historical data model, history of past states is kept, forming a sequence of snapshots. This requires adequate conceptual modeling, in particular for ensuring that in and across snapshots relations between entities remain consistent with the real world, despite data changes.

For example, at a given point in time, a person can be married to at most one other person (1:1 relationship between instances in a data model). If two people divorce, the marriage relationship is deleted in the current data model state. If a divorced person marries another person, a new relationship is introduced in the data model. This latter situation requires, in a historical data model, adequately modeling the 1:1 marriage relationship as a n:m relationship enriched with temporal information, in order to record all past relationships in a consistent manner (i.e., no concurrent marriages).

Reaching a conceptual model that is capable of capturing the temporal relations of instances is indeed an interesting problem. Like other models, the traditional, widely adopted, Entity-Relationship (ER) model lacks expressiveness for time-varying information. Several works have proposed temporal extensions to the ER model over the years: we refer the reader to [24] for an overview.

A possible solution allowing the modeling of time-varying information in an (extended) ER model is proposed in [6]. In their work, the authors propose an extension to the ER model that features the representation of advanced requirements on evolving data. In particular, the proposed model enables expressing temporal semantics for entity attributes, relationships, keys, and inheritance. A further advantage of the contribution is the support for the widely adopted relational model for the implementation of the data model.

4 Temporal Databases

Databases have been for decades the prime technology for implementing data models. As suggested in [40], a fundamental distinction can be drawn between database types based on their support to temporal data models. In particular, [40] identifies four types of databases: static, historical, static rollback, and temporal databases. We highlight their peculiarities here.

In a static database, the real world is represented through a snapshot at a particular point in time, which is considered to be the sole state of things over time. Of course, a snapshot may yield a state which does not necessarily reflect the current reality, as reality constantly mutates. Thus, as soon as any change in reality is recorded in such a static database, the past state, i.e., the information it yields about the real world, is discarded and lost forever. After a change is recorded, the new state is the one considered to correctly represent reality (currently and in the past as well as in the future), residing in the database timelessly.

To partially overcome the limitations of static databases, a historical database adds additional information about *when* in the real world there is a counterpart for the recorded information. This temporal information is also known as *valid time*. Technically, valid time is implemented as an additional field of database tables, and can be seen as an associated time axis to the data model, which becomes three-dimensional. The limitation of the valid time model is that it does not track any updates to a given tuple. Thus, the database stores information about a reality grounded in time (the valid time), which, however, has a timeless representation. Consequently, it is not possible to reconstruct the history of changes a tuple underwent.

The third type of database is the static rollback database, also known as *transaction time* database. In a static rollback database, records are extended with temporal information about when the record has been inserted, and possibly until when it is to be considered valid information. This temporal information is known as transaction time. Similar to valid time, transaction time is implemented as an additional field storing a time interval which records when a tuple was inserted and (potentially) when it was (logically) deleted. The limitation of this model is dual to the limitation of the valid time model. Indeed, the database stores information about a timeless reality, which, however, has a time-grounded representation (the valid time), so it does not track when in the real world facts are to be considered to hold.

Combining the previous approaches, i.e., storing both valid time and transaction time, allows database managers to overcome the respective limitations of databases which make use of either valid time or transaction time only. Storing both such times means associating tuples with two time axes, thus making the data model four-dimensional. The benefit of such an extension is that tuples can be seen as valid at any moment as of any moment in time, thus capturing any retroactive and proactive changes. Databases handling both valid and transaction time are better known with the designation *bitemporal databases*.

While temporal databases play a significant role in the storing of time-varying information, they are often insufficient for real-world problems requiring to analyze historical data, possibly of high dimension and resulting from aggregation of multiple data sources.

5   Temporal Multiversion Data Models

The multidimensional modeling of data warehouses [30] typically cared about representing time, as one of the features promised by data warehouse technology was always the analysis of data over time. One of the dimensions of a multidimensional data model is typically time, mostly in form of valid time or transaction time. The typical textbook examples show how facts like number of sales can be analyzed along dimensions like geographic area, product hierarchy, and, in particular, time.

The dimension time allows to represent the changes of facts indexed by the other dimensions over time, allowing the representation of time series. The hierarchical organization of the time dimension offers the possibility of analyzing these time series on different levels of temporal granularity.

So one could for an example ask queries like "which products have increased in sales in Europe over the last 2 years". OLAP operators help to analyze the data and achieve insights.

There is, however, a fundamental problem with this multidimensional approach, as the dimensions are considered orthogonal. Orthogonality with respect to the dimension time means time invariant, or in other words, it is not possible to represent changes of other dimension data and structure over the time. Inadequate treatment of these temporal and evolution issues lead to wrong analysis of the available data ([28]) and in consequence to bad decisions.

For an example, to answer a query like *"increase in the population of the European union in the last 40 years?"* one has to be aware of the following: First of all, the geopolitical entity "European Union" only exists since 1993, succeeding the "European Community", which itself was originally named "European Economic Community". Furthermore, in the considered period (1981 to 2021), the

European Union grew from 12 to 28 members and then reduced to 27 members in 2020, when the United Kingdom left the union. Finally, with the reunification of East- and West-Germany in 1990 one of the member countries had a massive internal reorganization. Comparing the numbers of 1990 and 1991, where the organization, the set of member states, itself did not change, may indicate a massive increase of inhabitants. In reality, the 1991 number also contains the 16.4 million people of former East-Germany. So if querying the number of inhabitants from 1983 to 2021, how can the resulting numbers be compared?

Temporal data warehousing [9,16,20] took up this issue of so called slowly changing dimensions [19] and developed a series of data models permitting both the multidimensional representation of facts changing over time as well as the changing master data and dimension structure data. Time series analysis can then be made by selecting reference time points for structures and by applying explicit mappings between the different versions of master and dimension data. A particular challenge for data warehousing and data mining of time series with changing schemas and master data is that the changes frequently are hidden and have to be detected when data is loaded into the data warehouse [15].

Not only data warehouses suffer from the problem of changing dimension data, but in general schemas of databases evolve over time due to changing requirements, changes in the universe of discourse, changes due to technological progress or changes in transaction profiles. The dominant way of dealing with these changes is to transform the already stored data to the new schema in a schema evolution process [10,38]. However, there are applications, where it is necessary to keep the old data in its original form and make it available for querying through the new schema. Databases with multi-version schemas have the ambition to support such applications [26,27].

6   Temporal Integrity Constraints

Data in many different areas, such as health, business, law and research, often have to fulfill, besides functional requirements, temporal requirements [13]. Such temporal requirements may be expressed as either descriptive or prescriptive temporal constraints. Descriptive constraints originate from given temporal properties of the environment (e.g., the law) relevant for the correctness or integrity of the data; prescriptive constraints originate from desirable properties (e.g., business goals) within the context of its use.

Independent from the origin, temporal constraints require data or events to be valid only in a certain time interval. We can express such constraints either as absolute (i.e., bound to a specific date), or as temporal relations between time points, e.g., in the form of *before* or *after* some event such as a transaction happened. In general, temporal constraints can be expressed as *lower-bound* or *upper-bound* constraints, or as combinations of them, between two time-points. A lower-bound constraint defines a minimum time span that needs to elapse between two time points, while an upper-bound constraint defines a maximum such time span. Lower-bound constraints are used in dynamic systems such as business processes to define *precedence constraints*, e.g., between actions, to define partial orderings. A pair of a lower- and an upper-bound constraint between two time points defines a *duration constraint*.

An example for the lower-bound constraint comes from the COVID-19 regulations in Austria, which requires that a vaccination is only valid if at least 22 days have elapsed since the administration of the first dose of the vaccine. Another COVID-19 regulation in Austria states that a certificate of anti-bodies (as verification of recovery) is valid at most three month after the registration. The protection against an infection provided by the vaccination, however, can be seen as duration, defined by the combination of a lower-bound constraint stating the minimum, and an upper-bound constraint stating the maximum time of protection.

A crucial aspect in the design phase of time-constrained data models is verifying whether a set of temporal integrity constraints may be contradicting, meaning that one constraint cannot be satisfied without violating another. The absence of such contradictions is something that we refer to as *temporal correctness*. Interestingly, the absence of contradictions in a set of temporal constraints is seen in the literature from different perspectives and according to different goals, which leads to a plethora of notions for temporal correctness [13]. Examples for such notions are the following:

– *Satisfiability*, also called *consistency*: the most relaxed notion for temporal correctness. A set of temporal constraints is satisfiable (consistent) if there exists at least one configuration, or setting of the involved variables, which satisfies all constraints. Satisfiability, however, does not prohibit the existence of possibly invalid configurations for which some temporal constraint is violated.

    For an example, a constraint "take this drug 2 days before you get an infection" would be satisfiable, as for each time point of an infection there would be a proper time-point for the application of the drug. However, the earlier time-point cannot be changed due to a later observation. Hence the constraint is satisfiable, but this does not prevent that time-failures, violations of temporal constraints, are avoidable.

– *Strong controllability*: a set of temporal constraints is strong controllable if for all possible configurations of the involved variables all constraints are satisfied. Strong controllability may be, however, too restrictive and not always possible, since it may happen that not all variables involved in temporal constraints may be controlled, but depend on external influences.

    The constraint in the example above, therefore, is satisfiable, but not controllable.

– *Conditional controllability*: it is meaningful only in presence of conditional constraints, i.e., constraints which must hold only under conditions attached to them. A set of temporal constraints is conditional controllable if there exists a configuration of the involved variables satisfying all constraints for each possible observed values of conditions.

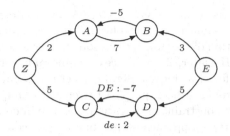

Fig. 1. Example of an STNU

- *Dynamic controllability*: it is meaningful only in dynamic systems, in which a controller is able to dynamically assign variables involved in temporal constraints. A set of temporal constraints is dynamically controllable if the controller can assign values in response to the past observations leading to the current state, so that all constraints can be fulfilled, despite uncertainties on future events.

How can we make sure that a data model meets a certain degree of temporal correctness such as the above? Depending on the context of the application, temporal aspects are encoded in models with different degrees of abstraction and expressiveness. For example, in the area of Business Process Management practitioners use process models and activity diagrams, while in the area of formal verification methods temporal logic is the most adopted formalism. Also integrity constraints can be of different complexity and of different types. Applications might state different requirements for the correctness of temporal models. And finally, models might be descriptive, represent some universe of discourse, or they might be prescriptive, representing goals or obligations.

Depending on the temporal data model type, the specification formalism, and the considered temporal correctness notion, different techniques are available for the verification of such temporal correctness.

For static systems, for instance, satisfiability may be sufficient while dynamic controllability may be not needed. In such a case, any SAT solver may be adopted. For a dynamic system such as a business process, dynamic controllability may be a more suitable notion to adopt. In this case, verification approaches frequently found in the literature are based on transformations to various types of Simple Temporal Networks (STNs) [11,29] or Timed Automata [46]. STNs are a formally grounded approach to represent temporal constraint satisfaction problems in the form of graphs, in which nodes encode time points and edges encode binary constraints between time points.

Considerable attention has been devoted to a particular STN type known as the STN with Uncertainty (STNU), due to its expressiveness for capturing real-world dynamic phenomena [43]. An example for an STNU is shown in Fig. 1.

An STNU includes semantics for uncontrollable assignments of time points, with the inclusion of *contingent* time points and edges. A contingent edge between two time points A and C (with C contingent) is expressed as (A, l, u, C), meaning that C occurs some uncontrollable time $\Delta \in [l, u]$ after A. Contingencies enable the representation of real-world situation characterized by uncontrollable, uncertain fragments. However, the presence of temporal uncertainty demands for advanced techniques for checking temporal correctness of an STNU.

Efficient algorithms have been proposed over the years, most of which are based on the notion of constraint propagation, i.e., the inference of implicit constraints [5]. Notably, these approaches have polynomial complexity, making the STNU a much suitable formalism for representing a broad number of temporal models and verifying their temporal correctness.

This glimpse on checking the controllability of sets of temporal constraints should highlight that checking temporal aspects in data models requires additional methods from the widely applied model checking approaches. As we showed, satisfiability which is typically considered in model checkers [4] is not sufficient for many applications, and more strict properties like controllability and dynamic controllability are necessary to ascertain that violations of temporal constraints are certainly avoidable.

7 Conclusions

Time is a peculiar phenomenon. Temporal aspects play a crucial role in many application domains of information technology. To capture temporal aspects correctly and to represent time properly in data models is inevitable for the development of information systems which yield accurate and useful answers for information needs. Hence, representing time and temporal aspects can be found in many technologies from conceptual modeling, information systems design, software development, to data science. We brought an overview about the perception of time in different data models and discussed techniques to represent time and deal with time, including the formulation of temporal constraints and checking fundamental properties of temporal models.

References

1. Allen, J.F.: Maintaining knowledge about temporal intervals. Commun. ACM **26**(11), 832–843 (1983)
2. Atluri, G., Karpatne, A., Kumar, V.: Spatio-temporal data mining: a survey of problems and methods. ACM Comput. Surv. (CSUR) **51**(4), 1–41 (2018)
3. Böhlen, M.H., Dignös, A., Gamper, J., Jensen, C.S.: Database technology for processing temporal data. In: 25th International Symposium on Temporal Representation and Reasoning (TIME 2018). Schloss Dagstuhl-Leibniz-Zentrum fuer Informatik (2018)
4. Bordeaux, L., Hamadi, Y., Zhang, L.: Propositional satisfiability and constraint programming: a comparative survey. ACM Comput. Surv. (CSUR) **38**(4), 12-es (2006)
5. Cairo, M., Rizzi, R.: Dynamic controllability made simple. In: 24th International Symposium on Temporal Representation and Reasoning (TIME 2017). LIPIcs, vol. 90, pp. 8:1–8:16 (2017)

6. Combi, C., Degani, S., Jensen, C.S.: Capturing temporal constraints in temporal ER models. In: Li, Q., Spaccapietra, S., Yu, E., Olivé, A. (eds.) ER 2008. LNCS, vol. 5231, pp. 397–411. Springer, Heidelberg (2008). https://doi.org/10.1007/978-3-540-87877-3_29

7. Combi, C., Galetto, F., Nakawala, H.C., Pozzi, G., Zerbato, F.: Enriching surgical process models by BPMN extensions for temporal durations. In: Proceedings of the 36th Annual ACM Symposium on Applied Computing, pp. 586–593 (2021)

8. Combi, C., Montanari, A.: Data models with multiple temporal dimensions: completing the picture. In: Dittrich, K.R., Geppert, A., Norrie, M.C. (eds.) CAiSE 2001. LNCS, vol. 2068, pp. 187–202. Springer, Heidelberg (2001). https://doi.org/10.1007/3-540-45341-5_13

9. Combi, C., Oliboni, B., Pozzi, G.: Modeling and querying temporal semistructured data. In: Kozielski, S., Wrembel, R. (eds.) New Trends in Data Warehousing and Data Analysis. Annals of Information Systems, vol. 3, pp. 1–25. Springer, Boston (2009). https://doi.org/10.1007/978-0-387-87431-9_14

10. Curino, C., Moon, H.J., Deutsch, A., Zaniolo, C.: Automating the database schema evolution process. VLDB J. **22**(1), 73–98 (2013)

11. Dechter, R., Meiri, I., Pearl, J.: Temporal constraint networks. Artif. Intell. **49**(1–3), 61–95 (1991)

12. Eder, J., Franceschetti, M.: Time and business process management: problems, achievements, challenges (invited talk). In: 27th International Symposium on Temporal Representation and Reasoning (TIME 2020). Schloss Dagstuhl-Leibniz-Zentrum für Informatik (2020)

13. Eder, J., Franceschetti, M., Lubas, J.: Time and processes: towards engineering temporal requirements. In: Proceedings of the 16th International Conference on Software Technologies (ICSOFT 2021), pp. 9–16 (2021)

14. Eder, J., Gruber, W.: A meta model for structured workflows supporting workflow transformations. In: Manolopoulos, Y., Návrat, P. (eds.) ADBIS 2002. LNCS, vol. 2435, pp. 326–339. Springer, Heidelberg (2002). https://doi.org/10.1007/3-540-45710-0_26

15. Eder, J., Koncilia, C., Mitsche, D.: Automatic detection of structural changes in data warehouses. In: Kambayashi, Y., Mohania, M., Wöß, W. (eds.) DaWaK 2003. LNCS, vol. 2737, pp. 119–128. Springer, Heidelberg (2003). https://doi.org/10.1007/978-3-540-45228-7_13

16. Eder, J., Koncilia, C., Morzy, T.: The COMET metamodel for temporal data warehouses. In: Pidduck, A.B., Ozsu, M.T., Mylopoulos, J., Woo, C.C. (eds.) CAiSE 2002. LNCS, vol. 2348, pp. 83–99. Springer, Heidelberg (2002). https://doi.org/10.1007/3-540-47961-9_9

17. Eder, J., Liebhart, W.: Workflow transactions. In: Workflow Handbook 1997, pp. 195–202. Wiley (1997)

18. Eder, J., Panagos, E., Rabinovich, M.: Workflow time management revisited. In: Bubenko, J., Krogstie, J., Pastor, O., Pernici, B., Rolland, C., Sølvberg, A. (eds.) Seminal Contributions to Information Systems Engineering, pp. 207–213. Springer, Heidelberg (2013). https://doi.org/10.1007/978-3-642-36926-1_16

19. Faisal, S., Sarwar, M.: Handling slowly changing dimensions in data warehouses. J. Syst. Softw. **94**, 151–160 (2014)

20. Golfarelli, M., Rizzi, S.: A survey on temporal data warehousing. Int. J. Data Warehous. Min. (IJDWM) **5**(1), 1–17 (2009)

21. Gonçales, L.J., Farias, K., Oliveira, T.C.D., Scholl, M.: Comparison of software design models: an extended systematic mapping study. ACM Comput. Surv. (CSUR) **52**(3), 1–41 (2019)

22. Graja, I., Kallel, S., Guermouche, N., Cheikhrouhou, S., Kacem, A.H.: Modelling and verifying time-aware processes for cyber-physical environments. IET Softw. **13**(1), 36–48 (2019)
23. Grandi, F.: Temporal databases. In: Encyclopedia of Information Science and Technology, 3rd edn., pp. 1914–1922. IGI Global (2015)
24. Gregersen, H., Jensen, C.S.: Temporal entity-relationship models-a survey. IEEE Trans. Knowl. Data Eng. **11**(3), 464–497 (1999)
25. Härer, F., Fill, H.-G.: Past trends and future prospects in conceptual modeling - a bibliometric analysis. In: Dobbie, G., Frank, U., Kappel, G., Liddle, S.W., Mayr, H.C. (eds.) ER 2020. LNCS, vol. 12400, pp. 34–47. Springer, Cham (2020). https://doi.org/10.1007/978-3-030-62522-1_3
26. Herrmann, K.: Multi-schema-version data management. Aalborg Universitetsforlag (2017)
27. Herrmann, K., Voigt, H., Pedersen, T.B., Lehner, W.: Multi-schema-version data management: data independence in the twenty-first century. VLDB J. **27**(4), 547–571 (2018). https://doi.org/10.1007/s00778-018-0508-7
28. Horner, J., Song, I.-Y.: A taxonomy of inaccurate summaries and their management in OLAP systems. In: Delcambre, L., Kop, C., Mayr, H.C., Mylopoulos, J., Pastor, O. (eds.) ER 2005. LNCS, vol. 3716, pp. 433–448. Springer, Heidelberg (2005). https://doi.org/10.1007/11568322_28
29. Hunsberger, L., Posenato, R.: Simple temporal networks: a practical foundation for temporal representation and reasoning. In: Combi, C., Eder, J., Reynolds, M. (eds.) 28th International Symposium on Temporal Representation and Reasoning (TIME 2021), volume 206 of Leibniz International Proceedings in Informatics (LIPIcs), pp. 1:1–1:5. Schloss Dagstuhl - Leibniz-Zentrum für Informatik, Dagstuhl (2021)
30. Jarke, M., Lenzerini, M., Vassiliou, Y., Vassiliadis, P.: Fundamentals of Data Warehouses. Springer, Heidelberg (2002). https://doi.org/10.1007/978-3-662-05153-5
31. Jensen, C.S., Snodgrass, R.T.: Temporal data management. IEEE Trans. Knowl. Data Eng. **11**(1), 36–44 (1999)
32. Kant, I.: Kritik der reinen Vernunft. BoD-Books on Demand (2020). http://odysseetheater.org/ftp/bibliothek/Philosophie/Kant/Kant%20Immanuel%20-%20Kritik%20der%20reinen%20Vernunft.pdf
33. Lanz, A., Weber, B., Reichert, M.: Time patterns for process-aware information systems. Requirements Eng. **19**(2), 113–141 (2012). https://doi.org/10.1007/s00766-012-0162-3
34. Li, X., Liu, Y.: Review of spatio-temporal data modeling methods. Data Anal. Knowl. Discov. **3**(3), 1–13 (2019)
35. Márquez-Chamorro, A.E., Resinas, M., Ruiz-Cortés, A.: Predictive monitoring of business processes: a survey. IEEE Trans. Serv. Comput. **11**(6), 962–977 (2017)
36. Parent, C., Spaccapietra, S., Zimanyi, E.: Spatio-temporal conceptual models: data structures+ space+ time. In: Proceedings of the 7th ACM International Symposium on Advances in Geographic Information Systems, pp. 26–33 (1999)
37. Pastor, O., Molina, J.C.: Model-Driven Architecture in Practice: A Software Production Environment Based on Conceptual Modeling. Springer, Heidelberg (2007). https://doi.org/10.1007/978-3-540-71868-0
38. Rahm, E., Bernstein, P.A.: An online bibliography on schema evolution. ACM SIGMOD Rec. **35**(4), 30–31 (2006)
39. Shahnawaz, M., Ranjan, A., Danish, M.: Temporal data mining: an overview. Int. J. Eng. Adv. Technol. **1**(1), 2249–8958 (2011)
40. Snodgrass, R., Ahn, I.: A taxonomy of time databases. ACM SIGMOD Rec. **14**(4), 236–246 (1985)

41. Snodgrass, R.T.: The TSQL2 Temporal Query Language, vol. 330. Springer, Heidelberg (2012)
42. Van Der Aalst, W.: Process mining: overview and opportunities. ACM Trans. Manage. Inf. Syst. (TMIS) **3**(2), 1–17 (2012)
43. Vidal, T.: Handling contingency in temporal constraint networks: from consistency to controllabilities. J. Exp. Theor. Artif. Intell. **11**(1), 23–45 (1999)
44. Wang, S., Cao, J., Yu, P.: Deep learning for spatio-temporal data mining: a survey. IEEE Trans. Knowl. Data Eng. 1 (2020). https://doi.org/10.1109/TKDE.2020.3025580
45. Young, H.D., Freedman, R.A., Sandin, T., Ford, A.L.: University Physics, vol. 9. Addison-Wesley, Reading (1996)
46. Zavatteri, M., Viganò, L.: Conditional simple temporal networks with uncertainty and decisions. Theoret. Comput. Sci. **797**, 77–101 (2019)

Big Data Analytics and Distributed Systems

Distributed Scalable Association Rule Mining over Covid-19 Data

Mahtab Shahin[1]($\boxtimes$) (iD), Wissem Inoubli[2] (iD), Syed Attique Shah[3] (iD),
Sadok Ben Yahia[2] (iD), and Dirk Draheim[1] (iD)

[1] Information Systems Group, Tallinn University of Technology,
Akadeemia tee 15a, 12618 Tallinn, Estonia
{mahtab.shahin,dirk.draheim}@taltech.ee
[2] Software Science Department, Tallinn University of Technology,
Akadeemia tee 15a, 12618 Tallinn, Estonia
{wissem.inoubli,sadok.ben}@taltech.ee
[3] Institute of Computer Science, University of Tartu, Tartu, Estonia
syed.shah@ut.ee

Abstract. The worldwide Covid-19 widespread in 2020 has turned into
a phenomenon that has shaken human life significantly. It is widely recognized that taking faster measurements is crucial for monitoring and
preventing the further spread of COVID-19. The advent of distributive
computing frameworks provides one efficient solution for the issue. One
method uses non-clinical techniques, such as data mining tools and other
artificial intelligence technologies. Spark is a widely used framework and
accepted by the big data community. This research used a cross-country
Covid-19 dataset to assess the performance of the Apriori and FP-growth
through different components of Spark (different numbers of cores and
transactions). This involves a scheme for classification and prediction by
recognizing the associated rules relating to Coronavirus. This research
aims to understand the difference between FP-growth and Apriori and
find the ideal parameters of Spark that can improve the performance by
adding nodes.

Keywords: Association rule mining · Big data · FP-growth · Spark ·
Apriori · Machine learning

1 Introduction

Coronavirus disease (COVID-19) belongs to a larger family of Coronaviruses
(CoV). This severe illness can be deadly as it assaults our respiratory cells and
causes an immune response that targets those infected cells, damages lung tissue,
and might finally shut off our supply of Oxygen by clogging our airways [4].
Countries worldwide have prioritized the early and automated diagnosis of this
disease to assign patients to quarantine and take further steps promptly. In some
severe cases, diagnosis has taken place in specialized hospitals to more efficiently

© Springer Nature Switzerland AG 2021
T. K. Dang et al. (Eds.): FDSE 2021, LNCS 13076, pp. 39–52, 2021.
https://doi.org/10.1007/978-3-030-91387-8_3

track disease transmission. Diagnosis is such a rapid procedure; therefore, the high expenses of further investigations have caused financial issues harming both states and patients, especially in areas where private health systems or economic issues can restrict one's access to medical care.

Classification, grouping, regression, and correlation are all aspects of data mining [21]. Data mining provides information about previously accurate independent itemsets and their relationship in extensive databases. Frequent Itemset Mining is the process of extracting frequent itemsets from transaction databases. It is crucial to look at association rules commonly utilized in real-world applications, including web data analysis, consumer behavior research, cross-marketing, catalog design, and medical records. In addition, Association Rule Mining (ARM) is involved in biological sciences, and researches illness detection and accurate classification prediction [31]. At its most basic level, the ARM entails analyzing patterns in data, or correlation, within a dataset using data mining tools. Every if-then association, also known as association rules, is defined by If-then statements that illustrate the potential of connections between itemsets in large databases of various sorts [32]. The support and confidence parameters are used to discover links between unrelated datasets or another data source, and ARM is created by looking for recurring data patterns. Support appears to reflect the regularity with which relationships occur in the database, whereas confidence indicates how often these associations have shown to be accurate [17,18,23,24]. All itemsets that fulfill such minimum support are generated for a given dataset. Within the second step, every frequent itemset is employed to develop all potential rules from the dataset; and rules that don't satisfy specified minimum confidence are removed. The main step of association rule mining is in distinctive frequent itemsets. Many ARM algorithms are presently in use: three typical classic representatives are Apriori [2], FP-growth [7], and Eclat [8].

This paper provides a design that supported Spark and association rule mining algorithms to seek an attention-grabbing relationship between Covid-19 data set. The findings would be gainful for patients, doctors, politics, and decision-makers in health informatics. This research addresses numerous contributions to the literature:

- It shows that applying an integrative k-NN/weighted k-NN algorithms with association rule mining improves prediction efficiency.
- It shows that the weighted k-NN has the highest accuracy compared to kNN for chronic disease data.
- It finds the ideal parameters that have positive effects on Spark jobs.
- It compares FP-growth and Apriori for the performance difference and how parameter tuning affects the results.

The remainder of the paper is organized as follows. In Sect. 2, we review the related works in this field. In Sect. 3, we explained the used algorithms briefly. In Sect. 4, we describe the details of the methodology, dataset, and pre-processing part. In Sect. 5, we provide the experimental results. Finally, In Sect. 6, we conclude the paper and present possible directions for future works.

2 Scrutiny of Related Work

One of the classical and well-known techniques of data mining is association rule mining [17]. Data mining determines a method to find out the relevant and gainful patterns in data [3]. This section will summarize prior works in the context of data mining techniques and association rule mining algorithms for finding frequent itemsets. Moreover, we will explain the tools that were used in this regard.

Kate and Nadig [15] proposed prediction models for breast cancer survival using the SEER dataset and machine learning approaches. They applied three different machine learning methods (naive Bayes, logistic regression, and decision tree) and discovered that the performance of the models varied greatly whenever evaluated independently at different phases. Soltani Sarvestani et al. [1] examined various research on the usage of other neural networks for accurate clinical detection of breast cancer in the largest and most active Hospital in South Iran; The idea was first assessed using publicly available statistics from throughout the world. They applied several neural network structures, and they functioned well. The PNN is the best-suited neural network model for categorizing WBCD and NHBCD data according to the overall results. This research also suggests that statistical neural networks can be utilized to aid doctors in breast cancer detection. Shukla et al. [25] proposed an unsupervised data mining creating patient cohort clusters. They applied a large dataset from the SEER program to recognize patterns associated with the survivability of breast cancer patients. These clusters, with associated patterns, were used to train the multilayer perceptron (MLP) model for enhanced patient survivability analysis. Examination of variable values in each cohort gives better insights into the survivability of a special subgroup of breast cancer patients. Wu and Zhou [27] developed two improved SVM methods to identify malignant cancer samples: support vector machine-recursive feature eliminate and support vector machine principal component analysis (SVM-PCA). Hinselmann, Schiller, Cytology, and Biopsy are four target variables that reflect the cervical cancer data. The three SVM-based methods diagnosed and categorized all four targets. They performed a comparison between these three approaches and compared the risk factor ranking result to the ground reality. The SVM-PCA technique is proven to be better than the others. Qiu et al. [20] proposed YAFIM (Yet Another Frequent Itemset Mining) and used it on real-world medical applications to discover the relationships in medicine. They concluded that the proposed method achieved 18 speedups for different benchmarks on average compared with the algorithms executed with MapReduce. It outperforms the MapReduce method about 25 times. To problem-solving of scanning the dataset in each iteration, Kumar Sethi and Ramesh [22] introduced Hybrid Frequent Itemset Mining (HFIM), which employs the vertical layout. The suggested algorithm was implemented over the Spark framework and comprised the concept of resilient distributed datasets to display in-memory processing to optimize the running time of operation. Their results showed that the HFIM performs better in terms of running time and memory consumption. Li and Sheu [19] proposed a divide-and-conquer-based scalable, highly parallelizable association rule mining heuristic (the SARL heuristic) that may reduce both time complexity and memory consumption while obtaining approximation results that are near to correct results. Comparative studies demonstrate that the suggested heuristic method outperforms algorithms by a substantial margin.

3   Preliminaries

A brief description of the algorithms used in the current study has been provided in the following.

3.1   Apriori

Agrawal et al. in 1993 proposed the AIS [2] as the first algorithm to generate all the frequent itemsets. Soon after, the developed version of AIS as the name of Apriori was introduced by Agrawal et al. Initially, association rule mining was utilized for market and sales data, where the function was to discover all the rules that would predict occurred items. This approach follows two steps [3]:

- In the *Join step*, calculate the union of two frequent itemsets of size n, assume taken A_n and B_n, which have a first $n-1$ element in common. $J_{n+1} = A_n \cup B_n$.
- In the *Prune step*, checked whether all the itemset of size n in j_{n+1} is frequent or not, and pruned those rules that do not satisfy the given condition (minimum support, confidence, and lift)

3.2   FP-Growth

Jiawei Han first introduced FP-growth in 2006 [11], where FP stands for frequent patterns. The strategy of FP-growth is based on the strategy of divide and conquer. Two scans have to do on the dataset. First, During the first scan of a database, find support for each item, and calculate a list of distributed frequent items in descending order (F-List). Second, it compresses the dataset into an FP-tree [16]. By using these steps, we can make FP-tree so that common prefixes can be provided.

3.3   k-Nearest Neighbours

According to Bank et al. [6], the general one percent of the data is futile; about one to five percent is manageable. Nevertheless, handling five to fifteen percent of missing data needs some advanced method. More than fifteen percent of missing data may significantly impact any characteristic of the data set. Missing value imputation techniques replace missing values from rows or specific classes with estimated ones, such as mean or mode values. The estimated values rely on various algorithms that return the outcome. Generally, missing values imputation often generates more effective results compared to other methods. kNN method was first proposed by Fix and Hodges [10] in 1951 and later developed by Thomas Cover [9]. It is one of the well-known imputation techniques for its ease of execution and provides fair output results. The principle of kNN is to fill missing values of the dataset according to different values of given k closest to missing items; applying distance function such as Euclidean distance function, evaluate the closeness or similarity between target instance and other instances in the data set. Then chose the top k closest instances as a candidate and determined weighted values as a replacement. The appealing advantages of this method including are:

- Appropriate for both quantitative and qualitative data.
- Avoid time consumption and computational cost, as it can make a predictive model for imputation.

3.4 Apache Spark

Apache Spark [14, 26] is known as a unified framework to analyze distributed big data processing. It was originally developed in 2009 at UC Berkeley University. The popularity of Spark is its ability to in-memory calculations that enable it to make faster 100 times compared to MapReduce. Apache Spark supports four basic libraries for machine learning, associated information mining, together with SparkSQL [5], Spark Streaming, Spark MLLib [30], and GraphX [28]. Spark deployment can be in three modes: standard mode, Mesos, and Hadoop Yarn. The principle of Spark is Resilient Distributed Datasets (RDDs). An RDD is a speeded immutable set of objects across a Spark cluster. According to master/slave architecture spark cluster contains of three main components [12, 13]:

- Driver Program: this component denotes the slave node in a Spark cluster. It maintains an object called Spark Context that manages running applications.
- Cluster Manager: this component can arrange the application's workflow since approved by Driver Program to workers. It also manages and controls every resource in the cluster and delivers its state to the Driver Program.
- Worker Nodes: every Worker Node denotes a container of one operation through a Spark program execution.

4 Methodology

4.1 Hardware and Software Configuration

The experiment of Hadoop and Spark were conducted on a high-performance computer by Python 3.7. It consisted of 11 nodes, and every single node was deployed with the same physical environment. Both Spark and Hadoop were configured on JDK version 8 and run the jobs on YARN. Also, HDFS is used to save intermediate data. The versions of Spark and Hadoop were 3.0.0 and 3.1.0, respectively. The details of nodes are shown in Table 1.

Table 1. System configuration.

Node type	Processor	Memory	OS	Docker version
Master	8	64	ubuntu 18.04	20.10.5
Slave	8	8		

In order to ssh the command line of the master node, we utilized PuTTY and access the HPC by its IP address.

4.2 Dataset Description

The Covid-19 data used in this experiment were taken from [29][1] – see Table 2 for details of the dataset.

Table 2. Properties of the used Covid-19 data set.

Size	# of Transaction	Time period
5.9 GB	3,048,576	December 2019 – January 2020

Among 31 attributes of this data set, we have selected 10 attributes to be included in our analysis, see Table 3, also compare with Fig. 1.

4.3 Data Pre-processing

Data preprocessing is one of the essential steps in the data mining process and is known as converting raw data into accurate data [2]. The main stages of data preprocessing are integration, cleaning, reduction, transformation, and discretization of the dataset. A preprocessing phase is developed with two goals, to optimize and speed up the process: (a) finding all sensitive transactions and determining weak rules; and (b) indexing different types of patterns affected by the sensitive transactions and the items. First, specify all sensitive transactions by scanning the whole database, by accomplishing the first purpose. Then, we removed the duplicated transactions to specify only everyday transactions instead of considering all database transactions. This process helps to reduce the size of the solutions and increase the speed of runtime. The second objective is to reduce database scanning by generating different index lists for sensitive transactions and items. Each transaction modification causes three different side effects: lost rule, hiding failure, and new rule.

---

[1] https://github.com/beoutbreakprepared/nCoV2019.

Table 3. Selected attributes

Attribute	Description
ID	Identify document for each reported case
Age	Age of the reported case
Gender	Male/female
City	Name of the reported city
Province	Name of the reported province
Country	Name of the reported country
Latitude	The latitude of the specific location
Longitude	The longitude of the specific location
Symptoms	List of reported symptoms in the case' description
Lives in Wuhan	0 the person does not live in Wuhan
	1 the person lives in Wuhan

	A	B	C	D	E	F	G	H	I	J
1	ID	age	sex	city	province	country	latitude	longitude	symptoms	livesInWuh.
2	1		male	Shek Lei	Hong Kon	China	22.36502	114.1338	anorexia-aching mu:	1
3	2	78	male	Vo Eugan∈	Veneto	Italy	45.29775	11.65838	fever	0
4	3	61	female			Singapore	1.35346	103.8151	headache	0
5	3		male	Zhengzho	Henan	China	34.62931	113.468	anorexia	1
6	4	32	female	Pingxiang	Jiangxi	China	27.51356	113.9029	caugh-chills	1
7	5	18	female	Yichun Cit	Jiangxi	China	28.30755	114.9732	chest discomfort	0
8	6	29	male	Shangrao	Jiangxi	China	28.77693	117.4692	backache-fever	0
9	7	52	male	Fuzhou Ci	Jiangxi	China	27.51128	116.4344	runny noise	0
10	8	76	male	Nanchang	Jiangxi	China	28.66149	116.0257	soreness	1

Fig. 1. First ten rows of the dataset

Filling Missing Values with k-Nearest Neighbours. The imputation of the missing values process comprises two main steps: The first step selects the set of attributes to the features with missing values as the target. Let the Covid-19 dataset be represented as a patient information expression matrix C with m columns and n rows corresponding to transactions and attributes, respectively. To impute the missing values of transaction X_c, $c \in \{1, \ldots, n\}$ and attributes X_c, $i \in \{1, \ldots, m\}$, it is to find k other transactions, each with a known value for attribute i and its features being the most similar to that of items.

$$d_{ij} = dist(x_i, x_j) = \sqrt{\sum_{p=1}^{n}(d_{ip} - d_{j_p})^2} \tag{1}$$

Where $dist\ (x_i,\ x_j)$ denotes the Euclidean distance between transaction x_i and x_j. n is the number of items, and x_{ip} is the pth of transaction x_i. The second step includes predicting the missing value using the observed values belonging to the selected item of transactions. At this stage, an average of values in experiment i from the k closest transactions is then used to estimate the missing value in transactions x_i. Can determine the estimated $\widetilde{x}_{ip}$ value of the missing x_{ip} as below:

$$\widetilde{x}_{ip} = \frac{\sum_{\forall x_a \in N_g} x_{ai}}{k} \tag{2}$$

In the above equations, x_i is the set of k nearest neighbors of transaction x_i. Moreover, in the weighted variation, the contribution of each transaction $x_a \in N_g$ is weighted by the similarity of its explanation to that of the transaction. Accordingly, higher weights are defined as a more similar transaction. A weighted average of values from k nearest transactions is then used to assess the missing value in the target transaction. This weight computation is as follow:

$$\widetilde{x}_{ip} = \sum_{\forall x_a \in N_g} x_{ai} w_i \tag{3}$$

where;

$$w_i = \frac{\frac{1}{d_i}}{\sum_k^{i=1} \frac{1}{d_i'}} \tag{4}$$

Table 4. Split-Validation results for the pre-processing

K#	KNN	WKNN
1	80.1	89.2
2	75	78.2
3	79.6	86.6
4	67.3	78.4
5	66.2	97.9
6	59.9	89.2
7	60.4	83.2

5   Results and Discussion

To implement the framework, we applied FP-growth and Apriori algorithms of the MLlib machine learning library. Then, the deployment and execution of the recommended system are done over a distributed computing environment formed of a different number of clusters, nodes, and transactions by Spark resources management. This section provides how to evaluate the performance of the Spark with three different experiments. We utilized the running time to present the efficiency because it can show the difference and efficiency directly (Fig. 2 and Table 4).

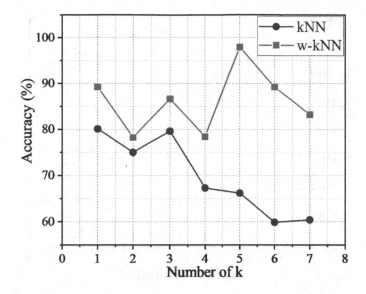

Fig. 2. Split-Validation results of kNN and wkNN

5.1 Core Utilization

This parameter determines the parallel computing ability for each executor. The executer-core is employed for configuring the number of CPU cores for each executor. As one CPU core can execute many tasks simultaneously, the more CPU cores assigned to the executors, the faster the Spark job. Experiments have been done for both algorithms in the same configuration to analyze the performance of Apriori and FP-growth. As shown in Fig. 3, we have applied the experiments in different numbers of core, from one to eight.

5.2 Node Utilization

Spark splits the work into multiple execute tasks on worker nodes. Thus Spark processors data in less time. Figure 4 shows that the running time strongly decreases as far as the number of nodes increases. Besides, we can find that the FP-growth curve is still sharper for all nodes than Apriori, which means FP-growth is more efficient than Apriori. The average running time of FP-growth and Apriori can be shown from Table 5.

5.3 Number of Transactions

We determined a comparative analysis between running time on FP-growth and Apriori to show the scalability. For that, we increased the number of transactions to take a sufficient database size. Later, we measured the speed of the

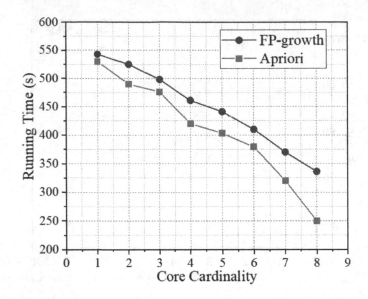

Fig. 3. Results of the number of cores

Table 5. Running time of Apriori and Fp-growth based on the variation of the number of nodes

Node #	Algorithms (s)	
	Apriori	FP-growth
1	553	2.1
3	437	11.6
5	297	281
7	238	230
9	119	116
11	59	41

FP-growth algorithm using the MLlib library compared to the Apriori in the same environment. The results are outlined in Fig. 5, which represents the line chart of execution times of different algorithms. As shown in the line chart, running the association rules with FP-growth is faster than Apriori. For example, it takes about 600 s to process 3 million transactions when Apriori takes 650 s. Furthermore, the FP-growth algorithm of Spark Mllib accomplishes good scalability because of the distributed computing on cluster nodes. As a result of the above analysis, FP-growth provided the most suitable environment to implement the Covid-19 dataset.

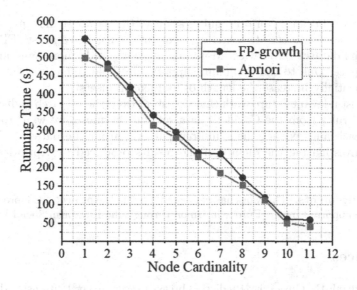

Fig. 4. Results of the number of nodes

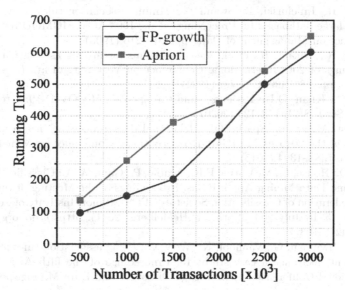

Fig. 5. The optimal performance for Apriori and FP-growth with different input splits

6 Conclusion

This paper aims to design a distributed framework for finding frequent itemsets of the Covid-19 dataset. We applied the Spark framework to expedite the parallel processing of data and decrease the calculation cost. Overall, the experiment results show that the performance of FP-growth is always superior to the Apriori algorithm. It is because Apriori requires scans of the database multiple times for generating candidate sets to generate frequent items; in contrast, FP-growth scans of the dataset only twice. Moreover, Apriori needs more time and large memory space. Due to the minimum support threshold reduction, the number and exponentially increase the length of frequent itemsets.

The most relevant future work that can stem from the research is discovering association rules from the Covid-19 data set. Furthermore, we want to expand this approach and extract the symptom patterns from the dataset. Hence, it is worth investigating the quality of the results produced by Apriori and FP-growth.

Acknowledgements. This work has been conducted in the project "ICT programme" which was supported by the European Union through the European Social Fund.

References

1. Abdelghani, B., Guven, E.: Predicting breast cancer survivability using data mining techniques. In: SIAM International Conference on Data Mining (2006)
2. Agrawal, R., Imieliński, T., Swami, A.: Mining association rules between sets of items in large databases. In: Proceedings of the 1993 ACM SIGMOD International Conference on Management of Data, pp. 207–216 (1993)
3. Agrawal, R., Srikant, R., et al.: Fast algorithms for mining association rules. In: Proceedings of 20th International Conference on Very Large Data Bases, VLDB, vol. 1215, pp. 487–499. Citeseer (1994)
4. Anwar, H., Khan, Q.U.: Pathology and therapeutics of COVID-19: a review. Int. J. Med. Stud. **8**(2), 113–120 (2020)
5. Armbrust, M., et al.: Spark SQL: Relational data processing in spark. In: Proceedings of the 2015 ACM SIGMOD International Conference on Management of Data, pp. 1383–1394 (2015)
6. Banks, D., House, L., McMorris, F.R., Arabie, P., Gaul, W.A.: Classification, Clustering, and Data Mining Applications: Proceedings of the Meeting of the International Federation of Classification Societies (IFCS), Illinois Institute of Technology, Chicago, 15–18 July 2004. Springer, Heidelberg (2011). https://doi.org/10.1007/978-3-642-17103-1
7. Brijs, T., Swinnen, G., Vanhoof, K., Wets, G.: Using association rules for product assortment decisions: a case study. In: Proceedings of the fifth ACM SIGKDD International Conference on Knowledge Discovery and Data Mining, pp. 254–260 (1999)
8. Chen, Y., Li, F., Fan, J.: Mining association rules in big data with NGEP. Clust. Comput. **18**(2), 577–585 (2015)
9. Cover, T., Hart, P.: Nearest neighbor pattern classification. IEEE Trans. Inf. Theory **13**(1), 21–27 (1967)
10. Fix, E., Hodges, J.L.: Discriminatory analysis. nonparametric discrimination: consistency properties. Int. Stat. Rev./Revue Int. Stat. **57**(3), 238–247 (1989)
11. Han, J., Pei, J., Yin, Y.: Mining frequent patterns without candidate generation. ACM SIGMOD Rec. **29**(2), 1–12 (2000)

12. Inoubli, W., Aridhi, S., Mezni, H., Maddouri, M., Nguifo, E.: A comparative study on streaming frameworks for big data. In: VLDB 2018–44th International Conference on Very Large Data Bases: Workshop LADaS-Latin American Data Science, pp. 1–8 (2018)
13. Inoubli, W., Aridhi, S., Mezni, H., Maddouri, M., Nguifo, E.M.: An experimental survey on big data frameworks. Futur. Gener. Comput. Syst. **86**, 546–564 (2018)
14. Inoubli, W., Aridhi, S., Mezni, H., Mondher, M., Nguifo, E.: A distributed algorithm for large-scale graph clustering (2019)
15. Kate, R.J., Nadig, R.: Stage-specific predictive models for breast cancer survivability. Int. J. Med. Inf. **97**, 304–311 (2017)
16. Kaur, G., Aggarwal, S.: Performance analysis of association rule mining algorithms. Int. J. Adv. Res. Comput. Sci. Softw. Eng. **3**(8), 856–58 (2013)
17. Kaushik, M., Sharma, R., Peious, S.A., Shahin, M., Ben Yahia, S., Draheim, D.: On the potential of numerical association rule mining. In: Dang, T.K., Küng, J., Takizawa, M., Chung, T.M. (eds.) FDSE 2020. CCIS, vol. 1306, pp. 3–20. Springer, Singapore (2020). https://doi.org/10.1007/978-981-33-4370-2_1
18. Kaushik, M., Sharma, R., Peious, S.A., Shahin, M., Yahia, S.B., Draheim, D.: A systematic assessment of numerical association rule mining methods. SN Comput. Sci. **2**(5), 1–13 (2021)
19. Li, H., Sheu, P.C.-Y.: A scalable association rule learning heuristic for large datasets. J. Big Data **8**(1), 1–32 (2021). https://doi.org/10.1186/s40537-021-00473-3
20. Qiu, H., Gu, R., Yuan, C., Huang, Y.: YAFIM: a parallel frequent itemset mining algorithm with spark. In: 2014 IEEE International Parallel & Distributed Processing Symposium Workshops, pp. 1664–1671. IEEE (2014)
21. Rasheed, J., et al.: A survey on artificial intelligence approaches in supporting frontline workers and decision makers for the COVID-19 pandemic. Chaos Solit. Fractals **141**, 110337 (2020). https://doi.org/10.1016/j.chaos.2020.110337. https://www.sciencedirect.com/science/article/pii/S0960077920307323
22. Senthilkumar, A., Hari Prasad, D.: An efficient FP-growth based association rule mining algorithm using hadoop MapReduce. Indian J. Sci. Technol. **13**(34), 3561–3571 (2020)
23. Shahin, M., et al.: Big data analytic in association rule mining: A systematic literature review. In: Proceedings of the International Conference on Big Data Engineering and Technology (2021). (in press)
24. Shahin, M., et al.: Cluster-based association rule mining for an intersection accident dataset. In: Proceedings of the IEEE International Conference on Computing, Electronic and Electrical Engineering (ICECUBE) (2021)
25. Shukla, N., Hagenbuchner, M., Win, K.T., Yang, J.: Breast cancer data analysis for survivability studies and prediction. Comput. Methods Program. Biomed. **155**, 199–208 (2018)
26. Spark, A.: Unified analytics engine for big data (2018). Accessed 5 Feb 2019
27. Wu, W., Zhou, H.: Data-driven diagnosis of cervical cancer with support vector machine-based approaches. IEEE Access **5**, 25189–25195 (2017)
28. Xin, R.S., Gonzalez, J.E., Franklin, M.J., Stoica, I.: GraphX: a resilient distributed graph system on spark. In: First International Workshop on Graph Data Management Experiences and Systems, pp. 1–6 (2013)
29. Xu, B., et al.: Epidemiological data from the COVID-19 outbreak, real-time case information. Sci. Data **7**(1), 1–6 (2020)
30. Zaharia, M., Chowdhury, M., Franklin, M.J., Shenker, S., Stoica, I., et al.: Spark: cluster computing with working sets. HotCloud **10**(10–10), 95 (2010)

31. Zaki, M.J.: Scalable algorithms for association mining. IEEE Trans. Knowl. Data Eng. **12**(3), 372–390 (2000)
32. Zhang, S., Webb, G.I.: Further pruning for efficient association rule discovery. In: Stumptner, M., Corbett, D., Brooks, M. (eds.) AI 2001. LNCS (LNAI), vol. 2256, pp. 605–618. Springer, Heidelberg (2001). https://doi.org/10.1007/3-540-45656-2_52

Threshold Benefit for Groups Influence in Online Social Networks

Phuong N. H. Pham[1(✉)], Bich-Ngan T. Nguyen[1], Quy T. N. Co[1], Canh-Pham[1(✉)], and Václav Snášel[2]

[1] Faculty of Information Technology, Ho Chi Minh City University of Food Industry, Ho Chi Minh, Vietnam
{phuongpnh,nganntb,canhpv}@hufi.edu.vn
[2] Technical University of Ostrava, Ostrava, Czech Republic
vaclav.snasel@vsb.cz

Abstract. The influence maximization (IM) is an optimization problem in the information propagation and social network analysis, which has the goal of finding a seed set that can influence largest number of users. There have been many studies on the IM problem, but most of them focus on maximizing influence effects based on individuals rather than on groups of users. In this paper, we studied a novel problem, the Threshold Benefit for Groups Influence (named TGI), defined as follows: given a social network G, a set of K target groups $\mathcal{C} = \{C_1, C_2, \ldots, C_K\}$ and a threshold $T > 0$, TGI asks us to find a seed set S in G with the minimum cost so that the benefit gained from the influence of the groups in C is at least T. We experimentally implement our proposed algorithm on real social network datasets diversity. It shows efficiency compared to other state-of-the-art IM algorithms in terms of cost and running time.

Keywords: Online social network · Viral marketing · Groups influence

1 Introduction

Influence Maximization is a popular concern in recent times, especially in social networks with millions of users such as Youtube, Facebook, Twitter, or the other online social networks like LinkedIn, Tumblr, etc. Users participating in social networks can connect with others easily, not only they are not only exchange and share information, but also update information quickly. Companies, organizations, and businesses have used to online social networks as a communication channel to advertise products, such as new products, online shopping, online training, spreading opinions or supply of human resources, etc. Online social networks (OSNs) has become an important platform in the field of marketing and advertising by leveraging the effectiveness of word-of-mouth. The primary aim of the IM problem is to discover the seed set of K users that the number of users in the online social network will be affected the most. Each user in the social network has two states: active and inactive, with the activation process switching the user's stable based on the influence of the seed set nodes. Kempe *et al.*

© Springer Nature Switzerland AG 2021
T. K. Dang et al. (Eds.): FDSE 2021, LNCS 13076, pp. 53–67, 2021.
https://doi.org/10.1007/978-3-030-91387-8_4

[9] are the first to consider the IM problem, which is as in many combinatorial optimization issues for the effect propagation problem with two classic models under named Independent Cascade (IC) and Linear Threshold (LT). Hence, the method inherits from Kempe's work, many variations on the IM problem have been introduced because of its important role in many practical applications, such as viral marketing [5,17], profit/revenue maximization [19,21], social recommendation, healthcare, rumor control, etc. However, the majority of the aforementioned existing works only focus solely on maximizing influence based on individual user nodes, ignoring the diffusion of influence to groups of individuals and communities. It can be shown that a small number of users who play a significant role in a group or community impact every user's behavior or choice in that group or community. Hence, the problem of maximum influence on a group or society can produce more beneficial results than the problem of maximum influence on an individual. As a result of the above practical significance, there have been several recent studies on groups influence by choosing a seed set with at least k users, such as groups influence [27], competitive groups influence, groups influence via network embedding [7]. In this article, we propose a method for optimizing the influence of groups instead of individuals in OSNs: assuming there is a social network under the information propagation model with $G = (V, E)$, where the set of groups C_i is the target group. To evaluate the role and importance of each node in the propagation process, we provide each node a cost $c(u)$ to pay for activating a node in the network, a benefit $b(u)$ if that node has influenced. Then, a fixed threshold t_i is assigned to each group C_i and for each node in group C_i, a fixed score $s(u)$ is assigned. As a result, our contents are presented in our work as follows:

- Firstly, we present widely known information propagation models such as IC and LT. Our algorithm is effectively implemented for the IC model in this paper, but it can be easily extended to other similar propagation models.
- Secondly, we propose a groups influence maximization algorithm that has more advantages than the influence of each node in the network, with a threshold t_i for each group.
- Thirdly, we present a group-based sampling (GBS) technique inherited from the RIS sampling method to estimate the epsilon group effect function.
- Finally, we conduct extensive experimentation of our proposed algorithm on real data social networks. The results show that our algorithm is efficient, providing quality solutions compared to other methods in terms of running time, memory usage, and total seed set cost.

Organization. The rest of the paper is organized as follows: In Sect. 1 and 2, we present an overview of issues to maximize influence and research works related to them. Section 3 introduces our information propagation model and the problem definition. The group sampling method and our proposed algorithm are then presented in Sect. 4. Section 5 presents experimental results, comparisons, and arguments. Finally, Sect. 6 concludes our paper.

2   Related Work

In recent years, the problem of IM has attracted the attention of many researchers from many different fields. Besides the early published works, many variations of the IM problem were introduced in turn. Domingos and Richardson [8] first introduced IM as a significant research problem related to the process of influence transmission in social networks. IM is a problem solving a small group of influential users that can maximize the spread of influence in the network. The influence propagation is simulated based on information propagation models. Kempe et al. [9] introduce the IM problem as a discrete optimization problem with two well-known widely used models, named IC and LT. They proved the IM problem to be NP-hard as well as compute the propagation of a seed set that is P-hard. Borg et al. [1] introduce a reverse sampling method, named RIS, which is the foundation for the development of linear-time algorithms with approximate solutions. Canh et al. [20] presented an extended version of Impact Maximization with Priority (IMP), a variant of the IM problem with priority constraints. The goal of the IMP issue is to choose a seed set with k nodes that can affect a certain priority set U larger than the threshold T in order to alter the seed set's effect on the priority set. The INCIM algorithm [3] estimates each node's propagation value as a combination of its local and global influences in order to measure each node's influence in its community as well as the influence of each community in the input graph. Beni et al. [2] proposes TI-SC, a survey-based community discovery method. The TI-SC algorithm chooses influential nodes by examining the relationships between core nodes and the scoring ability of other nodes. The score is updated after selecting each seed node to reduce overlap in seed node selection. Recently, Xuanhao Chen et al. [24] proposed a community-based Influence Maximization model to study the influence maximization problem in LBSN, taking both community structure and users' space-time behavior into user. Their work introduces two algorithms: one for detecting communities in LBSNs based on user mobility, and another for determining the most influential individuals based on the community.

Besides the problem of maximum influence based on the node, many researchers have been interested in the problem of maximum influence on the group in recent years. As we all know, each user in a social network frequently joins a specific group that shares characteristics such as similar interests, locations, or interests in specific topics. J. Zhu et al. [28] proposed a framework for selecting k seed users that combined the benefits of activated groups with the propagation costs of influence in order to maximize the expected return. They used the IC model to train an information diffusion model. Furthermore, they expressed their description as an optimization problem, proving that it is NP-hard and that the objective function is neither submodular nor supermodular. Similar manner, Nguyen et al. [18] investigated a set of k nodes that affect the greatest number of communities while also demonstrating that the influence function is not submodular or supermodular. Furthermore, the [10] algorithm, named GIN, generates various groups of graph nodes with more connections than others. It then chooses specific nodes from each group to narrow the search space to find the most influential nodes. It selects the seed nodes with the highest expected diffusion value using the greedy method. The test results showed that the GIN algorithm provided a high level of impact propagation along with low running times in comparison algorithms across all real-world data sets. Yuting Zhong *et al.* [26] recently proposed the Maximizing Group Coverage algorithm, which greedily chooses the best node based on evaluating node contributions to groups, ensuring success in estimating the maximum number of activated groups. The experimental results show that the MGC algorithm outperforms the base algorithm, Maximum Coverage, in terms of the number of activated group averages. The preceding studies on the problem of maximum influence on a group with various evaluation characteristics are different from the problem we studied. We propose a maximum effect algorithm for each group based on a given minimum cost, threshold constraint and present experimental results on various real-world datasets.

3   Information Diffusion Models

In Table 1, the frequently used notations are summarized.

Table 1. Table of notations

Notional	Description
n, m	The number of nodes and the number of edges in G, respectively
$N_{in}(v), N_{out}(v)$	The sets of incoming, and outgoing neighbor nodes of v
$\mathcal{C}$	$\mathcal{C} = \{C_1, C_2, \ldots, C_K\}$ is the set of target groups
C	$C = \bigcup_{i=1}^{K} C_i$
S	A seed set returned by our algorithms
S^*	An optimal solution of TGI problem
$C(u)$	The group contains node u
OPT	The optimal solution $\text{OPT} = c(S^*)$

3.1   Independent Cascade Model

We model an OSN as a directed graph $G = (V, E)$ where V is the set of nodes and E is the set of edges with $|V| = n$ and $|E| = m$. Let $N_{in}(v)$ and $N_{out}(v)$ be the set of in-neighbors and out-neighbor of node v, respectively. Given a seed set (initial influenced nodes) S, an information diffusion process happens in the network and hence more nodes can be activated. There have been a number of information diffusion models, among which Independent Cascade (IC) and Linear Threshold (LT) [9] are the two most basic and widely-used models in OSNs [1,4,14,15,17,18,22,23]. In this paper we focus on the IC model but our approach can be modified to handle the LT model as well.

In the IC model, each edge $e = (u, v) \in E$ has a propagation probability $p(e) \in [0, 1]$ representing the information transmission from a node u to a node v. The diffusion process from S happens in discrete time steps $t = 1, 2, \ldots$, as follows. At round $t = 0$, all nodes in S are *active* and other nodes in $V \backslash S$ are *inactive*. At step $t \geq 1$, for each node u activated at step $t - 1$, it has a single chance to activate each currently inactive node $v \in N_{out}(u)$ with a successful probability $p(e)$. If a node is activated it remains active till the end of the diffusion process. The propagation process ends at step t if no new node is activated in this step. Kempe *et al.* [9] show that the IC model is equivalent to a *live-edge* model defined as follows. From the graph $G = (V, E)$, we generate a random sample graph g by selecting edge $e \in E$ with probability $p(e)$ and not selecting e with probability $1 - p(e)$. We refer to g as a sample of G and write $g \sim G$. The probability that g is generated from G is:

$$\Pr[g \sim G] = \prod_{e \in E(g)} p(e) \cdot \prod_{e \notin E(g)} (1 - p(e)), \tag{1}$$

where $E(g)$ is the set of edges in the graph g. The influence spread from a set node S to a node u is:

$$\mathbb{I}(S, u) = \sum_{g \sim G} \Pr[g \sim G] \cdot r(S, u) \tag{2}$$

where $r(S, u) = 1$ if u is reachable from S in g and $r(S, u) = 0$ otherwise. The influence spread of S in network G is:

$$\mathbb{I}(S) = \sum_{u \in V} \mathbb{I}(S, u) \tag{3}$$

3.2   Problem Definition

We are given a social network $G = (V, E)$ under the IC model and a collection of K disjoint groups $\mathcal{C} = \{C_1, C_2, \ldots, C_K\}$ (called target groups), where $C_i \subseteq V, C_i \cap C_j = \emptyset$, for every pair of nodes (i, j) with $i \neq j$. Denote b_i is the benefit when C_i.

Denote by $C(u)$ the group that contains node u. To determine a group is influenced or not, we extend the influence group model in [18] by scoring each node in the group based on the fact that each user has a different role in his/her group. Thus, each node $u \in V$ has a *cost* $c(u)$ and a *score* $s(u)$. The weight $c(u)$ measures the cost or the price of the node u that has to pay if u is chosen as a seed node. The node score $s(u) > 0$ metrics the role of node u in group $C(u)$. Each group C_i assigns a threshold for t_i $(t_i > 0)$, which reflects the minimum total score that we must reach if we want to influence group C_i. We say that group C_i is influenced if the total score of influenced nodes in C_i is at least t_i. We define a *cost function* $c : 2^V \to \mathbb{R}_+$ and a *groups influence function* $\sigma : 2^V \to \mathbb{R}_+$ as follows. For a given seed set $S \subseteq V$,

- $c(S) = \sum_{u \in S} c(u)$ is total cost of S
- $\sigma_i(S)$ is the total score of nodes which influence by S in C_i

$$\sigma_i(S) = \sum_{v \in C_i} \mathbb{I}(S, v) s(v) \qquad (4)$$

- $\sigma(S)$ is (expected) the number of groups in $\mathcal{C}$ are influenced by the seed set S when the diffusion process ends, that is,

$$\sigma(S) = \sum_{C_i \in \mathcal{C}: \sigma_i(S) \geq t_i} b(v) \qquad (5)$$

In the special case where each group C_i has only one node the groups influence function $\sigma(\cdot)$ above becomes the influence spread function $\mathbb{I}(\cdot)$ of the IM problem. As a consequence, computing $\sigma(\cdot)$ is #P-hard. On the other hand, one can easily verify that the function $\sigma(\cdot)$ is neither submodular nor supermodular. The function $\sigma(\cdot)$ is submodular if for every pair of subsets $A, B \subseteq V$ it holds that $\sigma(A) + \sigma(B) \geq \sigma(A \cup B) + \sigma(A \cap B)$. If the inequality holds in the reversed direction we call $\sigma(\cdot)$ a supermodular function. For the completeness, we provide a counter-example in Example 1 below.

Example 1. Consider a graph G containing three nodes $\{a, b, c\}$, two directed edges (a, c), (b, c) with influence probability is 1. There is a target group $C = \{a, b\}$ with the threshold 2. We have $\sigma(\{a\}) - \sigma(\emptyset) = 0 < \sigma(\{a, b\}) - \sigma(\{a\}) = 1$, which means $\sigma(\{a\})$ is non-submodular. Also, we also have $\sigma(\{a, b, c\}) - \sigma(\{a, b\}) = 0 < \sigma(\{a, b\}) - \sigma(\{a\}) = 1$, which means $\sigma(\cdot)$ is non-suppermodular.

We now formally define the problem *Minimum cost for Influence Target Groups* (TGI), which will be studied in this paper.

Definition 1 (TGI problem). *An instance of* TGI *is given by* $(G, \mathcal{C}, T)$, *where* $G = (V, E)$ *is a social network under* IC *model, and* $\mathcal{C}$ *is a collection of disjoint target groups* $\{C_1, C_2, \ldots, C_K\}, C_i \cap C_j = \emptyset$. *The objective is to find a seed set* $S \subseteq V$ *of minimum total cost that the benefit function is at least* T, *i.e.,* $S = \arg\min_{S' \subseteq V, \sigma(S) \geq T} c(S')$.

4   Proposed Algorithm

4.1   Estimation of σ_i

We first introduce the Group Benefit Sample (GBS) concept, by modifying Reachable Reverse (RR) sample, to estimate $\sigma_i(S)$

Definition 2 (GBS sample). *Given a graph* G *and group* C_i, *a GBS sample is generated by the following steps:*

1. *Randomly select a node* u *with probability* $\frac{s(u)}{\Gamma}$ *(call* u *a source node), where* $\gamma_i = \sum_{u \in C_i} s(u)$

2. *Generate a sample graph g according to the live-edge model under* IC *model.*
3. *Return a node set R_i that is reachable from u in g. We call u a source node.*

Denote a random variable $X_i(S)$ as follows

$$X_i(S) = \begin{cases} 1, & \text{if } S \cap R_i = \emptyset \\ 0, & \text{otherwise} \end{cases} \tag{6}$$

We have the following lemma

Lemma 1. *Given a set node S, we have $\sigma_i(S) = \gamma_i \cdot \mathbb{E}[X_g(S)]$*

Given a set of GBSes $\mathcal{R}$, we have an estimation of $\hat{\sigma}_i(S)$ of $\sigma_i(S)$ as follows

$$\hat{\sigma}_i(S) = \frac{\gamma_i}{|\mathcal{R}|} \sum_{R_i \in \mathcal{R}} X_i(S) \tag{7}$$

We follow the method in [16] that uses the martingale theory for generating a sufficient number of samples to make a good approximation of the objective function. If the number of samples is at least $T \geq (2+\frac{2}{3}\epsilon)\frac{1}{\mu}\frac{1}{\epsilon^2}\ln(\frac{1}{\delta})$ for $\delta \in (0,1)$, $\hat{\sigma}_i(S)$ is an (ϵ, δ)-approximation of $\sigma(S)$, i.e.,

$$\Pr[(1-\epsilon)\sigma_i(S) \leq \hat{\sigma}_i(S) \leq (1+\epsilon)\sigma_i(S)] \geq 1 - \delta \tag{8}$$

We introduce an estimation of $\sigma(S)$

$$\hat{\sigma}(S) = \sum_{i \in [K]} \hat{\sigma}_i(S) \tag{9}$$

Algorithm 1: Generating a GBS for Group C_i

Input: Graph $G = (V, E)$, C_i, γ_i
Output: A GBS R_j

1. Pick a source node $u \in C_i$ with probability $\frac{s(u)}{\gamma_i}$
2. Initialize a queue $Q = \{u\}$ and $R_j = u$
3. **while** Q *is not empty* **do**
4. $v \leftarrow Q.pop()$
5.   **foreach** $x \in N_{in}(v)\backslash(R_j \cup Q)$ **do**
6.   With probability $p(x,v)$: $Q.push(x)$ and $R_j \leftarrow R_j \cup \{u\}$
7. **return** R_j

4.2 Main Algorithm

Among this subsection, we present our efficient algorithm for the TGI problem with input parameters, require a social network graph $G = (V, E)$, an influence threshold T, and a set of discrete groups C ($C_i \in C$, $C_i \subseteq V$), in which each C_i has a specific influence threshold t_i. The output of the algorithm is a set of seeds S provided that its influence estimate exceeds the influence threshold T so that resulting in the lowest total cost. We cover two search solutions in this algorithm.

Algorithm 2: RIS-based Heuristic Algorithm

Input: Graph $G = (V, E)$, set of group $\mathcal{C} = \{C_1, C_2, \ldots, C_K\}$, ϵ, δ
Output: A seed set S

1. $U \leftarrow V$
2. $S_i \leftarrow \emptyset, \forall i \in [K]$
3. $N_i = (2 + \frac{2}{3}\epsilon)|C_i|\frac{1}{\epsilon^2}\ln(\frac{1}{\delta})$
   /* Find the first candidate solution S_1 */
4. $S_1 \leftarrow \emptyset$
5. **for** $i = 1$ *to* K **do**
6.   Gennerate a set of $\mathcal{R}_i$ containning N_i GBSs for group C_i by using Algorithm 1
7.   **while** $\hat{\sigma}_i(S_i) < t_i$ **do**
8. $u \leftarrow \arg\max_{v \in U \backslash S_i} \frac{\hat{\sigma}_i(S_i \cup \{v\}) - \hat{\sigma}_i(S_i)}{c(v)}$
9. $S_i \leftarrow S_i \cup \{u\}$
10. $U \leftarrow U \backslash \{u\}$
11. $S_0 \leftarrow \bigcup S_i$
12. $S_1 = S_0$
13. **for** $i = 1$ *to* K **do**
14.   **while** $S_1 \neq \emptyset$ **do**
15. $u \leftarrow \arg\max_{v \in S_1} c(v)$
16.   **if** $\hat{\sigma}(S_1 \backslash \{u\}) \geq T$ **then**
17. $S_1 \leftarrow S_1 \backslash \{u\}$
18. $S_1 \leftarrow S_1 \backslash \{u\}$

   /* Find the second candidate solution S_2 */
19. $S_2 \leftarrow \emptyset$
20. **for** $i = 1$ *to* K **do**
21.   Gennerate a set of $\mathcal{R}_i$ containts N_i' GBSs R_i by using Algorithm 1
22. **while** $\hat{\sigma}(S_2) < T$ **do**
23. $u \leftarrow \arg\max_{v \in V \backslash S_2} \frac{\hat{\sigma}(S_2 \cup \{v\}) - \hat{\sigma}(S_2)}{c(v)}$
24. $S_2 \leftarrow S_2 \cup \{u\}$
25. $S \leftarrow \arg\min_{S' \in \{S_1, S_2\}} c(S')$
26. **return** S

In the first solution, we first search for the seed set S_i ($S_i \subseteq V$) that has the best effect on each group of C_i, so that the influence estimate of S_i exceeds t_i (as determined by the formula (7)), which is considered to have affected the entire C_i group, with the search condition mentioned in line 8 of the Algorithm 2. In this process, we use the *GBS* sample generation model (mentioned in Algorithm 1) to estimate the effect for each target group C_i, with the number of samples to be calculated as N_i (mentioned in line 3 Algorithm 2), we assign these parameters $\epsilon = 0.1$ and $\delta = 1/n$ as a default setting. After joining the sets S_i together, we consider the seed set S to perform non-optimal seed elimination. In here, we are searching for seeds that, if the influence estimate after eliminating them (by formula (9)) still exceeds the threshold T, should be discarded, with the highest cost seeds being preferred, in order to obtain the seed set with the lowest total cost. Besides, the second solution ensures IM search by the conventional method, allowing both solutions to be compared to find the best seed set. First, as solution 1, we generate GBS samples with the amount of N_i for the target groups C_i. Then, we perform a sequential search for seeds in the vertex set V of the G graph until we reach threshold T then stopped, so that the influence of this seed set is the largest, satisfying the search condition in line 23 of Algorithm 2. Finally, we compare the two solutions and choose the seed set with the lowest total cost.

5   Experiments

In this section, we conduct experiments to show the performance of the TGI algorithm and perform comparisons with other classic algorithms through metrics such as total cost, running time, and memory usage.

5.1   Experimental Settings

Datasets. In the experiment with the TGI algorithm, we experiment with 6 datasets for the problem of information propagation with different sizes. The datasets are described in Table 2 below. In which, the small and medium datasets calculated by the TGI algorithm are divided into several groups with a maximum of $K = 100$ groups (excluding discrete vertices), each group has from 1 to 10 vertices.

Table 2. Datasets

Dataset	# Nodes	# Edges	Avg.Deg	Directed	# K	Source
Email-Eu-core	1,005	20,777	3.3	Directed	30	[25]
Gnutella	6,301	20,777	3.3	Directed	100	[12]
Wiki-vote	6,301	20,777	3.3	Directed	100	[11]
Net-Hept	15,233	58,891	5.5	Undirected	100	[6]
Net-Phy	37,154	231,584	13.4	Undirected	100	[4]
Email-Enron	36,692	183,831	5.0	Undirected	100	[13]

Algorithms Compared. In this experiment, we use the High Degree algorithm, a basic algorithm often used in comparing problems related to information propagation. The DEGREE algorithm selects the vertices with the highest degree as the seed and propagates until the influence reaches the threshold then stops, where the Degree algorithm uses a Monte Carlo information propagation simulation to estimate the effect.

Parameters Setting. All experiments are propagated under the IC information propagation model with the given edge probability $p(u) = p(u, v) = 1/|N_{in}(v)|$.

In the formula to calculate the number of GBS samples (line number 3 in Algorithm 2), the parameters $\epsilon = 0.1$ and $\delta = 1/n$ are the default assignment values. And the Degree algorithm with the number of Monte Carlo iterations $t = 100,000$ times. The entire generating of GBS live-edge samples and the Monte Carlo propagation simulation iteration is time-optimized up to 4 threads.

For the TGI algorithm that aims to propagate the effect across groups, $\forall u \in C_i$ the score of u is assigned $s(u) = 1$, and the threshold of group C_i $t_i = \sum_{u \in C_i} s(u)/2$ for $i = 1 \ldots K$ Each node has its cost $c(u) = 1/|N_{out}(v)|$ with the support $(0, 1]$ and its benefit assigned $b(u) = 1$ if has entry degree, and vice versa.

We experimented all dataset with a Linux machine with a 2 x Intel(R) Xeon(R) CPU E5-2630 v4 @2.20 GHz, 64 GB RAM DDR4 @ 2400 MHz. The TGI algorithm and the High Degree algorithm are written in Python 3 language.

5.2   Experiment Results

In the experimental comparison between algorithms, we give the advantages and disadvantages of the TGI algorithm.

Comparision of the Total Cost for Seed Set. Regarding the seed set cost comparison, the datasets are compared in (Fig. 1), we evaluate the High Degree algorithm better in most of the experiments. The total cost at High Degree thresholds is more optimal as the thresholds are larger. To explain this reason, we notice that there is an inconsistency between the given number of K groups and the size of the dataset, as well as the magnitude of the cluster peaks belonging to the group. On this point, we will improve in future experiments with more stable group quality.

Comparision of Running Time. Regarding the running time comparison (Fig. 2), the running time of TGI will be better than the High Degree algorithm if the threshold T is predicted in the range that best fits the dataset. In general, the High Degree algorithm is a simple heuristic algorithm, using Monte Carlo simulation to propagate information, the algorithm will stop when the benefit estimate exceeds the threshold T, so when comparing the time running, High Degree really dominated. We have ensured that the Monte Carlo computation time of High Degree is roughly equivalent to the generation time of the proposed GBS samples with an iteration of 1000 times. However, the TGI algorithm uses the method of removing unnecessary seeds to ensure that the propagation estimate is closest to the threshold T, in order to obtain the lowest cost. Since then, the propagation estimation value of TGI not only stops when it exceeds T but also considers the search and elimination time, the time that takes up almost the majority of the time in the experiment with the TGI algorithm. We observe that this is inevitable when traversing the thresholds T to find the best seed set.

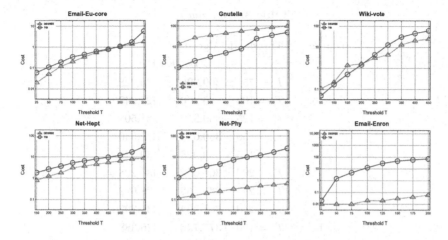

Fig. 1. Total cost compared of solutions

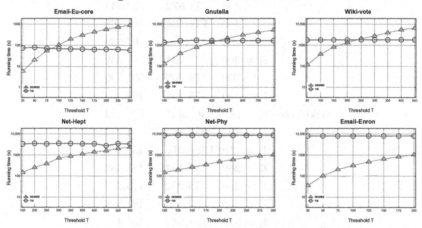

Fig. 2. Running time of algorithms

Comparision of Memory Usage. In terms of memory usage comparison, when comparing the two algorithms, the memory usage for the TGI algorithm is higher than High Degree but not significantly (Table 3). However, both algorithms have a rather high cost of memory usage.

Table 3. Memory usage of compared algorithms (MB)

Dataset	Threshold	Algorithm	
		DEGREE	TGI
Email Eu-core	$T_i = 25$	359,940	430,740
	$T_i = 50$	360,040	435,748

(*continued*)

Table 3. (*continued*)

Dataset	Threshold	Algorithm	
		DEGREE	TGI
	$T_i = 75$	360,296	430,228
	$T_i = 100$	360,296	428,948
	$T_i = 125$	360,312	431,508
	$T_i = 150$	360,632	437,540
	$T_i = 175$	360,924	430,228
	$T_i = 200$	361,032	430,228
	$T_i = 225$	360,568	428,692
	$T_i = 250$	361,364	434,292
Gnutella	$T_i = 100$	371,736	476,388
	$T_i = 200$	371,172	487,524
	$T_i = 300$	371,572	494,948
	$T_i = 400$	371,784	487,268
	$T_i = 500$	372,400	507,620
	$T_i = 600$	372,904	487,332
	$T_i = 700$	374,668	507,940
	$T_i = 800$	375,164	469,348
Wiki-Vote	$T_i = 50$	387,644	470,492
	$T_i = 100$	388,884	464,484
	$T_i = 150$	389,016	456,416
	$T_i = 200$	389,104	485,468
	$T_i = 250$	389,292	475,776
	$T_i = 300$	389,400	487,928
	$T_i = 350$	388,644	456,424
	$T_i = 400$	388,772	480,944
	$T_i = 450$	390,248	461,668
Net-Hept	$T_i = 150$	387,632	488,976
	$T_i = 200$	396,464	486,160
	$T_i = 250$	394,032	476,380
	$T_i = 300$	398,480	488,272
	$T_i = 350$	398,540	486,160
	$T_i = 400$	398,920	453,192
	$T_i = 450$	398,924	459,920
	$T_i = 500$	398,968	457,872
	$T_i = 550$	394,596	484,112
	$T_i = 600$	399,320	473,052
Net-Phy	$T_i = 100$	455,364	545,824
	$T_i = 125$	460,308	556,680

(*continued*)

Table 3. (*continued*)

Dataset	Threshold	Algorithm	
		DEGREE	TGI
	$T_i = 150$	463,116	556,000
	$T_i = 175$	467,540	543,264
	$T_i = 200$	469,600	544,032
	$T_i = 225$	472,404	553,444
	$T_i = 250$	475,424	549,412
	$T_i = 275$	479,536	528,048
	$T_i = 300$	481,404	528,576
Email-Enron	$T_i = 25$	454,092	558,012
	$T_i = 50$	445,024	572,864
	$T_i = 75$	459,980	533,628
	$T_i = 100$	465,308	539,912
	$T_i = 125$	458,180	571,372
	$T_i = 150$	469,012	547,356
	$T_i = 175$	463,532	573,584
	$T_i = 200$	476,108	543,588

6 Conclusion

In this paper, we introduced the TGI problem in Online Social Networks, which is a variant of the IM problem with a cost constraint arising from user influences in viral marketing. The goal of the TGI problem is to find a set of seeds with an effect on the greatest number of groups in the OSNs that is greater than the threshold T. The experimental results demonstrate the effectiveness of our algorithm in terms of cost and usage time when compared to other methods.

References

1. Borgs, C., Brautbar, M., Chayes, J.T., Lucier, B.: Maximizing social influence in nearly optimal time. In: Proceedings of the Twenty-Fifth Annual ACM-SIAM Symposium on Discrete Algorithms, SODA 2014, Portland, Oregon, USA, 5–7 January 2014, pp. 946–957 (2014)
2. Beni, H.A., Bouyer, A.: TI-SC: top-k influential nodes selection based on community detection and scoring criteria in social networks. J. Ambient Intell. Humaniz. Comput. **11**, 4889–4908 (2020)
3. Bozorgi, A., Haghighi, H., Zahedi, M.S., Rezvani, M.: INCIM: a community-based algorithm for influence maximization problem under the linear threshold model. Inf. Process. Manag. **52**, 1188–1199 (2016)
4. Chen, W., Lakshmanan, L.V.S., Castillo, C.: Information and Influence Propagation in Social Networks. Synthesis Lectures on Data Management. Morgan & Claypool Publishers (2013)

5. Chen, W., Wang, C., Wang, Y.: Scalable influence maximization for prevalent viral marketing in large-scale social networks. In: Proceedings of the ACM SIGKDD International Conference on Knowledge Discovery and Data Mining, New York, NY, pp. 1029–1038 (2010)
6. Chen, W., Wang, Y., Yang, S.: Efficient influence maximization in social networks. In: KDD 2009: Proceedings of the 15th ACM SIGKDD International Conference on Knowledge Discovery and Data Mining, pp. 199–208 (2009)
7. Dai, Y., Jiang, W., Li, K.: Group-based competitive influence maximization. In: 2018 IEEE SmartWorld, Ubiquitous Intelligence & Computing, Advanced & Trusted Computing, Scalable Computing & Communications, Cloud & Big Data Computing, Internet of People and Smart City Innovation (SmartWorld/SCALCOM/UIC/ATC/CBDCom/IOP/SCI), pp. 199–208 (2009)
8. Domingos, P.M., Richardson, M.: Mining the network value of customers. In: Proceedings of the Seventh ACM SIGKDD International Conference on Knowledge Discovery and Data Mining, San Francisco, CA, USA, 26–29 August 2001, pp. 57–66 (2001)
9. Kempe, D., Kleinberg, J.M., Tardos, É.: Maximizing the spread of influence through a social network. In: Proceedings of the Ninth ACM SIGKDD International Conference on Knowledge Discovery and Data Mining, Washington, DC, USA, 24–27 August 2003, pp. 137–146 (2003)
10. Aghaee, Z., Kianian, S.: Influence maximization algorithm based on reducing search space in the social networks. SN Appl. Sci. 2(2067) (2020)
11. Leskovec, J., Huttenlocher, D.P., Kleinberg, J.M.: Signed networks in social media. In: Proceedings of the 28th International Conference on Human Factors in Computing Systems, CHI 2010, Atlanta, Georgia, USA, 10–15 April 2010, pp. 1361–1370 (2010)
12. Leskovec, J., Kleinberg, J.M., Faloutsos, C.: Graph evolution: densification and shrinking diameters. TKDD 1(1), 2 (2007)
13. Leskovec, J., Lang, K.J., Dasgupta, A., Mahoney, M.W.: Community structure in large networks: natural cluster sizes and the absence of large well-defined clusters. Internet Math. 6(1), 29–123 (2009)
14. Li, Y., Zhang, D., Tan, K.-L.: Targeted influence maximization for online advertisements. PVLDB 8(10), 1070–1081 (2015)
15. Liu, B., Cong, G., Xu, D., Zeng, Y.: Time constrained influence maximization in social networks. In: 12th IEEE International Conference on Data Mining, ICDM 2012, Brussels, Belgium, 10–13 December 2012, pp. 439–448 (2012)
16. Nguyen, H.T., Thai, M.T., Dinh, T.N.: Stop-and-stare: optimal sampling algorithms for viral marketing in billion-scale networks. In: Proceedings of the 2016 International Conference on Management of Data, SIGMOD Conference 2016, San Francisco, CA, USA, 26 June–01 July 2016, pp. 695–710 (2016)
17. Nguyen, H.T., Thai, M.T., Dinh, T.N.: A billion-scale approximation algorithm for maximizing benefit in viral marketing. IEEE/ACM Trans. Netw. 25(4), 2419–2429 (2017)
18. Nguyen, L.N., Zhou, K., Thai, M.T.: Influence maximization at community level: a new challenge with non-submodularity. In: 39th IEEE International Conference on Distributed Computing Systems, ICDCS 2019, Dallas, TX, USA, 7–10 July 2019, pp. 327–337 (2019)
19. Pham, C.V., Duong, H.V., Thai, M.T.: Importance sample-based approximation algorithm for cost-aware targeted viral marketing. In: Proceedings of the Computational Data and Social Networks - 8th International Conference, CSoNet 2019, Ho Chi Minh City, Vietnam, 18–20 November 2019, pp. 120–132 (2019)

20. Pham, C.V., Ha, D.K.T., Vu, Q.C., Su, A.N., Hoang, H.X.: Influence maximization with priority in online social networks. Algorithms **13**, 183 (2020)

21. Tang, J., Tang, X., Yuan, J.: Profit maximization for viral marketing in online social networks: algorithms and analysis (2018)

22. Tang, Y., Shi, Y., Xiao, X.: Influence maximization in near-linear time: a martingale approach. In: Proceedings of the 2015 ACM SIGMOD International Conference on Management of Data, Melbourne, Victoria, Australia, 31 May–4 June 2015, pp. 1539–1554 (2015)

23. Tang, Y., Xiao, X., Shi, Y.: Influence maximization: near-optimal time complexity meets practical efficiency. In: International Conference on Management of Data, SIGMOD 2014, Snowbird, UT, USA, 22–27 June 2014, pp. 75–86 (2014)

24. Zhao, Y., Zhou, X., Zheng, K., Chen, X., Deng, L.: Community-based influence maximization in location-based social network. World Wide Web (2021)

25. Yin, H., Benson, A.R., Leskovec, J., Gleich, D.F.: Local higher-order graph clustering. In: Proceedings of the 23rd ACM SIGKDD International Conference on Knowledge Discovery and Data Mining, Halifax, NS, Canada, 13–17 August 2017, pp. 555–564 (2017)

26. Zhong, Y., Guo, L., Huang, P.: Maximizing group coverage in social networks. In: Zhang, Y., Xu, Y., Tian, H. (eds.) PDCAT 2020. LNCS, vol. 12606, pp. 274–284. Springer, Cham (2021). https://doi.org/10.1007/978-3-030-69244-5_24

27. Zhu, J., Ghosh, S., Weili, W.: Group influence maximization problem in social networks. IEEE Trans. Comput. Soc. Syst. **6**(6), 1156–1164 (2019)

28. Zhu, J., Ghosh, S., Wu, W., Gao, C.: Profit maximization under group influence model in social networks. In: Tagarelli, A., Tong, H. (eds.) CSoNet 2019. LNCS, vol. 11917, pp. 108–119. Springer, Cham (2019). https://doi.org/10.1007/978-3-030-34980-6_13

Application Based Cigarette Detection on Social Media Platforms Using Machine Learning Algorithms

Muhammad Umer Hashmi[1]([✉]), Ngoc Duy Nguyen[1], Michael Johnstone[1], Kathryn Backholer[2], and Asim Bhatti[1]

[1] Institute for Intelligent Systems Research and Innovation, Deakin University, Waurn Ponds Campus, Burwood, VIC, Australia
{uhashmi,n.nguyen,michael.johnstone,asim.bhatti}@deakin.edu.au
[2] Institute for Health Transformation, Deakin University, Waurn Ponds Campus, Burwood, VIC, Australia
kathryn.backholer@deakin.edu.au

Abstract. Cigarette and e-cigarette advertisements often portray positive images of smoking behaviour, especially amongst younger generations. It portrays a lifestyle in which smoking cigarettes or e-cigarettes are normal and an important part of human lives. Images of cigarette smoking on social media platforms have played an influential role in encouraging people to smoke. There is a growing need of advanced mathematical models and machine learning techniques to monitor the portrayal of cigarette and e-cigarette use on social media platforms, as well as other harmful products to human health. In this study, we have annotated a set of 1,333 smoking images collected from a wide array of communication media. In addition, we evaluated three state-of-the-art segmentation algorithms including Mask R-CNN, Cascade Mask-R-CNN and Hybrid Task Cascade (HTC) by using the MMDetection framework to detect smoking images within our annotated dataset. The study plays an important role towards developing a practical monitoring system, which can inform policy actions to restrict unhealthy advertisements on social media and other related platforms. Finally, our evaluation results show that Mask R-CNN outperforms Cascade Mask RCNN and HTC in terms of Average Precision and Average Recall.

Keywords: Machine learning · Deep learning · Computer vision · Advertising · Cigarette · Unhealthy ads · Data collection · Data annotation · Data analytics · Mask R-CNN · Cascade Mask R-CNN · Hybrid task cascade · Resnet-50

1 Introduction

The study aims to investigate machine learning based methods to detect cigarette (including e-cigarette) images that appear in social media and other related

T. K. Dang et al. (Eds.): FDSE 2021, LNCS 13076, pp. 68–80, 2021.
https://doi.org/10.1007/978-3-030-91387-8_5

platforms. These platforms have become a key avenue through which youth are exposed to cigarette advertising as they are largely unregulated and provide access to a large number of people though minimal promotional effort [1]. Whilst there is very little overt advertising of these products due to strong tobacco control policies, covert and unregulated promotion of cigarettes is highly prevalent [2]. Evidence shows that this type of advertising promotes positive sentiment among youth towards smoking and increases the likelihood of smoking initiation [3]. For example, these advertisements have shown to create a mindset that smoking cigarettes is normal and is considered as "cool stuff" [3].

Advertising through Facebook, Instagram, online gaming, and other social media platforms have been found out to be the most effective way of encouraging people to smoke. Surveys show that youth in the United States of America and United Kingdom may be particularly vulnerable [4]. These platforms have enabled smart marketing methodologies to encourage the young generation to smoke. Due to omni presence of these advertisements and market manipulation, people are exposed to cigarettes, even they do not want them [5]. Cigarettes have been culturally accepted everywhere in the world [6].

Because of the complexity of the ever-changing and complex online ecosystem, there is growing need for advanced mathematical models and machine learning techniques to closely monitor the excessive use of tobacco and e-cigarette. The data is easily accessible for researchers through smartphones, electronic medical records, public health data and industrial databases. With machine learning algorithms, we can create an adaptive model that can predict and analyse these products that are harmful to health [7]. In this study, we annotated a set of 1,333 smoking images collected from a variety of communication media. Subsequently, we used three state-of-the-art algorithms to conduct an evaluation analysis on our annotated dataset:

- Mask R-CNN [8].
- Cascade Mask R-CNN [9].
- Hybrid Task Cascade for Instance Segmentation (HTC) [10].

The paper is organised as follows. Section 2 highlights the related studies in the literature. Section 3 presents the background knowledge and our methodology. Section 4 shows the evaluation results of three state-of-the-art baselines on our annotated dataset. Finally, Sect. 5 concludes the paper.

2 Related Work

A study on tobacco consumption was first mentioned in [11], which is based on the collection and manual coding of video clips, recorded from Amazon Prime and Netflix, for the presence of tobacco and alcohol imagery. Like most studies examining unhealthy advertisements, this study relied on manual identification and content coding of images and other broadcast statistics [12]. The authors in [13] have collected data on children's exposure to alcohol marketing by using wearable cameras attached to children, which took a photo every 7 s across the day. The authors reported that the analysis of the 700,000 images collected took more than three months to complete.

On the other hand, the authors in [14–18] made a use of machine learning to effectively identify and classify inherent patterns in images that are collected from a wide array of communication media. Deep learning methods have shown tremendous results to analyse the exposure of social media images [19,20]. In this study, we also use machine learning methods, *i.e.* instance segmentation algorithms, to detect cigarettes and smoking images in social advertisements.

The authors in [21] made a database of beer, wine, champagne, and some non-alcoholic beverages' images from the Google's search engine. Non-alcoholic images (*e.g.*, water and coke) were added to make sure that the algorithm is capable of recognizing alcohol and non-alcohol images. Subsequently, the authors trained two deep learning algorithms based on Resnet and DenseNet [22–24] to conduct an evaluation analysis on their datasets.

3   Methodology

3.1   Dataset

We used a set of 1,333 smoking images collected from a wide array of communication media and Google's search engine to create a distinctive dataset of cigarette images. Together with the "cigarette" keyword, the following keywords are used during the search: advertisements, partying, movies, outdoors, cheerful, depressed, sad and celebrating. The dataset is divided into three subsets: training (55%), validation (15%), and testing dataset (30%). Finally, we used Hasty [25] to annotate the collected images and export them in COCO format [26].

3.2   MMDetection

In this paper, we have used MMDetection toolbox [27] as a benchmark to evaluate three state-of-the-art algorithms. Many studies use the framework to develop new algorithm and/or evaluate and compare it with existing algorithms [28–30]. These detectors may have different architectures but share common components including a backbone, a neck, an anchor head/anchor free head and RoI extractor. There are two type of detectors: single-stage and two-stage detectors as shown in Fig. 1. The single-stage detector such as "You Only Look Once" [31] has a faster inference time than two-stage detector, *e.g.* Faster-R-CNN, Mask R-CNN, Cascade Mask R-CNN. However, two-stage detectors outperform single-stage detectors with respect to Average Precision [32].

1. **Backbone:** This component is used to transform the input images into raw feature maps, *e.g.*, ResNet-50 excluding the last layer [33].
2. **Neck:** This component connects the "heads" and "backbone" by performing some reconfigurations/refinements on the feature maps transformed by backbone. Feature Pyramid Network is an example of this component.

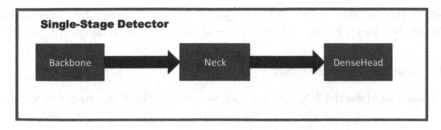

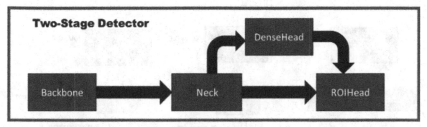

Fig. 1. One-stage and two-stage detector.

3. **Dense Head:** This component only processes on dense locations fed by Neck. These dense locations are mainly RetinaHead, Anchor Head, RPN-Head, FCOSHead and Anchor Free Head.
4. **ROI Extractor:** This component identifies the (ROI) Region of Interest and extracts its features from backbone or feature maps. The single ROIExtractor is the corresponding level of the Feature Pyramid.
5. **ROI Head:** The ROIHead takes feature maps as input and predicts the outcome based on that, *e.g.*, mask prediction or bounding boxes classification as per assigned task.

3.3   Mask R-CNN

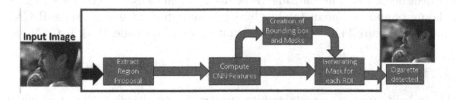

Fig. 2. The architecture of Mask R-CNN.

Mask R-CNN is deep learning algorithm, which aims to resolve object detection and instance segmentation problems. It separately identifies different objects in an image or video by creating bounding box, masks and classes. Firstly, it predicts the regions where there is an object in an image or video; and secondly predicts the tag/class of the object/objects; and finally generates a pixel-level mask of the object. Figure 2 shows an overall architecture of Mask R-CNN [8].

3.4   Cascade Mask R-CNN

Compared to Mask R-CNN, the cascade version includes three stages as follows:

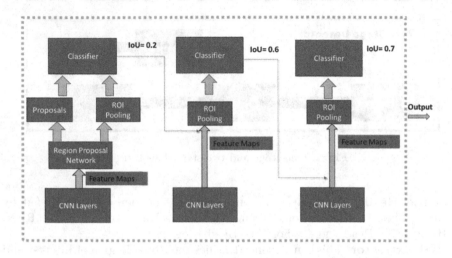

Fig. 3. Cascade Mark R-CNN.

- A prediction/mask head to initial stage of cascade R-CNN.
- A prediction/mask head to final stage of cascade R-CNN.
- A segmentation branch on the remaining stages of cascade framework.

Such stages are combined to predict the masks generated by the last stage of Object detection. The Cascade Mask R-CNN enhances the Average Precision of the mask and box predictability by some margin compared to Mask R-CNN framework. Figure 3 shows the basic architecture of Cascade Mask R-CNN [9].

3.5   HTC

HTC is built upon a cascade architecture with a multi-stage detector, which incorporates mask prediction and bounding box regression. It interleaves a pattern to boost the information flow amongst the branches of the mask by transferring the current mask features to the corresponding one. HTC also provides a separate semantic segmentation branch and merges it with the mask and bounding box. As a result, it significantly improves the information flow from the first stage to the last stage. Figure 4 shows the architecture of HTC [10].

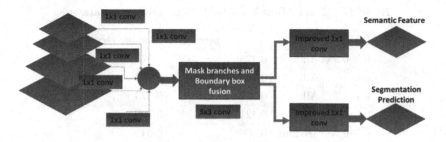

Fig. 4. The architecture of HTC.

4   Performance Evaluation

By using MMDetection framework, we applied the default configurations for the three algorithms with the following changes:

- Images were resized to 512 × 512 pixels.
- The number of training epochs was set to 12.
- The tag of "cigarette" was made and labelled on all images.

To reduce the training time, the network was initialized with a pre-trained model (COCO dataset) and subsequently was trained on our dataset. Our dataset is also divided into training, validation, and testing subset. We use the following metrics to conduct an evaluation analysis:

- Intersection over Union (IoU): measures the overlapped area between prediction bounding box with its ground truth. The value of 0.5 presents an average performance.
- Average Precision (AP): measures the accuracy of the algorithm among all detected objects.
- Average Recall (AR): measures the capability of detecting objects.
- Precision-Recall curve: provides a performance overview of the algorithm in both evaluation metrics, *i.e.*, AP and AR.

4.1   Mask R-CNN

Table 1 shows the validation performance of Mask R-CNN in terms of AP and AR. At IoU = 0.5, the AP achieves an average score of 79.8%. Similarly, Table 2 presents the testing performance of Mask R-CNN.

Table 1. AP and AR of Mask R-CNN on validation dataset.

IoU	Area	Average precision	Average recall
0.5–0.95	All	0.478	0.573
0.5	All	0.798	0.573
0.75	All	0.532	0.573
0.5–0.95	Small	0.421	0.525
0.5–0.95	Medium	0.576	0.64
0.5–0.95	Large	0.648	0.67

Table 2. AP and AR of Mask R-CNN on testing dataset.

IoU	Area	Average precision	Average recall
0.5–0.95	All	0.348	0.471
0.5	All	0.558	0.471
0.75	All	0.38	0.471
0.5–0.95	Small	0.339	0.452
0.5–0.95	Medium	0.449	0.627
0.5–0.95	Large	0	0

Figure 5 and Fig. 6 depict the Precision-Recall curve of Mask R-CNN on both the validation and testing datasets, respectively. Mask R-CNN achieves a well-performed result especially when IoU = 0.5. However, the performance of Mask R-CNN reduces significantly with the testing dataset because we deliberately select challenging smoking images where the area of the cigarette is small.

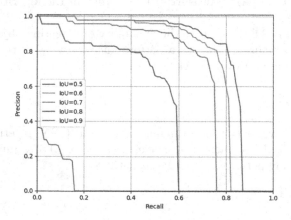

Fig. 5. Precision-Recall curve of Mask R-CNN on validation dataset.

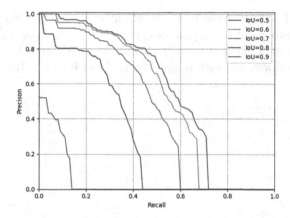

Fig. 6. Precision-Recall curve of Mask R-CNN on testing dataset.

4.2 Cascade Mask R-CNN

Table 3 presents the performance of Cascade Mask R-CNN on validation dataset. The performance is slightly lower than Mask R-CNN in terms of AP and AR. Table 4 shows the performance of Cascade Mask R-CNN on testing dataset.

Table 3. AP and AR of Cascade Mask R-CNN on validation dataset.

IoU	Area	Average precision	Average recall
0.5–0.95	All	0.469	0.543
0.5	All	0.793	0.543
0.75	All	0.516	0.543
0.5–0.95	Small	0.404	0.486
0.5–0.95	Medium	0.578	0.623
0.5–0.95	Large	0.609	0.65

Table 4. AP and AR of Cascade Mask R-CNN on testing dataset.

IoU	Area	Average precision	Average recall
0.5–0.95	All	0.343	0.422
0.5	All	0.523	0.422
0.75	All	0.386	0.422
0.5–0.95	Small	0.331	0.4
0.5–0.95	Medium	0.467	0.585
0.5–0.95	Large	0	0

Figure 7 and Fig. 8 show the Precision-Recall curve of Cascade Mask R-CNN on validation dataset and testing dataset, respectively. Although Cascade Mask R-CNN is worse than Mask R-CNN with respect to AP and AR. It achieves a slightly better performance with a high threshold of IoU, e.g. IoU = 0.9.

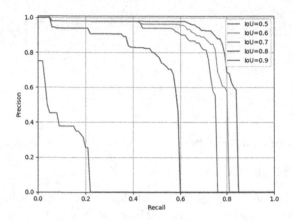

Fig. 7. Precision-Recall curve of Cascade Mask R-CNN on validation dataset.

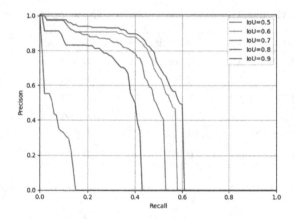

Fig. 8. Precision-Recall curve of Cascade Mask R-CNN on testing dataset.

4.3   HTC

Table 5 presents the AP and AR of HTC with different values of IoU and the area of smoking objects on validation dataset and testing dataset, respectively. HTC is worse than Mask R-CNN and Cascade Mask R-CNN in terms of precision and recall with different values of IoU. However, if we consider the area of smoking objects, HTC seems to perform well on large objects. Because cigarettes are small objects, it is reasonable that HTC cannot outperform Mask R-CNN and

Table 5. AP and AR of HTC on validation dataset.

IoU	Area	Average precision	Average recall
0.5–0.95	All	0.41	0.519
0.5	All	0.775	0.519
0.75	All	0.399	0.519
0.5–0.95	Small	0.334	0.453
0.5–0.95	Medium	0.535	0.61
0.5–0.95	Large	0.597	0.66

Table 6. AP and AR of HTC on testing dataset.

IoU	Area	Average precision	Average recall
0.5–0.95	All	0.243	0.373
0.5	All	0.381	0.373
0.75	All	0.279	0.373
0.5–0.95	Small	0.23	0.335
0.5–0.95	Medium	0.371	0.644
0.5–0.95	Large	0	0

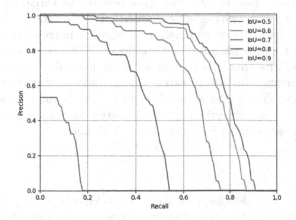

Fig. 9. Precision-Recall curve of HTC on validation dataset.

Cascade Mask R-CNN in overall. Similarly, Table 6 shows the performance of HTC on testing dataset.

Figure 9 and Fig. 10 show the Precision-Recall curve of HTC on validation dataset and testing dataset, respectively. Compared to Mask RCNN and it cascade version, HTC is a weaker detector with respect to both evaluation metrics although it performs well with large objects.

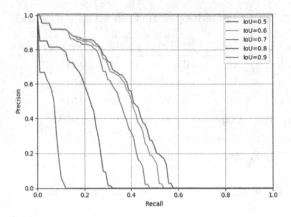

Fig. 10. Precision-Recall curve of HTC on testing dataset.

5   Conclusions

In this paper, we collected and annotated a set of smoking images from a variety of social sources. Subsequently, we used state-of-the-art instance segmentation algorithms to evaluate the capability of detecting smoking images. Evaluation analysis shows that Mask R-CNN performs best of the three algorithms when trained to our dataset, which can achieve an AP of 80% on the validation dataset. The study can inform the development of a broader range of real-world applications for detection of unhealthy advertisements. This is important to inform public health policies that can protect children from the harmful impacts of these advertisements. Our future extensions include an implementation of an algorithm that can detect inferred smoking behaviour in more detail. For example, several ads do not directly include cigarettes but instead use slogans or gesture to advertise smoking in a metaphorical manner.

References

1. Jenssen, B., Klein, J., Salazar, L., Daluga, N., Diclemente, R.: Exposure to tobacco on the internet: content analysis of adolescentsínternet use. Pediatrics **124**(2), e180-6 (2009)
2. Richardson, A., Ganz, O., Vallone, D.: Tobacco on the web: Surveillance and characterisation of online tobacco and e-cigarette advertising. Tob. Control **24**(4), 341–7 (2015)
3. Cavazos-Rehg, P.A., Krauss, M.J., Spitznagel, E.L., Grucza, R.A., Bierut, L.J.: The hazards of new media: youth's exposure to tobacco ads/promotions. Nicotine Tobacco Res. **16**, 437–44 (2014)
4. Anderson, P., de Bruijn, A., Hastings, G., Angus, K., Gordon, R.: Impact of alcohol advertising and media exposure on adolescent alcohol use: a systematic review of longitudinal studies. Alcohol Alcohol. **44**, 229–243 (2018)
5. Griffiths, R., Casswell, S.: Intoxigenic digital spaces? Youth, social networking sites and alcohol marketing. Drug Alcohol Rev. **29**, 525–530 (2010)

6. Roche, A., Bywood, P., Freeman, T., Pidd, K., Borlagdan, J., Trifonoff, A.: The social context of alcohol use in Australia. National Centre for Education and Training on Addiction, Adelaide, Australia (2009)
7. Figueroa, R.L., Flores, C.A.: Extracting information from electronic medical records to identify the obesity status of a patient based on comorbidities and bodyweight measures. J. Med. Syst. **40**(8), 191 (2016)
8. Gkioxari, G., He, K., Dollar, P., Girshick, R.: Mask R-CNN: Facebook AI research (FAIR). arXiv:1703.06870v3, vol. 3 (2018)
9. Cai, Z., Vasconcelos, N.: Delving into high quality object detection. arXiv:1712.00726v1, vol. 1 (2017)
10. Chen, K., et al.: Hybrid task cascade for instance segmentation. arXiv:1901.07518v2, vol. 2 (2019)
11. Barker, A.B., Smith, J., Hunter, A., Britton, J., Murray, R.L.: Quantifying tobacco and alcohol imagery in Netflix and Amazon Prime instant video original programming accessed from the UK. BMJ Open **9**, e025807 (2019)
12. Pinsky, I., Jundi, S.A.R.J.E., Sanches, M., Zaleski, M.J.B., Laranjeira, R.R., Caetano, R.: Exposure of adolescents and young adults to alcohol advertising in Brazil. Public Aff. **10**, 50–58 (2010)
13. Chambers, T., et al.: Quantifying the nature and extent of children's real-time exposure to alcohol marketing in their everyday lives using wearable cameras: children's exposure via a range of media in a range of key places. Alcohol Alcohol. **53**, 626–633 (2018)
14. Le, Y., Liu, J., Deng, C., Dai, D.Y.: Heart rate variability reflects the effects of emotional design principle on mental effort in multimedia learning. Comput. Hum. Behav. **89**, 40–47 (2018)
15. Ahmed, A.A.A., Donepudi, P.K., Choi, M.S.: Detecting fake news using machine learning: a systematic literature review (2020)
16. Dewey, C.: Facebook has repeatedly trended fake news since firing its human editors. In: The Washington Post (2016)
17. Nguyen, N.D., Nguyen, T.T., Creighton, D., Nahavandi, S.: A visual communication map for multi-agent deep reinforcement learning. arXiv preprint arXiv:2002.11882 (2020)
18. Nguyen, T.T., et al.: Genomic mutations and changes in protein secondary structure and solvent accessibility of SARS-CoV-2 (COVID-19 virus). Sci. Rep. **11**(1), 1–16 (2021)
19. Russakovsky, O., et al.: ImageNet large scale visual recognition challenge. Int. J. Comput. Vis. **115**, 211–252 (2015)
20. Grauman, K., Leibe, B.: Visual object recognition. CA, USA (2011)
21. Bonela, A.A., Kuntsche, E., Caluzzi, G., Miller, M., He, Z.: How much are we exposed to alcohol in electronic media? Development of the alcoholic beverage identification deep learning algorithm (ABIDLA). Drug Alcohol Dependence **208**, 107841 (2020)
22. He, K., Zhang, X., Ren, S., Sun, J.: Deep residual learning for image recognition. In: Proceedings of the IEEE Conference on Computer Vision and Pattern Recognition, Las Vegas, USA, pp. 77–778 (2016)
23. Huang, G., Liu, Z., Van Der Maaten, L., Weinberger, K.Q.: Densely connected convolutional networks. In: Proceedings of the 30th IEEE Conference on Computer Vision and Pattern Recognition, Honolulu, USA, pp. 2261–2269 (2017)
24. Kingma, D.P., Ba, J.: Adam: a method for stochastic optimization. In: Proceedings of the 3rd International Conference for Learning Representations, San Diego, USA, pp. 1–15 (2015)

25. (2021). https://hasty.ai/
26. Rostianingsih, S., Setiawan, A., Halim, C.I.: COCO (creating common object in context) dataset for chemistry apparatus. Proc. Comput. Sci. **171**, 2445–2452 (2020)
27. Chen, K., et al.: MMDetection: open MMLab detection toolbox and benchmark. arXiv preprint arXiv:1906.07155 (2019)
28. Liu, W., et al.: SSD: single shot multibox detector. In: Leibe, B., Matas, J., Sebe, N., Welling, M. (eds.) ECCV 2016. LNCS, vol. 9905, pp. 21–37. Springer, Cham (2016). https://doi.org/10.1007/978-3-319-46448-0_2
29. Goyal, P., Lin, T.-Y., Girshick, R., He, K., Dollar, P.: Focal loss for dense object detection. Presented at the IEEE International Conference on Computer Vision (2017)
30. Girshick, R., He, K., Dollar, P.: Rethinking imagenet pre-training. arXiv preprint arXiv:1811.08883 (2018)
31. Lakshmanamoorthy, R.: Guide to MMDetection: an object detection Python toolbox. https://analyticsindiamag.com/guide-to-mmdetection-an-object-detection-python-toolbox/
32. Carranza-García, M., Torres-Mateo, J., Lara-Benítez, P., García-Gutiérrez, J.: On the performance of one-stage and two-stage object detectors in autonomous vehicles using camera data. Remote Sens. **13**(1), 89 (2020)
33. Zhang, X., He, K., Ren, S., Sun, J.: Deep residual learning for image recognition. In: Proceedings of the IEEE Conference on Computer Vision and Pattern Recognition, pp. 770–778 (2016)

Efficient Brain Hemorrhage Detection on 3D CT Scans with Deep Neural Network

Anh-Cang Phan[1]($\boxtimes$), Ho-Dat Tran[1], and Thuong-Cang Phan[2]

[1] VinhLong University of Technology Education, Vĩnh Long, Vietnam
{cangpa,datth}@vlute.edu.vn
[2] CanTho University, Can Tho, Vietnam
ptcang@cit.ctu.edu.vn

Abstract. A brain hemorrhage is a type of stroke that can cause brain damage and can be life-threatening. The outcome of a brain bleed depends on its location inside the skull, the size of the bleed, and the duration between bleeding and treatment. Brain damage can be severe and results in physical and mental disability. Therefore, saving the lives of such patients completely depends on detecting the correct location of the hemorrhage in an early stage. U-Net is an architecture developed for fast and precise segmentation of biomedical images. Its success in medical image segmentation has been attracting much attention from researchers. In this paper, we propose a novel method for automatic brain hemorrhage detection on 3D CT images using U-Net with a transfer learning approach. The 3D CT images are preprocessed by slicing NIfTI files to 2D, splitting, filtering, and normalization to create input data for our model. We refine and pre-train the U-Net model to detect brain hemorrhage regions on the CT scans. Our proposed method is evaluated on a set of 3D CT-scan images and obtains an accuracy of 92.5%.

Keywords: Brain hemorrhage · 3D CT scans · Deep neural networks · U-Net

1 Introduction

Technology is being implemented in all possible aspects of daily life especially its use in healthcare. Technology trends in healthcare indicated that medical image segmentation and classification are completely potential. The demand for less time-consuming tasks in hospitals requires the automation of initiative diagnosis process, where artificial intelligence applications and appliances could support. Intracranial hemorrhage (ICH) is a type of stroke that occurs when blood vessels in the brain suddenly rupture and blood flows through the part of the brain. The Intracranial hemorrhage usually appears suddenly and deteriorates after the sign of intracranial hemorrhage initiates. MRI-contraindicated patients and its rapid acquisition time make CT scans a preferred diagnostic tool over Magnetic Resonance Imaging for the initial assessment of ICH. CT scans generate a sequence of

© Springer Nature Switzerland AG 2021
T. K. Dang et al. (Eds.): FDSE 2021, LNCS 13076, pp. 81–96, 2021.
https://doi.org/10.1007/978-3-030-91387-8_6

images using X-ray beams where brain tissues are captured with different intensities depending on the amount of tissue X-ray absorbency (Hounsfield units (HU)). CT scans are displayed using a windowing method. This method transforms the HU numbers into grayscale values ([0, 255]) according to the window level and width parameters. By selecting different window parameters, different features of the brain tissues are displayed in the grayscale image (e.g., brain window, stroke window, and bone window).

In CT scan images, the ICH regions appear as hyper-dense regions with a relatively undefined structure. The CT images are diagnosed by a radiologist to determine whether ICH has occurred. However, this diagnosis process relies on the availability of a subspecialty-trained neuroradiologist, and as a result, could be inefficient and even inaccurate, especially in rural areas where dedicated care is scarce [7]. The problem is solved by implementing the automatic segmentation of CT images. In this paper, we propose an effective method for detecting and segmenting brain hemorrhage regions in brain CT images using the pre-trained U-Net model and it is able to achieve a high detection rate of 92.5%. We make the following contributions: 1) Collect a dataset of brain hemorrhage from 3D CT images; 2) Propose the effective method for brain hemorrhage detection and segmentation using deep learning with transfer learning. We design and perfect the use of the adaptive deep neural network developed on U-Net model for hemorrhage detection from 3D CT images. It more efficient in terms of memory and complexity. The proposed network is very good feature extractor since it can capture and learn relevant features of brain hemorrhage automatically; 3) Use a transfer learning approach to well solve the problem of fast training, small training dataset and an accuracy improvement; 4) Compare and discuss the performance, accuracy of the proposed method with other methods. Experimental results show that the proposed method achieves an accuracy of 92.5%. It can be concluded that the proposed method using deep learning is efficient and fesible.

The remaining of this paper is organized as follows. Section 2 introduces and summarizes related work. In Sects. 3 and 4, we provide a detailed description of the main components of our approach. We describe the experimental setup to validate our approach along with a discussion of the results in Sect. 5. Finally, we conclude with a summary of our findings and comment on possible future directions in Sect. 6.

2   Related Works

In previous work, Phan et al. [13] proposed an approach for detection and classification of brain hemorrhage based on Hounsfield Unit and deep learning techniques. It not only determines the level and duration of hemorrhage but also automatically segments the brain hemorrhagic regions on MRI images. The study compared and evaluated three neural network models to select the most suitable model for classification. The results showed that the method using Hounsfield Unit and Faster R-CNN Inception is time-effective and has high accuracy with mean average precision (mAP) of 79%. Luong et al. [11] introduced a computer-aided ICH detection on CT images that combines a deep-learning model and typical image processing techniques. The deep-learning model based on MobileNet

V2 architecture was trained on Radiological Society of North America Intracranial Hemorrhage dataset. The model was validated on an ICH dataset collected from Vinh Long Province Hospital, Vietnam.

U-Net was developed by Ronneberger et al. [14] as a type of fully convolution network for biomedical image segmentation. The main idea was to supplement a usual contracting network by successive layers to increase the resolution of the output. They used excessive data augmentation to train the desired invariance and robustness properties for the model when the number of training samples is small. On the ISBI cell tracking challenge 2015, their segmentation results reached 92% on "PhC-U373" dataset and 77% on the "DIC-HeLa" dataset. Cicek et al. [2] used a network architecture like the standard U-Net. They introduced batch normalization (BN) before ReLU. They also made data augmentation by rotation, scaling, gray value augmentation and a smooth dense deformation field on both data and ground truth labels. They ran 70,000 training iterations on an NVIDIA TitanX GPU, which took approximately 3 days. Hssayeni et al. [7] collected head CT scans of subjects with traumatic brain injury between February and August 2018 from Al Hilla Teaching Hospital-Iraq of 82 subjects with an average age of 27.8 ± 19.5 years. Two radiologists annotated the non-contrast CT scans and recorded the ICH sub-types if an ICH was diagnosed. The dataset was released into the public domain in NIfTI formats. U-Net was shown to be effective on small training datasets, which they applied to ICH segmentation tasks. They did not perform any preprocessing on the original CT slices, except for removing 5 pixels from the image borders which were only the black regions with no important information. They trained the model with GeForce RTX2080 GPU with 11 GB of memory. The training took approximately 5 h in each cross-validation iteration. U-Net was also used for cell detection and shape measurements [3], segmentation of ambiguous images [9], and road extraction from aerial images [17]. Besides, there were architectures based on U-Net for medical image segmentation such as UNet++ [18] and Attention U-Net [12].

One of the challenges to deep learning model training is the lack of data. To solve this problem and optimize the accuracy, the studies use the transfer learning technique to optimize the training process. Transfer learning allows convolutional neural networks to pre-train on large datasets and then the model is fine-tuned by training on small datasets. The pre-trained model is performed on the same domain dataset and a different domain dataset. Firstly, in the same domain dataset, the dataset has been gathered from different sources that have microscopy images such as the erythrocytes DB dataset [5], the white blood dataset[1], the Wadsworth blood images[2], the ALL-IDB2 dataset [10], and the colorectal dataset [16]. Secondly, the different domain dataset is a collection of natural images that have been gathered from different sources. The first source

---

[1] https://github.com/dhruvp/wbc-classification/tree/master/Original_Images.
[2] https://www.wadsworth.org/.

is a dataset that consists of six classes [15]. The second source is a dataset that has 10 classes of animals[3]. The last source is a dataset with four classes [4].

Recently, the applications of convolutional neural networks (CNN) to identify and classify medical images are demonstrated that the method has an excellent performance. We implement the fully convolutional network, U-Net convolutional neural network, to detect and segment the area of brain damage. Dataset for the studies related to the medical field is a problem due to the privacy of patient information. Fortunately, a dataset contributed by the Physionet repository [6] are dedicated to the studies regarding ICH segmentation tasks. We build and train U-Net network on the dataset from Physionet. The CNN model is implemented, which has the ability to locate, and segment the brain hemorrhage regions in the 3D CT images. The research aims to improve the accuracy of the brain hemorrhage detection on the Physionet dataset.

3   Background

3.1   Architecture of U-Net Deep Neural Network

Convolutional neural networks (CNNs) are used to extract features for generating and learning the feature maps of images, and then use them to create the higher level feature maps. This work is well for classification because the image is converted to a vector. However, in image segmentation, we are not only converting the feature map to a vector but also reconstructing an segmented image from this vector. This is not a simple work and the U-Net architecture can solve this problem. The main idea of this network is to use the feature map for generating the feature vector and then extend this vector to obtain a segmented image. This will preserve the structural integrity of the image and reduce distortion greatly. U-Net is developed by Ronneberger et al. [14] and used mainly for image segmentation. The architecture of U-Net is illustrated in Fig. 1. U-Net architecture is built as a symmetric network consisting of two paths of the contraction and expansion. The contraction path, also known as the encoder in the left block, is similar to a regular convolution network and provides classification information. The expansion path, also known as the decoder in the right block, consists of deconvolution layers to learn and segment. In order not to lose information during convolution and deconvolution, short connections are used to link two blocks of the same level. The contraction does the task of feature extraction to find the features of the image for learning. On the other hand, it also decreases the size of the image by using convolutions and pooling layers to generate higher resolution feature maps. The expansion allows the network to learn localized classification information.

Additionally, it also increases the resolution of the output which can be passed to the convolutional layers to restore the original size of the image and passed to the last convolutional layer to create a fully segmented image. In the expansion,

---

[3] https://www.kaggle.com/alessiocorrado99/animals10#translate.py.
[4] https://www.kaggle.com/mbkinaci/chair-kitchen-knife-saucepan.

U-Net Architecture

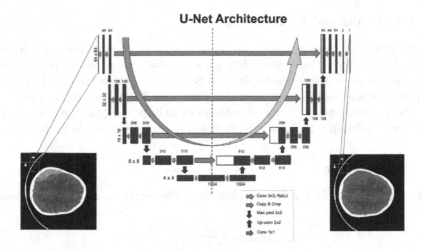

Fig. 1. The architecture of the U-Net convolutional neural network

in addition to the upsample, it also makes a symmetrical connection with the layers of the contraction until the last layer. Meaning that the feature map from the corresponding layer in the contraction is cropped and concatenated onto the upsampled feature map. Obviously, if we upsample in right block from the last layer in the contraction, the information of the original image is lost a lot. Therefore, a symmetrical connection with the contraction will help to recover information lost at the pooling layers.

The activation function is also called the transfer function. It determines the relationship between inputs and outputs of a node and a network. In particular, they are differentiable operators to transform input signals to outputs, while most of them add non-linearity. The commonly used activation functions in practice include Sigmoid function and Rectified linear unit (ReLU). The sigmoid function receives the outputs on the interval $(0, 1)$ from its inputs in the domain $\mathbb{R}$. It is often called a squashing function that squashes any input to some value in the range $(0, 1)$. The sigmoid function is a natural choice because of its smooth and differentiable approximation to a thresholding unit.

$$sigmoid(x) = \frac{1}{1 + \exp(-x)} \tag{1}$$

The rectified linear unit (ReLU) is the most popular choice in the deep learning community because of its simplicity of implementation and its good performance on a variety of predictive tasks. It provides a very simple nonlinear transformation. Given an element x, the function is defined as the maximum of that element or 0. In other words, this function discards all negative elements by setting the corresponding activations to 0.

$$\text{Re}LU(x) = \max(x, 0) \tag{2}$$

3.2   Overfitting and Underfitting

The network is overfitting when it is good at learning the training set but does not has the ability to generalize that gained knowledge to additional, unseen examples. We prevent the model from overfitting in many ways such as data augmentation and feature reduction. On the other hand, underfitting happens when the neural network is not able to accurately predict for the training set nor to the validation set. We add more training samples to avoid underfitting.

3.3   Loss Function

We used the dice coefficient as Loss function. The metric to assess performance is defined by [14].

$$Dice = 2\frac{|X \cap Y|}{|X| + |Y|} \tag{3}$$

where X is Predicted Volume and Y is Ground Truth Volume and the soft-max is defined as follows.

$$p_k(x) = \frac{\exp(a_k(x))}{(\sum\limits_{k'=1}^{K} \exp(a_{k'}(x)))} \tag{4}$$

where $a_k(x)$ denotes the activation in feature channel k at the pixel position x, K is the number of classes and $p_k(x)$ is the approximated maximum-function.

3.4   Xavier Initialization

There are several initialization methods. One of the most generally used is the Xavier initialization [4], named after the first author of its creators. Typically, the Xavier initialization samples weights from a Gaussian distribution with zero mean and variance $\sigma^2 = \frac{2}{n_{in}+n_{out}}$, where n_{in} is the number of input units in the weight tensor and n_{out} is the number of output units. We can also adapt Xavier's intuition to choose the variance when sampling weights from a uniform distribution. Note that the uniform distribution $U(-a, a)$ has variance $\frac{a^2}{3}$. Plugging $\frac{a^2}{3}$ into our condition on σ^2 yields the suggestion to initialize according to [4].

$$U\left[-\frac{\sqrt{6}}{\sqrt{n_{in} + n_{out}}}, \frac{\sqrt{6}}{\sqrt{n_{in} + n_{out}}}\right] \tag{5}$$

3.5   Measure of the Average Precision

The network models is evaluated by measure of the average precision (AP). In order to calculate it, we must first calculate Precision and Recall (Eq. 6). The Precision is calculated as the total number of correct predictions divided by the total number of predictions. Recall is calculated on the total number of correct predictions divided by the total number of correct cases. The accuracy of predicted cases is high giving high precision. High Recall means that the rate of missing actual positive cases is low.

$$Precision = \frac{TP}{TP + FP} \qquad\qquad Recall = \frac{TP}{TP + FN} \qquad (6)$$

In segmentation and classification models, AP [8] is the measure of the average precision computed by Eq. 7 with $\rho_{interp}(r)$ performing an 11-point interpolation to summarize the shape of the *Precision* x *Recall* curve by averaging the accuracy of a set of 11 evenly spaced points $[0, 0.1, 0.2,..., 1]$ and $\rho(\tilde{r})$ is the precision measured on $\tilde{r}$. To determine the average precision (AP), we use Eq. 7.

$$AP = \frac{1}{11} \sum_{r \in 0,0.1,...,1} \rho_{interp}(r) \qquad \rho_{interp}(r) = \frac{max\rho(\tilde{r})}{\tilde{r} : \tilde{r} \geq r} \qquad (7)$$

4   Proposed Method

The proposed method consists of the two phases of the training and testing as illustrated in Fig. 2. In the training phase, we perform three steps of the data pre-processing, augmentation, and features learning using U-Net model with transfer learning approach. Then, we re-train this model on our image dataset to be able to fit our study. This leads to improvements in training time, computational resources, and requirement of large datasets. After training process, we implement the test phase to segment the brain hemorrhage from the trained model.

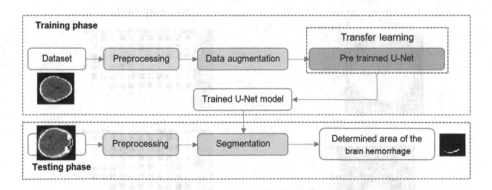

Fig. 2. The model overview of the proposed method.

The train-validation division is a technique for evaluating the performance of a machine learning algorithm. It is usually used for classification or regression problems and can be used for any supervised learning algorithm. The procedure involves taking a dataset and dividing it into two subsets. The first subset is used to fit the model and is referred to as the training dataset. The second subset is not used to train the model, instead, the input element of the dataset is provided to the model, then predictions are made and compared to the expected values. This second dataset is referred to as the test/validation dataset.

4.1 The Training Phase

Datasets are pre-processed and augmented to improve accuracy, reduce bias, and save time. In this section, we present the process of the data pre-processing, augmentation and transfer learning.

– **Pre-processing:** Since the 3D CT images usually have the large size, we make an improvement in this work by slicing them into thin slices and then converting into 2D images as illustrated in Fig. 3. Moreover, we continue to split these 2D images into smaller image patches stored in folders to improve the processing time for training U-Net model. We divide the 2D image into 64 smaller ones in the size of 64 × 64 with the help of the numpy library instead of saving the original image size of 512 × 512. As a result, 3D CT images in NIfTI files are transformed into 2D images in .png format files for training U-Net model. The filtering is referred as to down sampling the negative samples when they seemed to be too much noise. We utilize the filtering method to create the input dataset and improve the accuracy of transfer learning model. The whole process is conducted on 2D images for detection and segmentation of the brain hemorrhage.

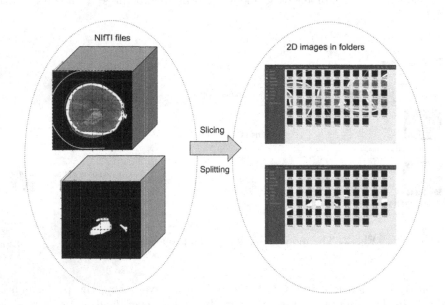

Fig. 3. Illustration of pre-processing 3D CT images

Statistics are crucial for machine learning developers to achieve the general characteristics of the data such as the relationships between features, the density of features, and the number of data points. The exploratory data analysis uses statistics combined with visualization to identify the characteristics.

- **Data augmentation:** In order to apply deep learning when we have the limited dataset, the number of images in the dataset is increased by approximately 10 times using the data augmentation technique. This is to improve the accuracy of the hemorrhage detection and segmentation. Two of the popular techniques of data augmentation used for our study are rotation and repetition.
- **Pre-training:** The U-Net model is trained with all positive samples. The key purpose of the pre-training is to optimize the accuracy efficiently, which helps us to gain good model weights.
- **Transfer learning for hemorrhage detection with U-Net:** Transfer learning is the process of exploiting and re-using knowledge learned by a pre-trained model to deal with a new problem without building a new model from scratch. In addition, this approach does not require large datasets and helps the model converge much faster than the untrained model. Taking advantage of this approach, we refine, re-train, and evaluate to locate areas of bleeding. Therefore, at this period, the model is continued to train with initialization weights by the pre-trained U-Net model. It is trained on the dataset with a wide variety of features, positive-negative samples.

4.2 The Testing Phase

The training process fine-tunes the parameters to achieve a good training model with the optimal measures of the Loss and accuracy. This process terminates when obtaining the good training model. We utilize this trained model to segment the ICH or non-ICH images in the testing phase. The images in the training and testing datasets are selected randomly and split in a ratio of 80:20.

5 Experimental Results

5.1 Description of Dataset and Environment

A dataset of 82 CT scans with 3D images was collected, including 36 scans for patients diagnosed with intracranial hemorrhage with the following types: Intraventricular, Intra-parenchymal, Subarachnoid, Epidural, and Subdural. Each CT scan for each patient includes about 30 slices with 5 mm slice-thickness. We convert nii extension files into png extension files. We divide CT images into 64 smaller ones in the size of 64×64 with the help of the numpy library instead of saving the original image size of 512×512.

Building systems with great performance requires well underlying hardware in addition to an understanding of the algorithms and models to capture the statistical aspects of the problem. The system for experiments run on Google Colab environment with the configuration of 16 GB RAM. The library used to support training network models is Tensorflow. We provide scenarios 9 and 10 to train models on the cloud with a configuration of 16 GB RAM and 4 CPUs.

5.2 Scenario Descriptions

An exploratory analysis of the data aims at discovering any kind of relevant insights on the dataset such as the correlations and variance during the creation of the different scenarios. After that, we make a comparison of the accuracy of the scenarios. In this paper, we study and train a brain hemorrhage segmentation model developed on the U-Net with the high accuracy. We design scenarios with an exploratory analysis of the input datasets, epochs, and initialization weights. Therefore, in the scenario from 1 to 10, we make changes on the number of input images, epochs and initialization weights corresponding to scenarios. The parameters of the scenarios should be fine-tuned to find the optimal training model that results the high segmentation accuracy.

Table 1 presents scenarios and parameters of the proposed method for experiments. For example, in scenario 1, we train the proposed model on a dataset of 2,181 images at 500 epochs with learning rate of $1e-5$ using Xavier initializer. In scenario 2, we use the initialization weights of scenario 1 and train on dataset of 2,181 images at 1,000 epochs with learning rate of $1e-6$. A similar description is for the remaining scenarios from 3 to 10 as shown in Table 1.

Table 1. Scenarios and parameters of the proposed method for experiments.

Scenario	Learning rate	Epochs	Training dataset	Initialization weights
1	$1e-5$	500	2,181 images	Xavier initializer [4]
2	$1e-6$	1,000	2,181 images	Scenario 1
3	$1e-6$	1,000	2,181 images	Scenario 2
4	$1e-7$	500	1,454 images	Scenario 3
5	$1e-8$	100	727 images	Scenario 4
6	$1e-7$	100	727 images	Scenario 5
7	$1e-5$	300	2,471 images	Scenario 4
8	$1e-5$	200	2,471 images	Scenario 1
9	$1e-5$	500	3,570 images	Xavier's initializer [4]
10	$1e-5$	500	2,471 images	Scenario 9

In order to get an good model, the learning rate should be set to 1e-5 constant. The number of the appropriate epochs is 500. The number of training images comprises 1,250 ICH slices. After data augmentation, the total training dataset contains 3,750 images for scenario 9. Besides, in order to avoid any class-imbalance issues between data with and without ICH, we applied a random under-sampling to the training data and reduced the number of non-ICH data to the same level as the data with ICH. Taking advantage of the fast learning ability on the positive dataset, the model used the pre-trained weights in scenario 9 to continue the training process. The same process as in the selection of the previous scenario is done. However, only the image rotation technique is applied to gain a variety of data. Consequently, the entire images for training and validation are 2,471 images including negative and positive images for scenario 10. The training takes approximately 4 days (less than 10 min each epoch). Despite the similarity of training procedures, the difference is found in the results of each scenario.

5.3   Scenario Evaluation

Once all scenarios are done and the possible values for the hyper-parameters are selected, is time to start with the development test. Therefore, with a fixed number of for loops, the scenarios learn to optimize the weights. We recognize that the scenarios have the complete ability to get high accuracy. The scenario that does better on the validation set. Learning curves are widely used in machine learning for algorithms that learn (optimize their internal parameters) incrementally over time, such as deep learning neural networks. It is common to create dual learning curves for a machine learning model during training on both the training and validation datasets. In terms of performance measure, learning curves are calculated on the metric by which the models will be evaluated and selected, e.g. accuracy [1]. In order to optimize and evaluate a segmentation model, we use Loss and AP measures. Based on these metrics, we fine-tune parameters of the training model to obtain the optimal parameters before applying for segmentation testing. The Loss values indicate the level of model error in the hemorrhagic segmentation. The error level of the model in brain hemorrhage detection is low if the value of Loss is low. In other words, during the training process, the accuracy of detecting brain hemorrhage will increase if Loss decreases. Therefore, we need to keep a Loss function which can penalize a model effectively while it is training on a dataset. It will improve the accuracy of ICH segmentation if we try to minimize the Loss value. A model optimization measure based on the Loss function will give a better insight for model selection. A good learning curve is identified by training and validation Loss values that decreases to a point of stability with a minimal gap between the two final Loss values. The Loss of the model will almost always be lower on the training dataset than the validation dataset. This means that we should expect some gap between the training and validation Loss learning curves. This gap is referred to as the "generalization gap". We split the dataset into 2 sub-datasets for training and validation. Focusing on the hyper-parameters, a tuple of hyper-parameters yields an optimal model which minimizes a predefined Loss function on given independent data. It can be crucial for maintaining accuracy and Loss stability. We search for the best values by trial and error (Table 2).

Table 2. The results of testing process for scenarios

Scenario	1	2	3	4	5	6	7	8	9	10
Accuracy (AP)	54%	64%	78%	86%	64%	81%	91%	76%	89%	**92.5%**

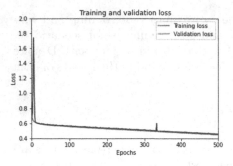
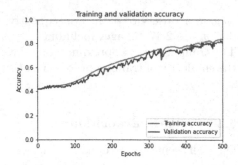

Fig. 4. The Loss values of scenario 9.   **Fig. 5.** The accuracy of scenario 9.

In Fig. 4, the curve of the Loss function values over the epochs for scenario 9 is represented. It shows how the performance of the network model is improving during the training. It is perceptible how the function continues decreasing after 500 epochs till the end and it would probably continue decreasing a bit more. However, the rate of improvement over time consuming and complexity is not worth it. Consequently, The Loss value of 0.42 at 500 epochs is appropriate to stop training process. Figure 5 contains the accuracy values for scenario 9. There is an improvement in the performance of each epoch. A consequence of the modifications is done to the network hyper-parameters, and to the increase of the size of the dataset used to fed the model. The time is less than 11 min per training and quite similar between them. Considering that the complexity of the models was practically the same, they have the same structure and similar inputs, similar times are expected. Obviously, the results prove that it is possible to get a better accuracy of a network by selecting the appropriate inputs. Additionally, from Figs. 4 and 5, we can see that, as the Loss value decreases, the detection accuracy is improved during training and validation. The curve of the Loss function values over the epochs for scenario 10 is presented in Fig. 6. It can be seen that Loss value is decreased fairly rapidly at the first point and ends with a fall to less than 0.4. A training and validation Loss remain the minimum gap that model has a rather good fit. Figure 7 shows that the accuracy fluctuates in the wide range at some points. However, the model still keeps on improving later. The main problem lies on both the imbalance of negative-positive values and some existing noise samples. We know that Loss functions provide more than just a static representation of how the model is performing. It means they show the model fits data in the first place. Most machine learning algorithms use some sort of Loss function in the process of optimization or finding the best parameters (weights) for data. Therefore, we continue to create the learning curve plot to observe the optimization.

Some plots are generated to compare both the predicted and the actual samples. The aim is to visually represent the accuracy of the prediction. The images show how prediction behaves on the test set. The three plots for scenario

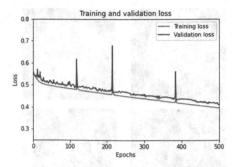
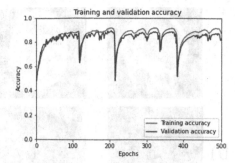

Fig. 6. The Loss values of scenario 10.   **Fig. 7.** The accuracy of scenario 10.

9 correspond to different data so that different annotated images are visualized in Fig. 8. We visualize some plots to evaluate the performance for scenario 10 in Fig. 9. Compared to previous scenarios, the accuracy of scenario 10 is acceptable not only regarding the achieving of predicted values, but also on the certainty over the false-positive region reduction.

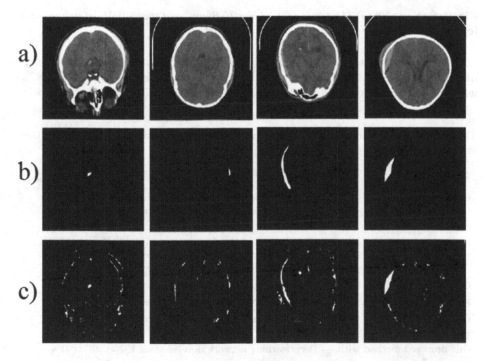

Fig. 8. Scenario 9: (a) The testing CT slices; (b) Samples of the radiologist delineation; (c) Brain hemorrhage detection using the proposed method.

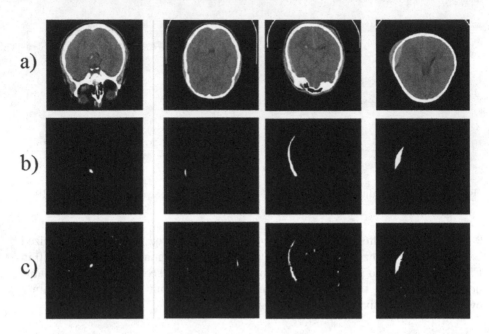

Fig. 9. Scenario 10: (a) The testing CT slices; (b) Samples of the radiologist delineation; (c) Brain hemorrhage detection using the proposed method.

Table 3 represents a comparison of the accuracy for brain hemorrhage segmentation methods. The accuracy of these methods achieves from 79% to 87% while the proposed method reaches the accuracy of 92.5%.

Table 3. Accuracy comparison of the different methods for brain hemorrhage detection and segmentation

Authors	Method	Accuracy
Phan et al. [13]	Detection and classification of brain hemorrhage based on Hounsfield Unit and deep learning techniques	79%
Luong et al. [11]	Combination of the deep-learning (MobileNet V2) model and typical image processing techniques	80.7%
Cicek et al. [2]	A network architecture like the standard U-Net	86.3%
Hssayeni et al. [7]	Based on U-Net with NifTI dataset	87%
Our proposed method with scenario 9	Deep learning network developed on U-Net	89%
Our proposed method with scenario 7	Deep learning network developed on U-Net	91%
Our proposed method with scenario 10	Deep learning network developed on U-Net	92.5%

6   Conclusion

In this paper, we propose an efficient method for brain hemorrhage detection on 3D CT scans with a deep neural network. The pre-trained U-Net model is refined, re-trained, and evaluated to detect brain hemorrhage. First of all, an exploratory analysis of the data is performed, with the main purpose of finding the initial insights on the dataset to build the network and experimental scenarios. In addition, during the exploratory analysis, the dataset has the difference of negative-positive samples, which needs to be balanced for the correct functioning of the network. The building process is done to choose the ideal selection of hyper-parameters for this problem. Probably the most work-loaded part of the paper is the development test, where the first scenarios are created. Moreover, the optimal hyper-parameters from the previous selection are chosen. The detections are performed after the training of the network. During the development of the scenarios, it is discovered that when training by the dataset consisting of multiple noises, the model is difficult to improve the accuracy, e.g. the learning curve remains unchanged. The solution to the problem is to train the model with positive samples in advance and then its weight parameters are trained with a positive-negative dataset. Our proposed method is evaluated on a set of 3D CT-scan images and obtains an accuracy rate of 92.5% for automatic brain hemorrhage detection.

References

1. Brownlee, J.: How to use learning curves to diagnose machine learning model performance. Mach. Learn. Mastery (2019)
2. Çiçek, Ö., Abdulkadir, A., Lienkamp, S.S., Brox, T., Ronneberger, O.: 3D U-net: learning dense volumetric segmentation from sparse annotation. In: Ourselin, S., Joskowicz, L., Sabuncu, M.R., Unal, G., Wells, W. (eds.) MICCAI 2016. LNCS, vol. 9901, pp. 424–432. Springer, Cham (2016). https://doi.org/10.1007/978-3-319-46723-8_49
3. Falk, T., et al.: U-net: deep learning for cell counting, detection, and morphometry. Nat. Methods **16**(1), 67–70 (2019)
4. Glorot, X., Bengio, Y.: Understanding the difficulty of training deep feedforward neural networks. In: Proceedings of the Thirteenth International Conference on Artificial Intelligence and Statistics, pp. 249–256. JMLR Workshop and Conference Proceedings (2010)
5. Gonzalez-Hidalgo, M., Guerrero-Pena, F., Herold-García, S., Jaume-i Capó, A., Marrero-Fernández, P.D.: Red blood cell cluster separation from digital images for use in sickle cell disease. IEEE J. Biomed. Health Inform. **19**(4), 1514–1525 (2014)
6. Hssayeni, M.: Computed tomography images for intracranial hemorrhage detection and segmentation. PhysioNet (2019)
7. Hssayeni, M.D., Croock, M.S., Salman, A.D., Al-khafaji, H.F., Yahya, Z.A., Ghoraani, B.: Intracranial hemorrhage segmentation using a deep convolutional model. Data **5**(1), 14 (2020)
8. Janani, R., Vijayarani, S.: An efficient text pattern matching algorithm for retrieving information from desktop. Indian J. Sci. Technol. **9**(43), 1–11 (2016)

9. Kohl, S.A., et al.: A probabilistic U-net for segmentation of ambiguous images. arXiv preprint arXiv:1806.05034 (2018)

10. Labati, R.D., Piuri, V., Scotti, F.: All-IDB: the acute lymphoblastic leukemia image database for image processing. In: 2011 18th IEEE International Conference on Image Processing, pp. 2045–2048. IEEE (2011)

11. Luong, K.G., et al.: A computer-aided detection to intracranial hemorrhage by using deep learning: a case study. In: Huang, Y.-P., Wang, W.-J., Quoc, H.A., Giang, L.H., Hung, N.-L. (eds.) GTSD 2020. AISC, vol. 1284, pp. 27–38. Springer, Cham (2021). https://doi.org/10.1007/978-3-030-62324-1_3

12. Oktay, O., et al.: Attention U-net: learning where to look for the pancreas. arXiv preprint arXiv:1804.03999 (2018)

13. Phan, A.-C., Cao, H.-P., Trieu, T.-N., Phan, T.-C.: Detection and classification of brain hemorrhage using hounsfield unit and deep learning techniques. In: Dang, T.K., Küng, J., Takizawa, M., Chung, T.M. (eds.) FDSE 2020. CCIS, vol. 1306, pp. 281–293. Springer, Singapore (2020). https://doi.org/10.1007/978-981-33-4370-2_20

14. Ronneberger, O., Fischer, P., Brox, T.: U-Net: convolutional networks for biomedical image segmentation. In: Navab, N., Hornegger, J., Wells, W.M., Frangi, A.F. (eds.) MICCAI 2015. LNCS, vol. 9351, pp. 234–241. Springer, Cham (2015). https://doi.org/10.1007/978-3-319-24574-4_28

15. Roy, P., Ghosh, S., Bhattacharya, S., Pal, U.: Effects of degradations on deep neural network architectures. arXiv preprint arXiv:1807.10108 (2018)

16. Sirinukunwattana, K., Raza, S.E.A., Tsang, Y.W., Snead, D.R., Cree, I.A., Rajpoot, N.M.: Locality sensitive deep learning for detection and classification of nuclei in routine colon cancer histology images. IEEE Trans. Med. Imaging $35(5)$, 1196–1206 (2016)

17. Zhang, Z., Liu, Q., Wang, Y.: Road extraction by deep residual U-net. IEEE Geosci. Remote Sens. Lett. $15(5)$, 749–753 (2018)

18. Zhou, Z., Rahman Siddiquee, M.M., Tajbakhsh, N., Liang, J.: UNet++: a nested U-net architecture for medical image segmentation. In: Stoyanov, D., et al. (eds.) DLMIA/ML-CDS -2018. LNCS, vol. 11045, pp. 3–11. Springer, Cham (2018). https://doi.org/10.1007/978-3-030-00889-5_1

Advances in Machine Learning for Big Data Analytics

Multi-class Bagged Proximal Support Vector Machines for the ImageNet Challenging Problem

Thanh-Nghi Do[1,2](✉)

[1] College of Information Technology, Can Tho University, Cantho 92000, Vietnam
dtnghi@cit.ctu.edu.vn
[2] UMI UMMISCO 209 (IRD/UPMC), Sorbonne University,
Pierre and Marie Curie University, Paris 6, France

Abstract. We propose the new multi-class of bagged proximal support vector machines (MC-Bag-PSVM) for handling the ImageNet challenging problem with very large number of images and a thousand classes. Our MC-Bag-PSVM trains in the parallel manner ensemble binary PSVM classifiers used for the One-Versus-All (OVA) multi-class strategy on multi-core computer with GPUs. The binary PSVM model is constructed by bagged binary PSVM models built in under-sampling training dataset. The numerical test results on ILSVRC 2010 dataset show that our MC-Bag-PSVM algorithm is faster and more accurate than the state-of-the-art linear SVM algorithm. An example of its effectiveness is given with an accuracy of 75.64% obtained in the classification of ImageNet-1000 dataset having 1,261,405 images in 2048 deep features into 1,000 classes in 29.5 min using a PC Intel(R) Core i7-4790 CPU, 3.6 GHz, 4 cores and Gigabyte GeForce RTX 2080Ti 11 GB GDDR6, 4352 CUDA cores.

Keywords: Large scale image classification · Support vector machines · Multi-class

1 Introduction

The classification of images is one of the most important topics in communities of computer vision and machine learning. The high performance classification algorithms help us find what we are looking for in very huge amount of images produced by internet users. The aim of image classification tasks is to automatically categorize the image into one of predefined classes. The performance of an image classification system largely depends on the feature extraction approach and the machine learning scheme. The most challenging is to classify images of extremely varied classes. ImageNet dataset [5,6] with more than 14 million images for 21,841 classes raises more challenges in training classifiers, due to the large scale number of images and classes. Many researches [7,9–11,26,32] proposed to use popular handcrafted features such as the scale-invariant feature

© Springer Nature Switzerland AG 2021
T. K. Dang et al. (Eds.): FDSE 2021, LNCS 13076, pp. 99–112, 2021.
https://doi.org/10.1007/978-3-030-91387-8_7

transform (SIFT [23,24]), the bag-of-words model (BoW [1,22,28]) and support vector machines (SVM [30]) for dealing with large scale ImageNet dataset. Recent deep learning networks, including VGG19 [27], ResNet50 [16], Inception v3 [29], Xception [4] are proposed to efficient classify ImageNet dataset with the prediction correctness more over 70%.

In this paper, we use the pre-trained deep learning network Inception v3 to extract invariant features from images. Followed which, we propose the new multi-class bagged proximal support vector machines (MC-Bag-PSVM) for classifying ImageNet challenging problem. We extend the PSVM learning algorithm [13] to create the MC-Bag-PSVM one for dealing with the large number of classes. Our MC-Bag-PSVM trains in the parallel manner ensemble binary PSVM classifiers used for the One-Versus-All (OVA [30]) multi-class strategy on multi-core computer with GPUs. The binary PSVM model is combined by bagged binary PSVM models built in bootstrap samples being under-sampling training dataset. The numerical test results on ILSVRC 2010 dataset show that our MC-Bag-PSVM algorithm is faster and more accurate than the state-of-the-art linear SVM algorithm such as LIBLINEAR [12]. MC-Bag-PSVM classifies ImageNet-1000 dataset having 1,261,405 images in 2048 deep features into 1,000 classes with an accuracy of 75.64% in 29.5 min using a PC Intel(R) Core i7-4790 CPU, 3.6 GHz, 4 cores and Gigabyte GeForce RTX 2080Ti 11 GB GDDR6, 4352 CUDA cores.

The remainder of this paper is organized as follows. Section 2 briefly introduces the SVM algorithm. The proximal SVM is presented in Sect. 3. The multi-class bagged proximal SVM algorithm is illustrated in Sect. 4 for handling large number of classes. Section 5 shows the experimental results before conclusions and future works presented in Sect. 6.

2   Support Vector Machines

We consider a simple binary classification task shown in Fig. 1, with the dataset $D = [X, Y]$ consisting of m datapoints $X = \{x_1, x_2, \ldots, x_m\}$ having correspond-

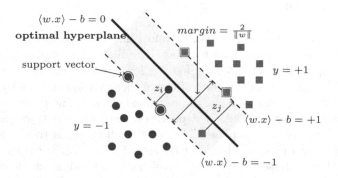

Fig. 1. SVM for the binary classification

ing labels $Y = \{y_1, y_2, \ldots, y_m\}$ being ± 1. The SVM algorithm proposed by [30] tries to find the best separating plane (denoted by the normal vector $w \in R^n$ and its bias $b \in R$) furthest from both class $+1$ and class -1. To pursue this aim, the training SVM algorithm tries to simultaneously maximize the margin between the supporting planes for each class and minimize errors.

The supporting plane $\langle w \cdot x \rangle - b = +1$ for class $+1$ satisfies (1):

$$\langle w \cdot x_i \rangle - b \geq 1 \quad for \ y_i = +1. \tag{1}$$

The supporting plane $\langle w \cdot x \rangle - b = -1$ for class -1 satisfies (2):

$$\langle w \cdot x_i \rangle - b \leq -1 \quad for \ y_i = -1. \tag{2}$$

The margin between two parallel supporting planes is $2/\|w\|$ (where $\|w\|$ is the 2-norm of the vector w). The maximization of the margin leads to minimize the norm $\|w\|$.

Any point x_i falling on the wrong side of its supporting plane is considered to be an error, denoted by z_i ($z_i \geq 0$, if the point x_i falling on the right side of its supporting plane then its corresponding $z_i = 0$).

Maximizing the margin and minimizing errors of the training SVM algorithm yield the quadratic programming (3).

$$min \ f(w, b, z) = (1/2)\|w\|^2 + \lambda \sum_{i=1}^{m} z_i$$

$$s.t. \begin{cases} y_i(\langle w \cdot x_i \rangle - b) + z_i \geq 1 \\ z_i \geq 0 \quad \forall i = 1, 2, \ldots, m \end{cases} \tag{3}$$

where the positive constant λ is used to tune errors and the margin size. Then, the classification function of a new datapoint x based on the plane is:

$$predict(x) = sign(\langle w \cdot x \rangle - b). \tag{4}$$

3   Proximal Support Vector Machines

Proximal SVM (PSVM [13]) illustrated in Fig. 2 tries to find the classification hyper-plane by maximizing margin between two parallel proximal planes and minimizing errors. It modifies the SVM formula (3) as follows:

- the inequality constraints are replaced by the equality ones
- maximizing the margin by minimizing $\|w, b\|$
- the least-squares error in the objective function f

Thus substituting the constraint in terms w, b for z into the objective function f of the quadratic program (3) gives an unconstraint optimization problem (5):

$$min \ f(w, b, [X, Y]) = \frac{1}{2}\|w, b\|^2 + \frac{\lambda}{2} \sum_{i=1}^{m} [1 - y_i(\langle w \cdot x_i \rangle - b)]^2 \tag{5}$$

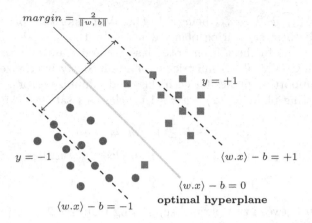

Fig. 2. Proximal SVM for the binary classification

The necessary and sufficient conditions of optimality for an unconstraint optimization problem (5) is the gradient with respect to w, b being zero. This yields the linear equation system as follows:

$$
\begin{pmatrix} w_1 \\ w_2 \\ \vdots \\ w_n \\ b \end{pmatrix} = (\frac{1}{\lambda}I + E^T E)^{-1} E^T \begin{pmatrix} y_1 \\ y_2 \\ \vdots \\ y_m \end{pmatrix} \tag{6}
$$

with matrix $E = \begin{pmatrix} x_{1,1} & x_{1,2} & \cdots & x_{1,n} & -1 \\ x_{2,1} & x_{2,2} & \cdots & x_{2,n} & -1 \\ \vdots & \vdots & \ddots & \vdots & \vdots \\ x_{m,1} & x_{m,2} & \cdots & x_{m,n} & -1 \end{pmatrix}$

and identity matrix $I = \begin{pmatrix} 1 & 0 & \cdots & 0 \\ 0 & 1 & \cdots & 0 \\ \vdots & \vdots & \ddots & \vdots \\ 0 & 0 & \cdots & 1 \end{pmatrix}$.

The training PSVM algorithm requires the linear equation system (6) instead of the quadratic programming (3) in the standard SVM. Therefore, the algorithmic complexity of PSVM is much lower than the one of standard SVM. PSVM training for the binary classification is described in Algorithm 1.

4   Multi-class Bagged Proximal Support Vector Machines

The binary SVM solver can be extended for handling multi-class problems (c classes, $c \geq 3$). Practical approaches for multi-class SVMs are to train series

---

Algorithm 1: PSVM(X, Y, λ) for the binary classification

---

   input:

        Training dataset $D = [X, Y]$

        Constant $\lambda > 0$ for tuning errors and margin size

   output:

        (w, b)

1 **begin**

2   Create matrix E

3   Solve linear equation system (6)

4   Return optimal plane (w, b)

5 **end**

---

of binary SVMs, including One-Versus-All (OVA [30]), One-Versus-One (OVO [20]). The OVA strategy shown in Fig. 3 builds c binary SVM models where the ith one separates the ith class from the rest. The OVO strategy illustrated in Fig. 4 constructs $c(c-1)/2$ binary SVM models for all binary pairwise combinations among c classes. The class of a new datapoint is then predicted with the largest distance vote. In practice, the OVA strategy is implemented in LIBLINEAR [12] and the OVO technique is also used in LibSVM [3].

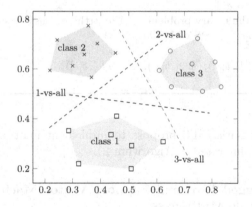

Fig. 3. Multi-class SVM (One-Versus-All)

4.1 Multi-class Proximal Support Vector Machines

When handling very large number of classes, e.g. ImageNet with $c = 1,000$ classes, the OVO strategy is too expensive because it needs training $499, 500$ of binary SVM classifiers and using them in the classification (compared to $1,000$ binary models learned by the OVA strategy). Therefore, the OVA strategy is suited for handling this case.

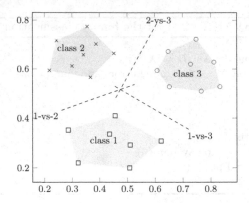

Fig. 4. Multi-class SVM (One-Versus-One)

Algorithm 2: MC-PSVM(X, Y, λ) for c classes

---

    input:
            Training dataset $D = [X, Y]$
            Constant $\lambda > 0$ for tuning errors and margin size
    output:
            $\{(w[1], b[1]), (w[2], b[2]), \ldots, (w[c], b[c])\}$

1 **begin**
2   **for** $c_i \leftarrow 1$ **to** c **do**
3   Create the binary problem for $Y[c_i]$ (the c_i class versus the rest)
4         $(w[c_i], b[c_i]) = \mathbf{PSVM}(X, Y[c_i], \lambda)$
5   **end**
6   Return $\{(w[1], b[1]), (w[2], b[2]), \ldots, (w[c], b[c])\}$
7 **end**

Fung and Mangasarian [14] propose to implement multi-class PSVM with the OVA strategy (illustrated in Algorithm 2).

4.2 Parallel Bagged Proximal Support Vector Machines for Large-Scale Multi-class

The multi-class PSVM algorithm using OVA leads to the two problems:

1. a drawback raised by OVA is the imbalanced datasets while building binary PSVM classifiers, it means that the ith class is very smaller than the rest, making poor performance in the classification,
2. the PSVM algorithm also takes very long time to train very large number of binary PSVM classifiers in sequential mode using the single processor.

Fung and Mangasarian [14] propose to use the Newton refinement for improving binary PSVM models. Parallel implementation on graphics processing units (GPUs) in [8,17] improves the training time for large datasets.

Due to these problems, we propose the parallel bagged PSVM algorithm (denoted by MC-Bag-PSVM) being able to efficiently handle the large number of datapoints and large-scale multi-class on standard multi-core computers (PCs) using GPUs. The first one is to build ensemble binary PSVM classifiers with under-sampling strategy in the parallel way using GPUs. The second one is to parallelize the training task of all binary PSVM classifiers on multi-core machines.

Bagging Binary PSVM Classifier

Our bagging binary PSVM classifier (Bag-PSVM) in the MC-Bag-PSVM algorithm is an adherent of the bagging framework [2] to separate the c_i class (positive class) from the $c - 1$ other classes (negative class).

Since the original bagging approach uses the bootstrap sample from the training dataset without regard to the class distribution, being the most intractable class separation for the extreme unbalance between the positive and the negative class in large-scale multi-class datasets.

Due to this problem, we propose to use the bootstrap sample with under-sampling the negative class for every bagging iteration of the Bag-PSVM algorithm.

Given the training dataset $D = [X, Y]$ consists of the positive class D_+ ($|D_+|$ is the cardinality of the positive class c_i) and the negative class D_- ($|D_-|$ is the cardinality of the negative class). Our bagging binary PSVM trains κ PSVM classifiers $\{(w, b)_1, (w, b)_2, \ldots, (w, b)_\kappa\}$ from bootstrap samples with under-sampling the negative class to separate the positive class c_i from the negative class.

At the ith iteration, the bootstrap sample B_i includes the positive class D_+ and $\sqrt{|D_+| \times |D_-|}$ datapoints sampling without replacement from the negative class D_-, and then learning Algorithm 3 learns $(w, b)_i$ from B_i. The Bag-PSVM averages all classifiers $\{(w, b)_1, (w, b)_2, \ldots, (w, b)_\kappa\}$ to create the final model $(w[c_i], b[c_i])$ for separating the class c_i from other ones. The Bag-PSVM is described in Algorithm 3.

Furthermore, we develop a parallel version of PSVM algorithm based on GPUs, as an inheritance of our work in [8]. The PSVM (at **line 6** in Algorithm 3) uses the CUBLAS library to perform matrix computations on the GPU massively parallel computing architecture. Therefore, it gains high training speeds.

Parallel Training of Multi-class Bag-PSVM

In the classification of c classes problem, the multi-class SVM algorithm independently trains c binary classifiers. To speed-up the training task, our multi-class Bag-PSVM (called MC-Bag-PSVM as described in Algorithm 4) that launches parallel Bag-PSVM algorithm to learn c binary classifiers for the very large number of classes, using the shared memory multiprocessing programming model OpenMP on multi-core computers.

Algorithm 3: Bag-PSVM($X, Y, \lambda, c_i, \kappa$) training algorithm for multi-class

---

input:

         Training dataset $D = [X, Y]$

         Positive constant $\lambda > 0$

         Positive class c_i versus other classes

         Number of bagging iteration κ

output:

         $w[c_i], b[c_i]$

1 **begin**

2   Split training dataset D into

3   the positive data D_+ (class c_i) and the negative data D_-

4   **for** $i \leftarrow 1$ **to** κ **do**

5   Create a bootstrap $B_i = \{X_i, Y_i\}$ including the positive data D_+

           and under-sampling $\bar{D}_-$ from D_- (with $|\bar{D}_-| = \sqrt{|D_+| \times |D_-|}$)

6        $(w, b)_i = \mathbf{PSVM}(X_i, Y_i, \lambda)$

7   **end**

8   Return $(w[c_i], b[c_i]) = \frac{1}{\kappa} \sum_{i=1}^{\kappa} (w, b)_i$

9 **end**

---

5   Experimental Results

We are interested in the experimental evaluation of MC-Bag-PSVM for handling large scale multi-class datasets. Therefore, it needs to assess the performance in terms of training time and classification correctness.

5.1   Software Programs

Parallel multi-class proximal SVM (denoted by MC-PSVM) and multi-class bagged proximal SVM (denoted by MC-Bag-PSVM) are implemented in C/C++, OpenMP [25] and the Automatically Tuned Linear Algebra Software (ATLAS [31]). We would like to compare with the best state-of-the-art linear SVM algorithm, LIBLINEAR [12] (the parallel version on multi-core computers).

    All experiments are conducted on a machine Linux Fedora 23, Intel(R) Core i7-4790 CPU, 3.6 GHz, 4 cores and 32 GB main memory and the Nvidia Gigabyte GeForce RTX 2080Ti 11 GB GDDR6, 4352 CUDA cores.

5.2   ILSVRC 2010 Dataset

The MC-Bag-PSVM algorithm is designed for dealing with the large amount number of data and large-scale multi-class, so we have evaluated its performance on the ImageNet challenging problems [5,6]. Full ImageNet dataset has more than 14 million images for 21,841 classes raises more challenges in training classifiers, due to the large scale number of images and classes.

Algorithm 4: Parallel MC-Bag-PSVM(X, Y, λ, κ) training algorithm for multi-class

---

input:
> Training dataset $D = [X, Y]$ with c classes
> Positive constant $\lambda > 0$
> Number of bagging iteration κ

output:
> $\{(w[1], b[1]), (w[2], b[2]), \dots, (w[c], b[c])\}$

1 **begin**
2   *#pragma omp parallel for schedule(dynamic)*
3   **for** $c_i \leftarrow 1$ **to** c **do**   /* class c_i */
4 | $(w, b)_{c_i} = $ **Bag-PSVM**$(X, Y, \lambda, c_i, \kappa)$
5   **end**
6   Return $\{(w[1], b[1]), (w[2], b[2]), \dots, (w[c], b[c])\}$
7 **end**

We propose to use a subset of ImageNet dataset (called ILSVRC 2010 dataset) to assess classification performances. This dataset is the most popular visual classification benchmark [4–7,9–11,16,27,29,32]. ILSVRC 2010 dataset contains 1,261,405 images with 1,000 classes.

Recent deep learning networks, especially Inception v3 [29] described in Fig. 5 is trained to efficiently classify ImageNet dataset with the prediction correctness more over 70% (significant improvement against ~20% accuracy obtained by the classical approach using linear SVM and handcrafted features). This pre-trained network model can be used as the feature extractor by getting the last **AvgPool** layer. Therefore, we use the pre-trained inception v3 to extract 2,048 invariant features from images.

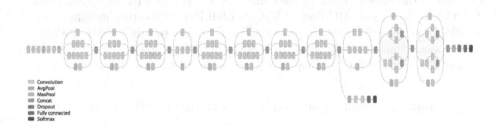

Convolution
AvgPool
MaxPool
Concat
Dropout
Fully connected
Softmax

Fig. 5. Architecture of Inception v3

ILSVRC 2010 dataset is randomly divided into training set with 1,009,124 rows and testing set with 252,281 rows (with random guess 0.1% due to 1,000 classes).

5.3   Tuning Parameter

For training linear SVM models, it needs to tune the positive constant C in SVM algorithms for keeping the trade-off between the margin size and the errors. We use the cross-validation (hold-out) protocol to find-out the best value $C = 100,000$.

LIBLINEAR uses L2-regularized Logistic Regression that is very closed to the softmax classifier used in deep learning networks, e.g. Inception v3 [29].

Furthermore, our MC-Bag-PSVM algorithm learns $\kappa = 5$ binary PSVM classifiers to separate c_i class from other ones.

Due to the PC (Intel(R) Core i7-4790 CPU, 4 cores) used in the experimental setup, the number of OpenMP threads is setting to 4 for all training tasks.

5.4   Classification Results

We obtain classification results of MC-PSVM, MC-Bag-PSVM and LIBLINEAR in Table 1, Fig. 6 and Fig. 7. The fastest training algorithm is in bold-faced and the second one is in italic. The same presentation format is used for classification accuracy.

MC-PSVM classifies ILSVRC 2010 dataset in 0.46 minutes with 66.94% accuracy. LIBLINEAR learns to classify ILSVRC 2010 dataset in 9,813.58 min with 73.66% accuracy. Our MC-Bag-PSVM achieves 75.64% accuracy with 29.5 minutes in the training time.

In the comparison of training time among multi-class classification SVM approaches, we can see that original MC-PSVM is fastest training algorithm with the lowest accuracy. MC-PSVM is 64.22, 21,364.84 times faster than MC-Bag-PSVM and LIBLINEAR, respectively. Our MC-Bag-PSVM is 332.66 times faster than LIBLINEAR.

In terms of overall classification accuracy, MC-Bag-PSVM gives the highest accuracy in the classification. In the comparison, algorithm by algorithm, shows that the superiority of MC-Bag-PSVM on LIBLINEAR corresponds to 1.98%. MC-Bag-PSVM significantly improves 8.7% compared to original MC-PSVM.

The classification results allow to believe that our proposed MC-Bag-PSVM algorithm is efficient for handling such large-scale multi-class datasets.

Table 1. Overall classification accuracy for ILSVRC 2010 dataset

No	Algorithm	Time (min)	Accuracy (%)
1	LIBLINEAR	9,813.58	_73.66_
2	MC-PSVM	**0.46**	66.94
3	MC-Bag-PSVM	_29.5_	**75.64**

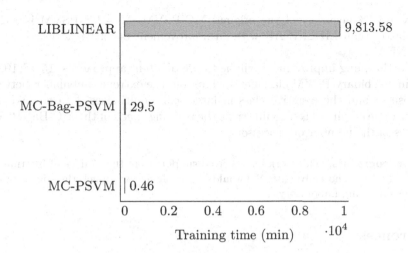

Fig. 6. Training time (min) for ILSVRC 2010 dataset

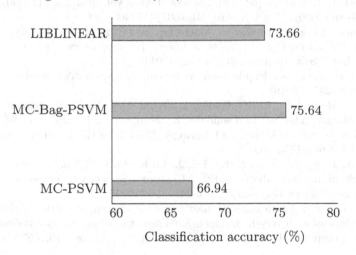

Fig. 7. Overall classification accuracy for ILSVRC 2010 dataset

6   Conclusion and Future Works

We have presented the new MC-Bag-PSVM training algorithm for dealing with
the ImageNet challenging problem having very large number of images and a
thousand classes. MC-Bag-PSVM is to parallelize training tasks of binary PSVM
classifiers on multi-core computer with GPUs. Bagged binary PSVM models
built in under-sampling training dataset forms the binary PSVM model used
for the OVA multi-class strategy. The numerical test results on ILSVRC 2010
dataset show that our MC-Bag-PSVM algorithm is faster and more accurate
than the state-of-the-art linear SVM algorithm, LIBLINEAR. MC-Bag-PSVM

improves 8.7% accuracy of the original MC-PSVM. MC-Bag-PSVM is 332.66 times faster than LIBLINEAR with an improvement of 1.98% accuracy versus LIBLINEAR.

A forthcoming improvement will be to use different approaches [15,18,19,21] to train the binary PSVM classifier for handing the extreme unbalance between the positive and the negative class in large-scale multi-class datasets. We will provide more empirical test to illustrate the training speed of the MC-Bag-PSVM in terms of the number of processors.

Acknowledgments. This work has received support from the College of Information Technology, Can Tho University. We would like to thank very much the Big Data and Mobile Computing Laboratory.

References

1. Bosch, A., Zisserman, A., Muñoz, X.: Scene classification Via pLSA. In: Leonardis, A., Bischof, H., Pinz, A. (eds.) ECCV 2006. LNCS, vol. 3954, pp. 517–530. Springer, Heidelberg (2006). https://doi.org/10.1007/11744085_40
2. Breiman, L.: Bagging predictors. Mach. Learn. **24**(2), 123–140 (1996)
3. Chang, C.C., Lin, C.J.: LIBSVM: a library for support vector machines. ACM Trans. Intell. Syst. Technol. **2**(27), 1–27 (2011)
4. Chollet, F.: Xception: deep learning with depthwise separable convolutions. CoRR abs/1610.02357 (2016)
5. Deng, J., Berg, A.C., Li, K., Fei-Fei, L.: What does classifying more than 10,000 image categories tell us? In: Daniilidis, K., Maragos, P., Paragios, N. (eds.) ECCV 2010. LNCS, vol. 6315, pp. 71–84. Springer, Heidelberg (2010). https://doi.org/10.1007/978-3-642-15555-0_6
6. Deng, J., Dong, W., Socher, R., Li, L.J., Li, K., Li, F.F.: ImageNet: a large-scale hierarchical image database. In: IEEE Computer Society Conference on Computer Vision and Pattern Recognition, pp. 248–255 (2009)
7. Do, T.-N.: Parallel multiclass stochastic gradient descent algorithms for classifying million images with very-high-dimensional signatures into thousands classes. Vietnam J. Comput. Sci. **1**(2), 107–115 (2014). https://doi.org/10.1007/s40595-013-0013-2
8. Do, T.-N., Nguyen, V.-H., Poulet, F.: Speed up SVM algorithm for massive classification tasks. In: Tang, C., Ling, C.X., Zhou, X., Cercone, N.J., Li, X. (eds.) ADMA 2008. LNCS (LNAI), vol. 5139, pp. 147–157. Springer, Heidelberg (2008). https://doi.org/10.1007/978-3-540-88192-6_15
9. Do, T.-N., Poulet, F.: Parallel multiclass logistic regression for classifying large scale image datasets. In: Le Thi, H.A., Nguyen, N.T., Do, T.V. (eds.) Advanced Computational Methods for Knowledge Engineering. AISC, vol. 358, pp. 255–266. Springer, Cham (2015). https://doi.org/10.1007/978-3-319-17996-4_23
10. Do, T.-N., Tran-Nguyen, M.-T.: Incremental parallel support vector machines for classifying large-scale multi-class image datasets. In: Dang, T.K., Wagner, R., Küng, J., Thoai, N., Takizawa, M., Neuhold, E. (eds.) FDSE 2016. LNCS, vol. 10018, pp. 20–39. Springer, Cham (2016). https://doi.org/10.1007/978-3-319-48057-2_2

11. Doan, T.-N., Do, T.-N., Poulet, F.: Large scale classifiers for visual classification tasks. Multimed. Tools Appl. **74**(4), 1199–1224 (2014). https://doi.org/10.1007/s11042-014-2049-4

12. Fan, R.E., Chang, K.W., Hsieh, C.J., Wang, X.R., Lin, C.J.: LIBLINEAR: a library for large linear classification. J. Mach. Learn. Res. **9**(4), 1871–1874 (2008)

13. Fung, G., Mangasarian, O.L.: Proximal support vector machine classifiers. In: Proceedings of the seventh ACM SIGKDD International Conference on Knowledge Discovery and Data Mining, San Francisco, CA, USA, 26–29 August 2001, pp. 77–86 (2001)

14. Fung, G., Mangasarian, O.L.: Multicategory proximal support vector machine classifiers. Mach. Learn. **59**(1–2), 77–97 (2005)

15. He, H., Garcia, E.A.: Learning from imbalanced data. IEEE Trans. Knowl. Data Eng. **21**(9), 1263–1284 (2009)

16. He, K., Zhang, X., Ren, S., Sun, J.: Deep residual learning for image recognition. CoRR abs/1512.03385 (2015)

17. Herrero-Lopez, S., Williams, J.R., Sanchez, A.: Parallel multiclass classification using SVMs on GPUs. In: Proceedings of the 3rd Workshop on General-Purpose Computation on Graphics Processing Units, pp. 2–11. ACM, New York (2010)

18. Hido, S., Kashima, H.: Roughly balanced bagging for imbalanced data. In: SIAM International Conference on Data Mining, pp. 143–152 (2008)

19. Japkowicz, N. (ed.): AAAI'Workshop on Learning from Imbalanced Data Sets. No. WS-00-05 in AAAI Tech Report (2000)

20. Kreßel, U.H.G.: Pairwise classification and support vector machines. In: Schölkopf, B., Burges, C.J.C., Smola, A.J. (eds.) Advances in Kernel Methods, pp. 255–268. MIT Press, Cambridge (1999)

21. Lenca, P., Lallich, S., Do, T.-N., Pham, N.-K.: A comparison of different off-centered entropies to deal with class imbalance for decision trees. In: Washio, T., Suzuki, E., Ting, K.M., Inokuchi, A. (eds.) PAKDD 2008. LNCS (LNAI), vol. 5012, pp. 634–643. Springer, Heidelberg (2008). https://doi.org/10.1007/978-3-540-68125-0_59

22. Li, F., Perona, P.: A Bayesian hierarchical model for learning natural scene categories. In: 2005 IEEE Computer Society Conference on Computer Vision and Pattern Recognition (CVPR 2005), 20–26 June 2005, San Diego, CA, USA, pp. 524–531 (2005)

23. Lowe, D.: Object recognition from local scale invariant features. In: Proceedings of the 7th International Conference on Computer Vision, pp. 1150–1157 (1999)

24. Lowe, D.: Distinctive image features from scale invariant keypoints. Int. J. Comput. Vis. **60**, 91–110 (2004)

25. OpenMP Architecture Review Board: OpenMP application program interface version 3.0 (2008). http://www.openmp.org/mp-documents/spec30.pdf

26. Perronnin, F., Sánchez, J., Liu, Y.: Large-scale image categorization with explicit data embedding. In: IEEE Computer Society Conference on Computer Vision and Pattern Recognition, pp. 2297–2304 (2010)

27. Simonyan, K., Zisserman, A.: Very deep convolutional networks for large-scale image recognition. CoRR abs/1409.1556 (2014)

28. Sivic, J., Zisserman, A.: Video Google: a text retrieval approach to object matching in videos. In: 9th IEEE International Conference on Computer Vision (ICCV 2003), 14–17 October 2003, Nice, France, pp. 1470–1477 (2003)

29. Szegedy, C., Vanhoucke, V., Ioffe, S., Shlens, J., Wojna, Z.: Rethinking the inception architecture for computer vision. CoRR abs/1512.00567 (2015)

30. Vapnik, V.: The Nature of Statistical Learning Theory. Springer, Heidelberg (1995). https://doi.org/10.1007/978-1-4757-2440-0
31. Whaley, R., Dongarra, J.: Automatically tuned linear algebra software. In: Ninth SIAM Conference on Parallel Processing for Scientific Computing (1999). cD-ROM Proceedings
32. Wu, J.: Power mean SVM for large scale visual classification. In: IEEE Computer Society Conference on Computer Vision and Pattern Recognition, pp. 2344–2351 (2012)

Selective Combination and Management of Distributed Machine Learning Models

Takeshi Tsuchiya[1,2](✉), Ryuichi Mochizuki[1], Hiroo Hirose[1],
Tetsuyasu Yamada[1], Keiichi Koyanagi[3], and Quang Tran Minh[4,5]

[1] Suwa University of Science, Chino, Japan
{tsuchiya,hirose,yamada}@rs.sus.ac.jp
[2] Institute for Data Science Education, Tokyo International University,
Kawagoe, Japan
ttsuchi@tiu.ac.jp
[3] Waseda University, Tokyo, Japan
keiichi.koyanagi@waseda.jp
[4] Faculty of Computer Science and Engineering, Ho Chi Minh City University of
Technology, 268 Ly Thuong Kiet, District 10, Ho Chi Minh City, Vietnam
quangtran@hcmut.edu.vn
[5] Vietnam National University Ho Chi Minh City, Linh Trung Ward,
Thu Duc District, Ho Chi Minh City, Vietnam

Abstract. This study presents a method for selecting and combining
feature models constructed by the machine learning on the processing
task capability. The evaluation of combining the feature models shows
that the processing task capability can be improved by selecting and
reaching feature models based on their similarity to the vector of queries
without combining all feature models. Then, we discuss a method for
constructing logical the R-Tree algorithm on the distributed fog nodes.
For future work, we will implement the proposed method on various
types of data.

Keywords: Fog computing model · Distributed future model ·
Similarity of future models

1 Introduction

Recently, it has become possible to analyze user activities and interests using various types of information, such as user activities on Web services, Web browsing time, domain navigational history, and input information on Web sites. These analyses have significantly contributed to the advancement of Web services [1]. However, the information used in these analyses can be acquired by not just the Web service provider and user, but also third parties, such as Web-advertising providers who display advertisements on several websites. Therefore, there are privacy issues, such as leakage of sensitive information (e.g., interests and preferences), and in some cases, individual's identity [2].

© Springer Nature Switzerland AG 2021
T. K. Dang et al. (Eds.): FDSE 2021, LNCS 13076, pp. 113–124, 2021.
https://doi.org/10.1007/978-3-030-91387-8_8

In the future, the number of applicable services, analysis advancement, and the number of available data is expected to increase. However, because of the aforementioned privacy issues, it is difficult to accumulate user information across Web services and use them as secondary data. Thus, it is necessary to establish an information infrastructure that utilizes distributed technology without accumulating learning data.

Currently, federated learning of cohorts (FLOC) [3] is proposed to solve the user privacy problem. This method sends the browser history of the user to servers through the browser and the server assigns group IDs based on the tendency of these histories. It means that users are classified according to the characteristics of their similar Web activities. Thus, it can provide the same functions as conventional services using cookies, such as providing interest information based on the characteristics of these groups and distributing Web advertisements. However, this method does not solve the privacy problem; instead, it accumulates information with the third party by providing the browser history and performs user classification by determining similarity.

We have proposed a distributed machine learning platform for constructing learning models without accumulating training data and combining these feature models into arbitrary learning models [4]. The proposed method constructs a learning model with a group of arbitrarily sized nodes called fog nodes. Combining the feature models indicates the possibility of not directly learning the training data but only taking in the results after learning. In this study, we compare the effectiveness of performance by the difference of combining feature models. Then, we discuss the feature model selection algorithm for the processing task. Additionally, we clarify the efficient management method for combining feature models among fog nodes.

2   Related Research and Approaches

In this section, we describe related research works and introduce a novel approach constituting the main contribution of this study to solve the research problem raised in related publications.

2.1   Related Research

Current big data analysis is based on accumulating all data in a cloud environment managed by the same policy, and analysis is conducted using statistical methods and machine learning methods. As a result, the load on processing increases with an increase in the amount of training data and complexity of processing; thus, it is crucial to look for alternative data analysis methods. and many proposals are being considered.

Fog computing has been proposed to improve the throughput of cloud services by distributing cloud functions to the edge nodes of a network [5,6]. In this model, fog nodes at the network edge provide the functionality of a conventional cloud network, for example, data management and specific data processing functions.

A fog node works as a service gateway and local server for users. Therefore, it enables the transparent implementation of data processing in cloud networks and middleware. Currently, load-balancing methods for cloud networks based on fog nodes are widely discussed. In this study, we propose a method for ensuring the privacy of the information managed in user groups by employing fog nodes as a private proxy function.

A machine learning method without accumulating training data has been proposed in [8]. This method distributes a basic feature model with trained data (Fig. 1). The user can optimize the model for each environment by training additional features with their training data, and the difference in information between the feature model and the one obtained by the additional training is accumulated. Additionally, the basic feature model is updated from the average of these differences. Since the basic feature model is based on learning data from other domains, it must be qualitatively similar to the actual learning data in terms of distribution characteristics to be applicable to processing tasks. Thus, the target tasks for which the updated basic feature model can be applied are limited. To solve this problem, it is essential to acquire the universality of the basic model (improvement of the generalization performance) by processing complex data on a large amount of data [10]. However, since the training data's universality has several applications and is not absolute, there are limitations for using training data from other domains. Therefore, it is necessary to develop a method for constructing a feature model that efficiently learns a small number of training data.

The above methods adopt an approach of simplifying the data learning phase by providing pre-trained feature models based on large learning data. Therefore, the learning data acquired from the user are updated only with small differences in the feature models. However, this study adopts an approach of combining feature models constructed from distributed data to adapt to the performance of the processing task. Therefore, it is possible to adapt to the processing task by arbitrarily selecting the target of combining feature models by the user. Therefore, this study focuses on adapting to the processing task and discusses methods for selecting feature models to be combined.

2.2 Approach

In this study, we assume feature analysis using machine learning word2vec [7] for feature analysis of text information, such as Web contents and server logs. The following compares the differences between the performance task methods.

Our previous study [4] showed that the performance of a processing task changes depending on the combining feature model. When we assume a practical environment, the issue occurs as to what feature models should be combined for a processing task. Previous researchers [4] combined all distributed feature models and only showed the effectiveness of their combination by combining the performance with the common method, data accumulation type. Here, it does not mention the efficiency and performance change of the processing task in the model-combining process. This study discusses a method for combining

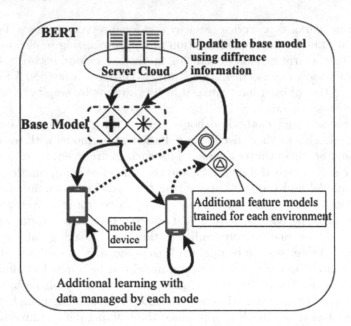

Fig. 1. Outline of the Base model Approach

feature models better than conventional methods by adaptively combining feature models to processing tasks (Fig. 2). First, we discuss the selection method for combining feature models by comparing the changes in performance and characteristics of combining feature models. Then, we discuss a feature model management approach effective for this feature model selection method.

The feature model is managed by a fog node selected from a group of nodes that can share the training data (lower part of Fig. 2). When training data cannot be shared, except for a specific node, a single node is defined as a fog node. All fog nodes are assigned a fog node ID for identification, enabling the identification of nodes and access to specific nodes. A fog node constructs a feature model using the training data managed by this node. This study discusses a structural model management method based on the learning model features on a logical space by collaboration among fog nodes.

Feature models for text data have a smaller and fixed dimensionality of vectors, which are the entities of the feature model, than images. Therefore, it can be assumed that there is no effect of the "curse of dimensionality," which increases the management cost with the dimensionality. Therefore, we discuss a distributed algorithm for indexing feature models in a logical space.

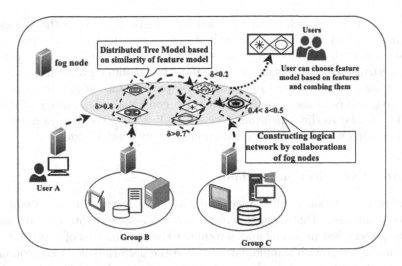

Fig. 2. Outline of the proposal model

3 Combining Feature Models

This section discusses the selection of feature models in combining distributed feature models constructed from a corpus and clarifies the effect of differences in the feature models that are selected on the basis of their performance.

3.1 Distributed Feature Models

The combined feature model uses 100, 000 Wikipedia [11] articles as the training data and divides the data into 100 nodes. The corpus comprises sufficient number and diversity of categories; thus, there is no bias toward a particular category. Each node constructs a feature model for the deployed data. Here, each fog node synchronizes the statistics of words by executing word2vec [12] of the method proposed in the previous study [4], and each of them constructs a feature model. Thus, it is possible to combine up to 100 feature models.

The feature models are constructed using word2vec as high-dimensional vectors having the occurrence characteristics of words in the training data. First, combining feature models is conducted by averaging the dimension values of each feature model. Then, the performance of the combined feature models is evaluated for each combined feature model using evaluation data assuming a processing task.

3.2 Sequential Combining Method

In reference [4], the feature models managed by each fog node are combined sequentially in the order of their fog node IDs as a simple method for combining feature models. Here, feature models are combined sequentially regardless of the characteristics of each feature model. This method is called the sequential combining method. However, this method cannot be adapted to the processing task. We found that the performance of this method is similar to the conventional method for accumulating training data by combining all feature models. However, we have no metric to guarantee the performance of this process.

At first, the candidate feature models are combined in the order of the fog node IDs, similar to the sequential method. Additionally, the feature models are evaluated using sample data for evaluation at every combination.

3.3   Adaptive Selection Method

The adaptive selection method focuses on selecting feature models adapted to the processing task. Therefore, this method combines candidate feature models with the current feature model and determines the combination of feature models by evaluating them with sample data. In case the performance of the combined feature model is better than the previous one, the combined feature model is adopted. However, if the performance is reduced, this combined feature model is discarded. Then, the previous feature model is combined with the next feature model to be evaluated. These steps are repeated.

This method must have a process of combining all feature models similar to the above method. However, this method is inefficient in a distributed environment because it will take a long time to construct a feature model with good performance. The adaptive selection method is expected to increase the time required for its establishment since it requires an evaluation process.

3.4   Similarity Model Retrieved Method

In contrast to the above methods for combining future model, we discuss a similarity model retrieval method that retrieves and selects feature models adapted to the processing task or having specific characteristics. This method retrieves and combines a list of feature models that have similarities to any feature model designated by the user. Here, the similarity is used as an index for acquiring feature models, and each feature model is managed structurally based on its characteristics in advance, enabling us to select and combine only those feature models satisfying the requirements without having to search for all of them. Therefore, it is necessary to consider and evaluate the threshold value to determine the degree of similarity of the feature models to be combined.

Figure 3 shows the performance of each feature model using the above methods. Additionally, it shows the relationship between the number of feature model combinations (x-axis) and each method's processing task performance (y-axis).

3.5   Comparison of Performance by Combining

The performance of each feature model is shown in Fig. 3 using the methods described above. Figure 3 shows the relationship between the number of feature model combinations (x-axis) and the processing task performance (y-axis) for each of the methods described above.

The performance change using the basic sequential combination method is shown by "Sequential Combining" in Fig. 3. Here, the methods performance is increased in the first part up to about 30 nodes proportionally to the number of feature model combinations. Then, the performance of the processing task slowly decreases with the combined feature models. Moreover, the performance of the task processing is characterized by the change and unstable performance for each combination based on the sequential ordering of fog node IDs. For instance, the first half improves performance according to the number of combinations, but the performance of the second half decreases since the combinations generated in the second half are not adapted to the processing task.

The adaptive selection method is indicated by "Adaptive Selection" in Fig. 3. This method stably shows better performance by combining feature models than the sequential combination method. The reason is that combining feature models are limited to models that improve performance. The effectiveness of the selecting feature models for combining is clear. However, the problem is that all feature models must be retrieved and sequential selection of combining and evaluation using sample data must be conducted repeatedly. Therefore, there are some issues to be solved, such as the distributed scalability of the learning models to be combined; the determination of the threshold to decide the combination; the selection method of the evaluation sample data.

As an example, we retrieve feature models similar to the feature model with the average value of all feature models and combined them satisfying the condition. Figure 3 shows the relationship between the number of combined feature models and processing task performance. Compared to other methods, the performance difference caused by combining similarity feature models is small, indicating a low effectiveness of our method.

The "Similarity Model Retrieved (Reverse)" shows the case of combining the feature models with the lowest similarity having the most different characteristics. Therefore, the method performs the processing task with fewer feature model combinations than the other methods. It shows a more efficient approach to reach the feature models than all feature model queries. Additionally, combining feature models with high similarity is analogous to the overfitting occurrence, where certain dimensions are enhanced. If the characteristics generated by overfitting do not represent the characteristics of query data used in the evaluation, it is difficult to achieve good performance. However, combining feature models with different characteristics is effective since feature models with similar overall characteristics are retrieved and combined without emphasizing any certain dimension. Moreover, this method enables combining feature models without retrieving all of them.

3.6   Combining Policy

From the results of Sect. 3.5, selecting feature models to combine by retrieval of similar feature models is effective from the performance and efficiency viewpoint.

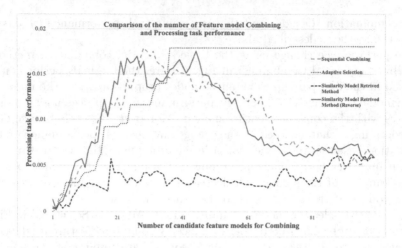

Fig. 3. Process of combining feature models

However, it is necessary to manage the feature models using a unique metric, such as similarity, to retrieve arbitrary feature models efficiently. Thus, we consider a method for managing feature models in the logical space constructed by cooperating fog nodes in the following sections.

4   Feature Model Management

This section discusses a method for users to reach arbitrary feature models by managing them based on their features on the logical space constructed by fog nodes.

4.1   Management Policy

For feature-based retrieval, the spatial indexing method (a method for partitioning the vector space) and locality-sensitive hashing method (an approximation method based on the randomness and probabilistic processing of hash functions) are used in conventional research. Since our feature model is based on textual information as learning data, it does not have a high dimensionality as image data. Additionally, the number of dimensions is uniquely determined by the synchronization among fog nodes when converted into numerical values (vectorization). The effect of the "curse of dimensionality," which is the increase in management cost proportional to the number of vector dimensions, is insufficient. We propose to apply logical R-Tree [9] algorithm constructed on the distributed fog nodes, which enables the indexing of feature models on the logical space.

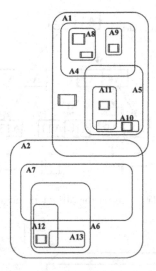

Fig. 4. Allocation of feature model on logical space

4.2   Similarity Retrieval of Feature Models

R-Tree [9] divides the logical space using the minimum bounding rectangle (MBR) and enables access to data in the neighborhood. An example is illustrated in Fig. 4. Each dimension of the feature model represents the features acquired by data learning at each fog node. Combining feature models requires the similarity of the entire feature model (vector), i.e., the approximation of the feature model in the vector space instead of the approximation of the values of each dimension. The R-Tree constructs a logical tree structure based on the range of MBR and can manage the distances that indicate the approximations among feature model objects (Fig. 4). For instance, it is possible to reach an arbitrary feature model, such as the feature model with the lowest similarity, as a candidate for combining ones.

Applying the R-Tree requires placing the feature models managed by each fog node in the logical space in advance. Thus, the query is retrieved for similarity to each feature model. All feature models are based on a vector consisting of the average of each dimension as the reference vector. Additionally, the feature models are managed based on their similarity to this vector. Figure 5 shows the construction of a tree structure. At the time of retrieval, it derives the similarity between the query vector and this reference vector. Then, it decides the candidates to combine the feature models based on whether the similarity is high or low. Since the feature model on the R-Tree is the similarity with the reference vector, similarity with the query is finally derived and determines the candidates of feature model comminating or not.

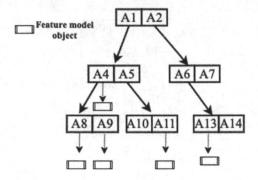

Fig. 5. Tree model for object management

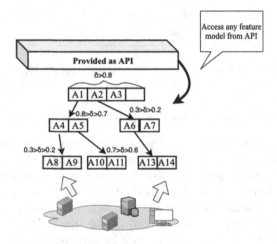

Fig. 6. Implementation of feature model management

4.3 R-Trees in a Distributed Environment

The R-Tree algorithm can be configured flexibly for the number of objects to be managed and the division range. The current implementation constructs the tree structure managed under 20 future models per node. The tree structure is constructed so that the range space of similarity managed by the nodes is $\frac{1}{2}$.

4.4 Evaluation

The implementation of the proposed R-Tree is still in progress, and we could not evaluate it in this study. However, it is implemented as an upper layer of the already implemented distributed machine learning infrastructure with fog nodes (Fig. 6), and it will be evaluated in the future. The performance is expected to vary greatly in the evaluation, depending on the choice of feature models to be combined (Fig. 3). We would like to define a metric for selecting these combinations by evaluating the selection patterns of various feature models.

5   Issues and Summary

In this paper, we focused on the process of combining distributed feature models and clarified the change in performance according to different combination methods. From this result, we discussed a method for managing feature models by the similarity to enable their combination with arbitrary feature models. Unfortunately, we could not reach the evaluation phase but discussed a method for managing them by applying the R-Tree algorithm. The proposed method enables the selection of combined feature models by the similarity between them and the query. In future studies, we will evaluate these methods and clarify a flexible method for combining feature models adapted to the processing task.

Acknowledgement. This research was partially supported by the Ministry of Education, Science, Sports and Culture, Grant-in Aid for Scientific Research (C), 2021–2023 21K11850, Takeshi TSUCHIYA.

Work by Tran Minh Quang acknowledges the support of time and facilities from Ho Chi Minh City University of Technology (HCMUT), VNU-HCM.

References

1. Tsuchiya, T., Hirose, H., Yamada, T., Yoshinaga, H., Koyanagi, K.: Predicting user interests based on their latest web activities. In: Proceedings of 13th International Conference on Mobile Ubiquitous Computing, Systems, Services and Technologies (UBICOMM), September 2019
2. Bessis, N., Dobre, C.: Big Data and Internet of Things: A Roadmap for Smart Environments. Studies in Computational Intelligence, pp. 169–186. Springer, Cham (2014). https://doi.org/10.1007/978-3-319-05029-4
3. https://github.com/WICG/floc
4. Tsuchiya, T., Mochizuki, R., Hirose, H., Yamada, T., Koyanagi, K., Minh Tran, Q.: Distributed data platform for machine learning using the fog computing model. SN Comput. Sci. **1**(3), 1–9 (2020). https://doi.org/10.1007/s42979-020-00171-6
5. Bonomi, F., Milito, R., Natarajan, P., Zhu, J.: Fog computing: a platform for internet of things and analytics. In: Bessis, N., Dobre, C. (eds.) Big Data and Internet of Things: A Roadmap for Smart Environments. SCI, vol. 546, pp. 169–186. Springer, Cham (2014). https://doi.org/10.1007/978-3-319-05029-4_7
6. Alrawais, A., Alhothaily, A., Hu, C., Cheng, X.: Fog computing for the internet of things: security and privacy issues. IEEE Internet Comput. **21**(2), 34–42 (2017)
7. Mikolov, T., Chen, K., Corrado, G., Dean, J.: Efficient estimation of word representations in vector space, CoRR, Vol. abs/1301.3781 (2013)
8. McMahan, H.B., Moore, E., Ramage, D., Hampson, S., Arcas, B.A.: Communication-efficient learning of deep networks from decentralized data. In: Proceedings of the 20th International Conference on Artificial Intelligence and Statistics, JMLR: W&CP, vol. 54, pp. 169–186 (2014)

 9. Guttman, A.: R-trees: a dynamic index structure for spatial searching, vol. 14, no. 2. ACM (1984)
10. Devlin, J., Chang, M.-W., Lee, K., Toutanova, K.: BERT: pre-training of deep bidirectional transformers for language understanding, CoRR, abs/1810.04805, October 2018
11. Wikipedia: https://en.wikipedia.org/
12. Mikolov, T., Chen, K., Corrado, G., Dean, J.: Efficient estimation of word representations in vector space, CoRR, Vol. abs/1301.3781 (2013). word2vec

Improving ADABoost Algorithm with Weighted SVM for Imbalanced Data Classification

Vo Duc Quang[1,2]([✉]), Tran Dinh Khang[1]([✉]), and Nguyen Minh Huy[1]

[1] Hanoi University of Science and Technology, Hanoi, Vietnam
khangtd@soict.hust.edu.vn, huy.nm170773@sis.hust.edu.vn
[2] Vinh University, Vinh City, Vietnam
quangvd@vinhuni.edu.vn

Abstract. Recently, different boosting algorithms have been proposed in order to improve the performance of classification for imbalanced data. In this paper, we present an improved ADABoost algorithm, called Im.ADABoost, for imbalanced data including two main improvements: *(i)* initializing different error weights adapted to the imbalance rate of the datasets; *(ii)* calculating the confidence weights of the member classifier that is sensitive to the total errors caused on the positive label. Additionally, we combine Im.ADABoost with Weighted-SVM to enhance classification efficiency on imbalanced datasets. Our experimental results show some promising potential of the proposed algorithm.

Keywords: Imbalanced dataset · ADABoost · Support vector machine

1 Introduction

The analysis of imbalanced datasets has received much attention from researchers, especially in practical problems such as diagnosing diseases in medicine, detecting environmental problems, financial transaction fraud, and network attacks,.... However, there are a number of the major challenge remains that need further study in the field of machine learning and data mining [12]. In this study, we consider classification problems on the imbalanced data of two class labels, in which the class label with the majority of data samples is called the negative label (-1) and the other with the minority of data sample is called the positive label ($+1$).

Usually, traditional classification algorithms always consider data samples as equal. Algorithmic improvement studies are often geared towards trying to train the classification model with the highest accuracy rate. However, when applying these algorithms to imbalanced dataset problems, the classification model will be biased towards prioritizing class label recognition as (-1). In the case of a highly imbalanced dataset, the trained classification model tends to classify all label samples ($+1$) as labels (-1). These results give a very high accuracy but fail to

© Springer Nature Switzerland AG 2021
T. K. Dang et al. (Eds.): FDSE 2021, LNCS 13076, pp. 125–136, 2021.
https://doi.org/10.1007/978-3-030-91387-8_9

properly classify any of the positive label samples. In the context of problems that need to accurately detect all labels (+1), this becomes meaningless. According to [10], approaches to improve classification on imbalanced datasets usually are:

- Using preprocessing methods on the dataset: (*i*) reduce the number of samples on the label (−1); (*ii*) generate additional artificial data samples label (+1) or (*iii*) combine both techniques. The above methods aim to reduce the imbalance of datasets in order to make traditional machine learning algorithms work efficiently [3,19,26].
- Improving algorithms: adjust the traditional algorithms so that they are more suitable for (+1) labels. The most popular method is error weight assignment, cost-based learning [1,5,9,14,18,21,22]. Some other studies apply deep learning models to imbalanced datasets [4,11,13,25].

Among the above approaches, the cost-based learning algorithm assigns a higher cost weight when the model misclassifies the sample label (+1) to the label (−1). Accordingly, this algorithm has many outstanding advantages such as: keeping the original characteristics of datasets, having many ways to improve the training parameters, and minimizing the error cost function by using loops in conjunction with turning the parameters. However, using a particular algorithm may not fully consider the attributes of datasets. Therefore, many studies combine member classifiers to produce a better composite classifier [7,8]. In particular, the ADABoost algorithm proposed by Freund [6] has been improved by many researchers, notably the study of combining ADABoost with Support Vector Machine (SVM) [2,15–17,20,23,24]. These improvements are intended to take advantage of ADABoost's adaptive iterability and SVM's scalability on datasets with different characteristics. In [15], the authors proposed a method that combines ADABoost with Weight SVM (W-SVM) to improve classification efficiency on imbalanced data.

However, the algorithm in [15] and other studies using ADABoost on imbalanced data initialize equal error weights for each data sample. This algorithm is not suitable when being applied to problems that need to prioritize the correct classification of labels (+1). In addition, ADABoost calculates the confidence weight of the membership classification algorithm based on the total error in the entire dataset without considering the details of each label type (+1) and (−1). Based on these observations, we propose an algorithm, called Im.ADABoost, by making two major improvements to the original ADABoost. Our improvements consist of: (*i*) initializing the set of different error weights adapted to the imbalance rate of the datasets; (*ii*) calculating the confidence weights of the member classifiers based on sensitivity to the total error caused on positive labels (+1), i.e., if the member classifier misclassifies more samples (+1), the lower its confidence weights will be.

We also combine Im.ADABoost with W-SVM into Im.ADABoost.W-SVM algorithm to classify imbalanced datasets. We used the recall, accuracy and fscore measures in our experiments to compare the performance with other classification algorithms on different imbalanced datasets. Experimental results show that the Im.ADABoost.W-SVM algorithm gives better classification performance on imbalanced datasets, especially when datasets have a high imbalance ratio. The rest of the paper is structured as follows: Sect. 2 recalls the description of the original ADABoost algorithm and related ones; Sect. 3 presents improvements to the ADABoost algorithm combined with the W-SVM membership classifier; Sect. 4 describes the experimental results; Some discussion of the results and directions for future work are presented in Sect. 5.

2   Preliminaries

In classification problems on imbalanced datasets, using only a particular classification algorithm may not be efficient. ADABoost is an iterative algorithm that combines membership classification algorithms. This allows detailed testing of each sample in the dataset space by assigning each data sample an error weight. Through each iteration, the ADABoost re-evaluates the classification results of each membership classification algorithm, thereby calculating better parameters to use for the next iteration. The ADABoost algorithm is presented in Algorithm 1.

Algorithm 1: ADABoost

Input: A dataset with N samples $X = \{(x_1, y_1), \ldots, (x_N, y_N)\}$ with
$\quad\quad y = \{-1, +1\}$, T: maximum iteration, h_i: the member classifier
Output: H: Ensemble classifier

1 **initialize:** the error weight $D^1 = \{\omega_i^1\}$ on each sample x_i with
$\quad i = 1, \ldots, N$
2 **for** $t = 1$ **to** T **do**
3 $\quad$ Set $h_t \leftarrow Training(X)$ with the error weight D^t;
4 $\quad$ Calculate total error traning: $\varepsilon_t = \sum_{t=1}^{N} \omega_i^t, y_t \neq h_t(x_i)$;
5 $\quad$ Calculate confident weight of h_t: $\alpha_t = \frac{1}{2} ln \frac{1-\varepsilon_t}{\varepsilon_t}$;
6 $\quad$ Update the error weight for the next iteration:
$\quad\quad \omega_i^{t+1} = \frac{\omega_i^t . exp[-\alpha_t y_i h_t(x_i)]}{L_t}$, where L_t the normalization constant and
$\quad\quad \sum_{i=1}^{N} \omega_i^{t+1} = 1$;
7 **return** $H = sign(\sum_{t=1}^{T} \alpha_t h_t)$.

The inputs of the algorithm include: (i) X is a dataset with N samples (x_i, y_i), where x_i is the attribute vector and y_i is the class label $y_i \in \{-1, +1\}$; (ii) D^1 is a set of equal error weights for each sample $w_i^t = \frac{1}{N}$; (iii) T is the maximum number of iterations; (iv) h_t is the membership classification algorithms.

In each loop, the classifier h_t classifies the dataset X at Step 3 of Algorithm 1. The classification quality of h_t is evaluated through the sum of error ε_t at Step 4 and confidence weight α_t at Step 5. Then the algorithm updates the error weight distribution ω_i^{t+1} using the formula at Step 6.

The synthetic classification model is calculated according to the formula H at Step 7. The classification label of the sample is determined based on the sign function: label $(+1)$ when $H > 0$ and label (-1) when $H < 0$. When the total error ϵ_t on the dataset is equal to 0.5, then $\alpha_t = 0$, then the classifier h_t does not contribute to the decision of the composite classifier H.

We can see that the error weight ω_i^t assigned to each data sample is initialized equally. At each iteration, ADABoost analyzes the classification result of each member learner and updates the weights for each sample, as follows: increase the error weight assigned to the sample if it is misclassified and decrease the error weight if the sample is correctly classified. However, in the case of an imbalanced dataset, we need to adjust the algorithm to take a closer look at the positive labels, that is, to assign a higher error weight to the positive label samples. In addition, ADABoost calculates the confidence weight of the membership classification algorithm based on the total error on the entire dataset without considering the details of each label type $(+1)$ and (-1). These observations are the basis for us to develop ADABoost improvement methods for imbalanced data in Sect. 3, in which the algorithm prioritizes increasing the error weights when the member classifier misclassifies a positive label sample $(+1)$.

Wonji Lee et al. [15] proposed to combine ADABoost with W-SVM as a membership classification algorithm. This algorithm uses parameters z_i^t to adjust the weights on samples in W-SVM. The value of z_i^t is calculated based on the number of samples x_i distributed in the SVM marginal space. The formula to calculate z_i^t is :

If the sample x_i is in the border subclass BSV:

$$z_n^t = \begin{cases} \frac{N_{BSV}}{2N_{BSV-}} & if \quad y = -1, \\ \frac{N_{BSV}}{2N_{BSV+}} & if \quad y = +1. \end{cases} \tag{1}$$

If the sample x_i is in the border subclass SV:

$$z_n^t = \begin{cases} \frac{N_{SV}}{2N_{SV-}} & if \quad y = -1, \\ \frac{N_{SV}}{2N_{SV+}} & if \quad y = +1. \end{cases} \tag{2}$$

If the sample x_i is noise:

$$z_n^t = exp(\frac{N_{noisy}}{N_+}). \tag{3}$$

In the above formulas, N_{BSV} is the number of samples in the SVM boundary, N_{SV} is the number of samples generating the support vector and N_{Noisy} is the noise sample; Positive class label samples are represented by $(+)$, negative class label samples are represented by $(-)$; N_+ is the number of positive label samples. Based on the ADABoost.W-SVM algorithm, in Sect. 3, we propose two improvements to ADABoost algorithm, then combine our improved algorithm with W-SVM to classify the imbalanced datasets.

3 Proposed Method

3.1 Initialize Adaptive ADABoost Weights

In Sect. 2, we analyzed the limitation of ADABoost in initializing error weights when it is applied to an imbalanced dataset. In this section, we propose a new method to initialize error weights that adapts to the class-label imbalance ratio of the dataset. This method aims to assign a higher error weight on the positive label sample, $+1$. Assume N_{min} and N_{maj} are the number of samples of the minority and majority class, respectively, where $N_{min} \leq N_{maj}$. In ADABoost, each sample is assigned an error weight of $\frac{1}{N_{min}+N_{maj}}$ and therefore, ADBoost is inefficient for imbalanced datasets. We adjust the error weights by adding a Δ_{min} value to error weights of positive samples and subtracting a Δ_{maj} value from error weights of negative samples. This means the error weights on each positive sample will be $\frac{1}{N}+\Delta_{min}$, and on each negative sample will be $\frac{1}{N}-\Delta_{maj}$, where Δ_{min} and Δ_{maj} must satisfy the following conditions:

- Error weights are greater than 0, that is: $0 < \Delta_{min}, \Delta_{maj} < \frac{1}{N}$;
- The total error on the samples is equal to 1, that is:
  $\frac{N_{min}}{N} + N_{min} * \Delta_{min} + \frac{N_{maj}}{N} - N_{maj} * \Delta_{maj} = 1$;
- When the number of positive samples is equal to the number of negative samples, the dataset is balanced, then the error weight on each sample is equal to $\frac{1}{N}$.

If the ratio of the number of samples of positive to negative labels is set to $\delta = \frac{N_{min}}{N_{maj}}$, with $0 < \delta \leq 1$, then the above expression becomes:

$$\begin{cases} 0 < \Delta_{min}, \Delta_{maj} < \frac{1}{N} \\ \Delta_{maj} = \delta * \Delta_{min}. \end{cases} \tag{4}$$

We propose to choose $\Delta_{maj} = \frac{1-\delta}{N}$, thus $\Delta_{min} = \frac{1-\delta}{\delta*N}$. Accordingly, the set of bias weights is $D^t = \omega_i^t$ with $i = 1, 2, \ldots N$, and

$$\omega_i^t = \begin{bmatrix} \frac{1}{N} + \frac{1-\delta}{\delta*N}, & if & y_i = +1 \\ \frac{1}{N} - \frac{1-\delta}{N}, & if & y_i = -1. \end{bmatrix} \tag{5}$$

It can be seen that, when applying the formula (5) to datasets with different imbalance rates, the error weights on positive and negative labels will increase and decrease respectively depending on δ. When the dataset is balanced, meaning that $\delta = 1$, then Δ_{min} and $\Delta_{maj} = 0$, and therefore the initialization weights D^1 return to ADABoost's default (error weights on all samples are equal to $1/N$). In addition, in order to dynamically adjust the Δ_{min} and Δ_{maj} values according to the individual characteristics of the datasets, we propose a more general formula using the exponential parameter θ as follows:

$$\begin{aligned} \Delta_{maj} &= \frac{(1-\delta)^\theta}{N}, \\ \Delta_{min} &= \frac{(1-\delta)^\theta}{\delta*N}. \end{aligned} \tag{6}$$

For each particular dataset, we can find the most suitable exponential parameter value through the process of testing on a given set of values. This improvement makes ADABoost more generalizable on datasets with different imbalance rates. Moreover, if the algorithm uses a threshold that eliminates unnecessary membership classifiers, then it converges faster, meaning that it also reduces the number of iterations.

3.2 Positive Label Sensitive Confidence Weights of the Membership Classifier

At Step 5 of Algorithm 1, the confidence weight α_t of the member classifier h_t is calculated by a function that is inversely proportional to the total error ε_t. We find that this total error is considered equally across the misclassified samples. For the classification problem on the imbalanced dataset, the algorithm should give priority to assigning a high error weight when it misclassifies many positive-label samples $(+1)$. We propose a new total error ε_t^* instead of ε_t, which is calculated by the total error of positive labels, denoted by ε_t^+, and that of negative labels, denoted by ε_t^-, i.e. $\varepsilon_t^* = \varepsilon_t^- + \varepsilon_t^+$ where $\varepsilon_t^+ = \sum_{i=1}^{N} \omega_i^t, y_i \neq h_t(x_i), y_i = +1$ and $\varepsilon_t^- = \sum_{i=1}^{N} \omega_i^t, y_i \neq h_t(x_i), y_i = -1$. Obviously, ε_t^* depends on ε_t^+ and ε_t^-, and if we want our model to classify precisely on positive labels, then we need to increase ε_t^+ and therefore, we redefine ε_t^* as follows:

$$\varepsilon_t^* = \varepsilon_t^- + \gamma * \varepsilon_t^+, \text{ subject to } \gamma > 1. \tag{7}$$

Since $0 < \varepsilon_t^- + \varepsilon_t^+ < 1$, we choose $\gamma = 2 - (\varepsilon_t^- + \varepsilon_t^+)$. Then, the confidence weight of the model is

$$\alpha_t^* = \frac{1}{2} ln \frac{1 - \varepsilon_t^*}{\varepsilon_t^*}. \tag{8}$$

Obviously, the total error value ε_t^* in (7) of the model increases with the total error on the positive label ε_t^+, resulting in the confidence weight value α_t^* being adjusted down accordingly. This means the algorithm will try to correctly classify as many positive-label data samples as possible.

3.3 Im.ADABoost.W-SVM Algorithm

We call ADABoost with two improvements in Sects. 3.1 and 3.2 an Im.ADABoost. Accordingly, we propose an algorithm using Im.ADABoost combined with W-SVM, called Im.ADABoost.W-SVM, where W-SVM is used as a member classifier. The Im.ADABoost.W-SVM scheme is described in Fig. 1 and Im.ADABoost.W-SVM algorithm is shown in Algorithm 2.

Our Im.ADABoost algorithm initializes the inputs $z_i^1 = 1$ and D^1, the set of adaptive error weights, calculated by (5) and (6) in Sect. 3.1 (i.e. the input in Fig. 1). The algorithm runs for T iterations and in each loop it performs as follows. First, the algorithm uses W-SVM to classify the dataset X by using the parameters z_i^1 and the error weights on the samples $D_t = \omega_i^t$ with $i = 1, 2, \ldots, N$ (i.e. Step W-SVM in Fig. 1). Then, the algorithm computes and updates the

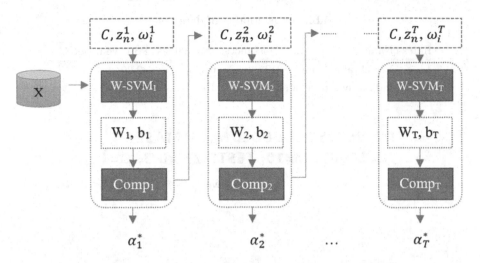

Fig. 1. Scheme of Im.ADABoost algorithm combined with W-SVM.

Algorithm 2: Im.ADABoost.W-SVM

Input: A dataset with N samples $X = (x_1, y_1), \ldots, (x_N, y_N)$, T: maximum iteration, h_i: the member classifier, C: W-SVM control parameters

Output: H: Ensemble classifier

1 **initialize:** $z_i^1 = 1, i = 1, 2, \ldots, N$ and ω_i^1 (using Eq.(5),(6));
2 **for** $t = 1$ **to** T **do**
3       $h_t \leftarrow$ Training W-SVM(X) with the error weight D^t and $z_i^t * \omega_i^t$, $i = 1, 2, \ldots, N$;
4   Calculate z_i^{t+1} (using Eq.(1),(2),(3));
5   Calculate total training error: ε_t^* (using Eq.(7));
6   Calculate the confidence weight of the h_t classifier: α_t^* (using Eq.(8));
7   Calculate error weight allocation for next loop: $\omega_i^{t+1} = \frac{\omega_i^t e^{-\alpha_t v_i h_t(x_i)}}{L_t}$, L_t is normalization constant, $\sum_{i=1}^{N} \omega_i^{t+1} = 1$;
8 **return** $H(x) = sign(\sum_{t=1}^{T} \alpha_t^* h_t(x))$.

---

parameters z_i^2 and D_2 for the next loop based on the *true/false* classification results on the samples (i.e. Step Comp$_1$ in Fig. 1). It should be noted that the reliability α_t^* is calculated by (8) in Sect. 3.1. After the completion T loops, the aggregate classifier H predicts the class labels for the samples using the formula: $H(x) = sign(\sum_{t=1}^{T} \alpha_t^* h_t(x))$.

4 Experiments

In this section, we present experiments to evaluate the performance of our Im.ADABoost algorithm. To do so, we compared the classification results of Im.ADABoost.W-SVM with that of ADABoost.W-SVM [15] and ADABoost

Table 1. Description of datasets

Index	Name	#instances	#variables	Positive ratio(%)
1	Vertebral column	310	6	32.26
2	Indian liver patient	583	10	28.64

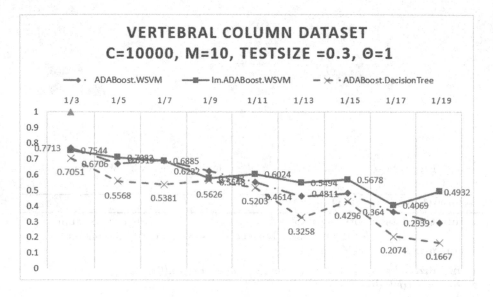

Fig. 2. Classification results for Vertebral column dataset.

DecisionTree [7] on two datasets with different imbalance rates[1] detailed in Table 1. On each dataset, we randomly select samples of (+1) labels, cut some positive labels (+1) to create a new sub-dataset with increasing imbalance ratio 1:5, 1:7, 1:9, 1:11, 1:13, 1:15, 1:17, 1:19. Our experimental scenarios use the following parameters: TestSize is 0.3 and 0.4; the number of Boosting iterations $T = 10$; the input parameter for SVM C is tested from 50:10:15000 to get the best value is $C = 10000$; the parameter θ of the Im.ADABoost is tested from 20:-1:1 to choose the most suitable for the dataset. We compare the classification efficiency of the algorithms by using the measures: *precision, recall, F1-Score*, taking the average results from the 10-fold method.

Using the *F1-Score* measure, $\theta = 1$, the classification results of positive labels (+1) of the algorithms on the *Dataset Vertebral column* dataset are presented in Fig. 2. The classification results of the Im.ADABoost.W-SVM algorithm on the *Dataset Vertebral column* dataset with different θ values are illustrated in Fig. 3. These results show that with datasets have positive label ratio, from 1:3 to 1:19, the ADABoost.W-SVM and Im.ADABoost.W-SVM algorithms always gives better results than ADABoost.DecisionTree. In

---

[1] https://www.kaggle.com/datasets.

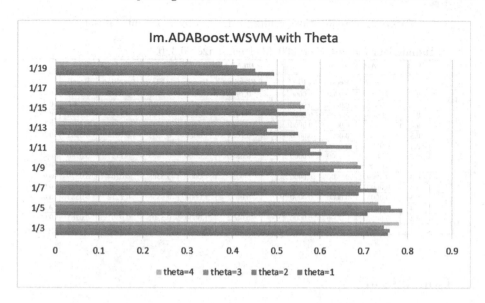

Fig. 3. Classification results for Vertebral column dataset with Theta.

addition, when the dataset has a low imbalance ratio, positive label ratio from 1:3 to 1:9, ADABoost.W-SVM and Im.ADABoost.W-SVM give approximately the same results. However, we can always find a suitable value θ that makes Im.ADABoost.W-SVM better than ADABoost.W-SVM. When the dataset has a high imbalance ratio, positive label ratio from 1:11 to 1:19, the Im.ADABoost.W-SVM algorithm gives a much better classification performance than ADABoost.W-SVM and ADABoost.DecisionTree with most experimental values θ.

Classification results on the *Indian liver patient* datasets are presented in Table 2. In all parameters used, these results show that the ADABoost.W-SVM algorithm does not correctly classify any positive label, and ADABoost.DecisionTree gives progressively worse results according to the decrease in the percentage of positive labels while Im.ADABoost.W-SVM correctly classifies all positive labels. This is very meaningful in the prediction problems of rare events. Moreover, when observing the values of θ, we see that, for different imbalance ratios of datasets, we can always choose a suitable θ value to help Im.ADABoost.W-SVM give the best performance. For example, given that dataset has a positive label ratio is 1:7, when testing the value $\theta = 20 : -1 : 1$, if $\theta \leq 4$, the Im.ADABoost.W-SVM algorithm gives results $recall = 1$, i.e. correctly classifies all positive labels (+1). It should also be noted that the adjusted values for error weights in the Im.ADABoost algorithm decrease as the θ exponent increases accordingly. When $\theta > 4$, the adjusted value for the error weights of the samples is too small, and this impact is not enough to prioritize the correct classification of positive labels (+1) in the Im.ADABoost algorithm.

Table 2. Classification results for Indian liver patient dataset.

Dataset **Indian liver patient**, C=10000, M=10, testsize =0.3, 0.4

positive label rate (+1)	Test size	ADABoost .WSVM			Im.ADABoost .WSVM				ADABoost .ID3		
		Precision	Recall	F1	θ	Precision	Recall	F1	Precision	Recall	F1
1:5	0.3	0.0000	0.0000	0.0000	3	0.1975	1.0000	0.5987	0.2852	0.1902	0.2377
	0.4	0.0000	0.0000	0.0000		0.2019	1.0000	0.6010	0.2749	0.1980	0.2365
1:7	0.3	0.0000	0.0000	0.0000	4	0.1438	1.0000	0.5719	0.2377	0.1328	0.1853
	0.4	0.0000	0.0000	0.0000		0.1443	1.0000	0.5722	0.2195	0.1447	0.1821
1:9	0.3	0.0000	0.0000	0.0000	6	0.1064	1.0000	0.5532	0.1816	0.0982	0.1399
	0.4	0.0000	0.0000	0.0000		0.1070	1.0000	0.5535	0.1748	0.1316	0.1532
1:11	0.3	0.0000	0.0000	0.0000	7	0.0870	1.0000	0.5435	0.1197	0.0746	0.0971
	0.4	0.0000	0.0000	0.0000		0.0870	1.0000	0.5435	0.0846	0.0592	0.0719
1:13	0.3	0.0000	0.0000	0.0000	8	0.0735	1.0000	0.5368	0.0704	0.0579	0.0642
	0.4	0.0000	0.0000	0.0000		0.0778	1.0000	0.5389	0.1130	0.0789	0.0960
1:15	0.3	0.0000	0.0000	0.0000	10	0.0672	1.0000	0.5336	0.1212	0.0585	0.0898
	0.4	0.0000	0.0000	0.0000		0.0674	1.0000	0.5337	0.1127	0.0746	0.0936
1:17	0.3	0.0000	0.0000	0.0000	11	0.0602	1.0000	0.5301	0.0301	0.0263	0.0282
	0.4	0.0000	0.0000	0.0000		0.0565	1.0000	0.5282	0.1066	0.0684	0.0875

5 Conclusion

In this paper, we have proposed an algorithm, called Im.ADABoost, by making two major improvements to the original ADABoost. The main modifications include *(i)* initializing the set of different error weights adapted to the imbalance rate of the datasets, which is adjusted by a parameter θ; and *(ii)* calculating the confidence weights of the member classifiers based on sensitivity to the total error caused on positive labels (+1), i.e., if the member classifier misclassifies more samples (+1), the lower its confidence weights will be. We also combine Im.ADABoost with W-SVM to classify the imbalanced datasets. Preliminary experimental results on two datasets named Vertebral Column and Indian Liver Patient show that Im.ADABoost achieves high classification efficiency compared to the original ADABoost algorithm with DecisionTree and ADABoost.W-SVM. Especially, when the datasets have small numbers of positive labels +1, the Im.ADABoost algorithm gives superior classification results. In addition, for each specific dataset, it is always possible to find the adjustable exponential parameter θ by experiment to help Im.ADABoost achieve the best classification results. We will further improve the calculation of confidence weights of the member classifier based on sensitivity total errors on positive labels and do more experiments on other datasets.

References

1. Akbani, R., Kwek, S., Japkowicz, N.: Applying support vector machines to imbalanced datasets. In: Boulicaut, J.-F., Esposito, F., Giannotti, F., Pedreschi, D. (eds.) ECML 2004. LNCS (LNAI), vol. 3201, pp. 39–50. Springer, Heidelberg (2004). https://doi.org/10.1007/978-3-540-30115-8_7
2. Benjamin, X.W., Nathalie, J.: Boosting support vector machines for imbalanced data sets. Knowl. Inf. Syst. **21**, 1–20 (2010). https://doi.org/10.1007/s10115-009-0198-y

3. Chawla, N., Bowyer, K., Hall, L., Kegelmeyer, W.: SMOTE: synthetic minority over-sampling technique. J. Artif. Intell. Res. **16**, 321–357 (2002)
4. Dong, X., Gao, H., Guo, L., Li, K., Duan, A.: Deep cost adaptive convolutional network: a classification method for imbalanced mechanical data. IEEE Access **8**, 71486–71496 (2020). https://doi.org/10.1109/ACCESS.2020.2986419
5. Elkan, C.: The foundations of cost-sensitive learning. In: Proceedings of the Seventeenth International Conference on Artificial Intelligence: 4–10 August 2001, Seattle, vol. 1, pp. 973–978 (2001)
6. Freund, Y.: Boosting a weak learning algorithm by majority. Inf. Comput. **121**(2), 256–285 (1995). https://doi.org/10.1006/inco.1995.1136
7. Freund, Y., Schapire, R.E.: A decision-theoretic generalization of on-line learning and an application to boosting. J. Comput. Syst. Sci. **55**(1), 119–139 (1997)
8. Galar, M., Fernández, A., Barrenechea, E., Bustince, H., Herrera, F.: Ordering-based pruning for improving the performance of ensembles of classifiers in the framework of imbalanced datasets. Inf. Sci. **354**, 178–196 (2016). https://doi.org/10.1016/j.ins.2016.02.056
9. Guo, H., Viktor, H.: Learning from imbalanced data sets with boosting and data generation: the databoost-im approach. SIGKDD Explor. **6**(1), 30–39 (2004). https://doi.org/10.1145/1007730.1007736
10. Hilario, A., Garcia Lopez, S., Galar, M., Prati, R., Krawczyk, B., Herrera, F.: Learning from Imbalanced Data Sets, Artificial Intelligence. Springer, Cham (2018). https://doi.org/10.1007/978-3-319-98074-4_9
11. Johnson, J., Khoshgoftaar, T.: Survey on deep learning with class imbalance. J. Big Data **6**, 1–54 (2019). https://doi.org/10.1186/s40537-019-0192-5
12. Jordan, M., Mitchell, T.: Machine learning: trends, perspectives, and prospects. Science (New York N.Y.) **349**, 255–60 (2015). https://doi.org/10.1126/science.aaa8415
13. Khang, T.D., Tran, M.K., Fowler, M.: A novel semi-supervised fuzzy c-means clustering algorithm using multiple fuzzification coefficients. Algorithms **14**(9), 258 (2021)
14. Khang, T.D., Vuong, N.D., Tran, M.K., Fowler, M.: Fuzzy c-means clustering algorithm with multiple fuzzification coefficients. Algorithms **13**(7), 1–11 (2020)
15. Lee, W., Jun, C.H., Lee, J.S.: Instance categorization by support vector machines to adjust weights in AdaBoost for imbalanced data classification. Inf. Sci. **381**, 92–103 (2016). https://doi.org/10.1016/j.ins.2016.11.014
16. Li, X., Wang, L., Sung, E.: AdaBoost with SVM-based component classifiers. Eng. Appl. Artif. Intell. **21**, 785–795 (2008). https://doi.org/10.1016/j.engappai.2007.07.001
17. Lima, N.H.C., Neto, A.D.D., Dantas de Melo, J.: Creating an ensemble of diverse support vector machines using AdaBoost. In: 2009 International Joint Conference on Neural Networks, pp. 1802–1806 (2009)
18. Lin, C.F., Wang, S.D.: Fuzzy support vector machines. IEEE Trans. Neural Netw. **13**(2), 464–471 (2002). https://doi.org/10.1109/72.991432
19. Liu, X.Y., Wu, J., Zhou, Z.H.: Exploratory undersampling for class-imbalance learning. IEEE Trans. Syst. Man Cybern. Part B (Cybern.) **39**(2), 539–550 (2009). https://doi.org/10.1109/TSMCB.2008.2007853
20. Rengasamy, S., Punniyamoorthy, M.: Performance enhanced boosted SVM for imbalanced datasets. Appl. Soft Comput. **83**, 105601 (2019). https://doi.org/10.1016/j.asoc.2019.105601
21. Sun, Y., Kamel, M.S., Wong, A.K., Wang, Y.: Cost-sensitive boosting for classification of imbalanced data. Pattern Recogn. **40**(12), 3358–3378 (2007)

22. Tao, X., et al.: Affinity and class probability-based fuzzy support vector machine for imbalanced data sets. Neural Netw. **122**, 289–307 (2020)
23. Tharwat, A., Gabel, T.: Parameters optimization of support vector machines for imbalanced data using social ski driver algorithm. Neural Comput. Appl. **32**(11), 6925–6938 (2019). https://doi.org/10.1007/s00521-019-04159-z
24. Turki, T., Wei, Z.: Boosting support vector machines for cancer discrimination tasks. Comput. Biol. Med. **101**, 236–249 (2018). https://doi.org/10.1016/j.compbiomed.2018.08.006
25. Yan, Y., Chen, M., Shyu, M.L., Chen, S.C.: Deep learning for imbalanced multimedia data classification. In: 2015 IEEE International Symposium on Multimedia (ISM), pp. 483–488. IEEE, Miami (2015). https://doi.org/10.1109/ISM.2015.126
26. Zeng, M., Zou, B., Wei, F., Liu, X., Wang, L.: Effective prediction of three common diseases by combining SMOTE with Tomek links technique for imbalanced medical data. In: 2016 IEEE International Conference of Online Analysis and Computing Science (ICOACS), pp. 225–228 (2016). https://doi.org/10.1109/ICOACS.2016.7563084

Feature Learning and Data Generative Models for Facial Expression Recognition

Vu Thanh Nguyen[1(✉)], Mai Viet Tiep[2(✉)], Luong The Dung[2], Vu Thanh Hien[3],
Tuan Dinh Le[4], Ton Quang Toai[5], My Nguyen[6], and Phan Trung Hieu[7]

[1] Ho Chi Minh City University of Food Industry, Ho Chi Minh City, Vietnam
nguyenvt@hufi.edu.vn
[2] Academy of Cryptography Techniques, Ho Chi Minh City, Vietnam
[3] Ho Chi Minh City University of Technology (HUTECH), Ho Chi Minh City, Vietnam
vt.hien@hutech.edu.vn
[4] Long An University of Economics and Industry, Tân An, Long An Province, Vietnam
le.tuan@daihoclongan.edu.vn
[5] Ho Chi Minh City University of Foreign Languages - Information
Technology, Ho Chi Minh City, Vietnam
tonquangtoai@huflit.edu.vn
[6] Deakin University, Burwood, Australia
nguyenvie@deakin.edu.au
[7] University of Information Technology, Ho Chi Minh City, Vietnam
hieupt@uit.edu.vn

Abstract. During communication process, people's emotions are usually shown on the face (we call it expression). Capturing expression signals is very important to communication among individuals. In this article, we develop a neuron network architecture for automatic recognition of seven facial expressions: angry, disgust, fear, happy, sad, surprise and neutral. Furthermore, we also developed an CycleGAN architecture used to learn the data allocation of the training data set to solve the problem of data imbalance of the problem need solving. The efficiency of the architectures was evaluated on the FER2013 data set with an accuracy of 72.24%, which is higher than the result of the winner in the FER2013 competition (71.16%).

Keywords: CycleGAN · Facial expression recognition · Handling imbalanced data

1 Introduction

During daily communication, we usually use two basic means of communication: verbal communication and non-verbal communication. Non-verbal communication includes: facial expressions, eyes, tone, body posture and expressions of communicators. David Lapakko [1] argues that we usually express our expressions and attitudes in a non-verbal way rather than verbal way. Therefore, developing a facial expression recognition system

© Springer Nature Switzerland AG 2021
T. K. Dang et al. (Eds.): FDSE 2021, LNCS 13076, pp. 137–151, 2021.
https://doi.org/10.1007/978-3-030-91387-8_10

is one of the key conditions in the development of intelligent Human-Machine interaction systems.

When doing study of facial expression recognition, psychologist Ekman and co-authors [2] divided facial expressions into seven main types of expressions: (1) angry, (2) disgust, (3) fear, (4) happy, (5) sad, (6) surprise, (7) neutral and he also suggested a system named Facial Action Coding System (FACS) to list areas on human face that show expressions. Thus, a facial expression is a combination of many Action Units (AUs), each AUs corresponding to each Facial Action group.

However, developing an algorithm to detect AUs appearing on face is not easy. Therefore, in 2013, to promote the study of facial expression recognition, Kaggle organized the facial expression recognition contest on FER2013 data set [3]. This is a difficult data set for a number of reasons such as: the data set has an imbalance between classes, the data set contains many noisy labels, the data set has both images and drawings,... The first prize winner of the contest can only recognize 81.16% on the FER2013 data set. In this article we will research and develop facial expression recognition algorithm based on FER data set 2013. The main contributions of this article include two parts as follow:

1. Solving the data imbalance problem in the FER2013 dataset by developing a data generation model. This model uses data in the neutral class and converts it to data of the disgust class. This model helped generate 2700 more data points for the disgust class.
2. Developing a neural network architecture to learn and extract features of facial expressions. Through the experimental process, we have determined the appropriate set of parameters to solve the problem. The results of the model achieved an accuracy of 72.24%.

The remaining contents of this study include the following parts: Sect. 2 presents typical works related to facial expression recognition and data generation models. Section 3 presents FER2013 database, in which, we will see some difficulties when using this database to develop models. The fourth part presents a method of solving the problem by developing a data generation model to solve the imbalance problem in FER2013 data and developing the model to learn the features of facial expressions.

2 Related Works

2.1 Expression Recognition on Face Images

The first work on FER2013 data set is "Deep Learning using Linear Support Vector Machines" [4] by Yichuan Tang in 2015. In this work, Yichuan Tang uses the loss function L2-SVM to train neural networks for data classification. Lower layer weights are learned by backpropagating the gradients from the top layer linear SVM.

Yichuan Tang compares performances of softmax with deep learning using L2-SVMs (DLSVM). Both models tested and compared with each other are given in Table 1.

In 2018, Author Lei Chen and co-authors [5] carried out the work "Attention Map Comparison between Support Vector Machines and Convolutional Neural Networks". In this work, the authors have built 7 different models: ResNet-34, ResNet-18, VGG-13,

Table 1. Comparison of models

	Softmax	DLSVM
Public leaderboard	69.3%	69.4%
Private leaderboard	70.1%	71.2%

Adaboost, Random Forest, SVM-PCA, SVM. Among these models, the model ResNet-34 gives the best 64.2% accuracy.

In 2019, Author Shervin Minaee and co-authors [10] proposed a deep learning approach based on attentional convolutional network. The result of the model achieved an accuracy of 70.2%.

With the great success of deep learning, especially convolutional neural networks (CNNs), many features can be extracted and learned for a decent facial expression recognition system [11, 12].

2.2 Generative Adversarial Network Model

A generative model is a model that describes how data is generated. The initial assumption of the generating model is that the data set has an unknown p_data distribution. The task of the generation model is to construct a distribution p_model $\approx$ p_data. With the p_model built, we can create new data models by sampling from the p_model model.

In generation models, GANs (Generative Adversarial Networks) [6] are commonly used. GANs were invented by Goodfellow and co-authors in 2014. GANs' main idea is to develop two networks: generator network and discriminator network. The generator network takes as input a random noisy vector and tries to convert it to as real data as possible. Whereas the discriminator network is responsible for distinguishing whether input data is generated by the generator or from a given dataset. In GANs, these two models are trained alternately, competing with each other, until the generator network can generate data that the discriminator network cannot distinguish whether it is the generator generated or not.

Zhu and his colleagues proposed CycleGAN [7] in 2017, which is a form of GANs model but is used to convert an image from one domain to an image in another domain without using pairs of images of the same size and shape.

3   FER2013 Data Set Analysis

3.1   Overview of FER2013 Data Set

We use FER2013 [3] data set to solve facial expression recognition problem. The FER2013 database was created by Pierre-Luc Carrier and Aaron Courville for their study. Then they provided preliminary data sets for using in the 2013 Kaggle competition on facial expression recognition.

The FER2013 dataset contains grayscale images, containing 48 $\times$ 48 pixel face image and is labeled with one of seven expressions, respectively: angry, disgust, fear, happy, sad, surprise, and neutral. Some images of the seven expressions are shown in Fig. 1.

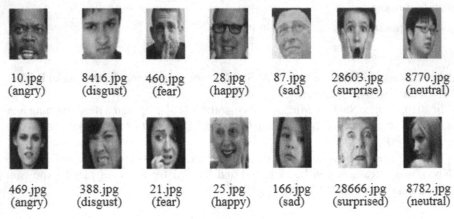

| 10.jpg (angry) | 8416.jpg (disgust) | 460.jpg (fear) | 28.jpg (happy) | 87.jpg (sad) | 28603.jpg (surprise) | 8770.jpg (neutral) |

| 469.jpg (angry) | 388.jpg (disgust) | 21.jpg (fear) | 25.jpg (happy) | 166.jpg (sad) | 28666.jpg (surprised) | 8782.jpg (neutral) |

Fig. 1. Seven facial expressions

All data is in the FER2013.csv file with a capacity of 287 MB. Each image is on a line, consisting of three columns: expression, pixels, and Usage, as shown in Table 2.

Table 2. FER2013 data set structure

Emotion	Pixels	Usage
0	70 80 82 72 58 58 60 63 54 58 60 48 89 115 121 119 115 110 98 91 84…	Training
0	151 150 147 155 148 133 111 140 170 174 182 154 153 164 173 178…	Training
2	231 212 156 164 174 138 161 173 182 200 106 38 39 74 138 161 164…	Training
4	24 32 36 30 32 23 19 20 30 41 21 22 32 34 21 19 43 52 13 26 40 59 65…	Training

The "emotion" column describes the facial expressions, the values in this column have values from 0 to 6 representing seven expressions. The "pixels" column contains a set of pixel values, each row containing 48×48 pixel values. The "Usage" column indicates the image is in Training, Validation or Test. For convenience of processing, these data points will be converted to image files and numbered from 1 onwards.

3.2 Statistics of Data Set FER2013

Statistics on All Data. There are a total of 35,527 images in the data set FER2013. The number of images and the ratio of each expression type are statistically descriptive as shown in Table 3.

Table 3. Statistics of the number of images by each expression

Expression name	Number of image	Percentage
Angry	4.593	13,8%
Disgust	547	1,5%
Fear	5.121	14,3%
Happy	8.989	25%
Sad	6.077	16,9%
Surprise	4.002	11,2%
Neutral	6.198	17,3%
Total	35.527	100%

As shown in Table 3. We see that the number of images of the "disgust" class is quite small compared to other expressions.

During the competition, two image sets: 28,709 images and 3,589 image sets were provided As shown in Table 3. We see that the number of images of the "disgust" class is quite small compared to other expressions.

To the contestants to train and self-assess their models, so we call these 2 sets training and validation. And another set of 3,589 images are kept private to check and find out the winner of the contest. After the competition, this entire dataset is made available to everyone for study. Thus, in order to ensure fairness, in the process of Developing the model to solve the problem, we divide the data set into three sets as given in the contest as described in Table 4.

Table 4. Number of images in each dataset

Data set	Number of image	Percentage
Training	28.709	80%
Validation	3.589	10%
Test	3.589	10%

Data rate of three sets Training: Validation: Test respectively 80%: 10%: 10%.

Statistics on Training Set. The total number of images in the training set is 28,709 images. The number of images distributed for each class is illustrated in Fig. 2.

Through the graph in Fig. 2, we can see: happy class has the most number of images (7215 images, accounting for 25.13%). Disgust class had the lowest number of images (436 images, accounting for 1.52%). The remaining classes have a relatively even number of images (from 3000 to nearly 5000 images).

From this, we conclude as follows: classes such as angry, fear, happy, sad, surprise, neutral have quite a lot of data (all over 1000 images) so they are suitable for applying

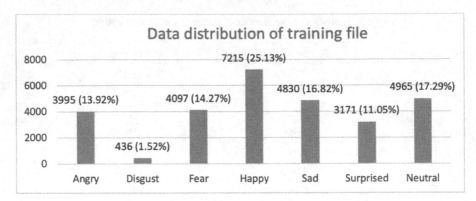

Fig. 2. Statistics on training set

deep learning methods. Disgust class has quite little data (436 < 1000) and the difference is quite large compared to other classes (disgust class accounts for 1.52% compared to other classes with 10% or more) so it creates an unbalanced data set and will make it difficult to train the model.

Statistics on Validation Set. In FER2013, the validation dataset is called the public test set. The total number of images in the validation set is 3,589 images. The number of images distributed for each class is illustrated in the Fig. 3:

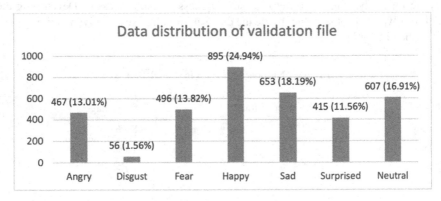

Fig. 3. Statistics on validation set

Figure 3 shows that data of disgust class is quite small compared to other classes, accounting for only 1.56%.

3.3   Challenges on the FER2013 Data Set

The images in the FER2013 dataset are collected in the natural and diverse environment such as male and female genders (see Fig. 4), have different ages (see Fig. 5), and have many shooting angles (see Fig. 6), there are different levels of brightness and contrast (see Fig. 7). There may be accessories or other elements on the face that overlap the face (see Fig. 8).

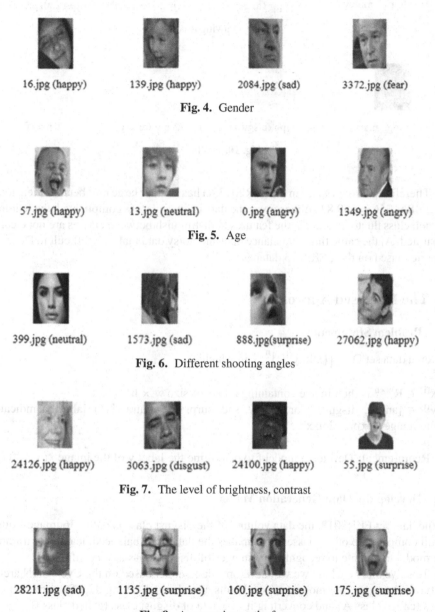

16.jpg (happy)   139.jpg (happy)   2084.jpg (sad)   3372.jpg (fear)

Fig. 4. Gender

57.jpg (happy)   13.jpg (neutral)   0.jpg (angry)   1349.jpg (angry)

Fig. 5. Age

399.jpg (neutral)   1573.jpg (sad)   888.jpg(surprise)   27062.jpg (happy)

Fig. 6. Different shooting angles

24126.jpg (happy)   3063.jpg (disgust)   24100.jpg (happy)   55.jpg (surprise)

Fig. 7. The level of brightness, contrast

28211.jpg (sad)   1135.jpg (surprise)   160.jpg (surprise)   175.jpg (surprise)

Fig. 8. Accessories on the face

Additionally, in the data set, in addition to the images, there are some drawings (see Fig. 9) and some noisy labels (see Fig. 10).

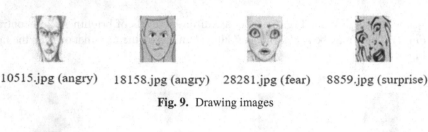

10515.jpg (angry)   18158.jpg (angry)   28281.jpg (fear)   8859.jpg (surprise)

Fig. 9. Drawing images

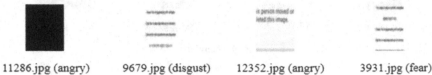

11286.jpg (angry)   9679.jpg (disgust)   12352.jpg (angry)   3931.jpg (fear)

Fig. 10. No face

Therefore, the data space in the FER2013 set has a rather large number of dimensions, $48 \times 48 \times 256 = 589{,}824$ dimensions, the data in each class is complicated, the features in each class fluctuate greatly, the features to distinguish between classes are not clearly separated. At the same time, imbalance data and noisy data makes it difficult to Develop a model based on the FER2013 database.

4 The Proposed Approach

4.1 Problem Statement

Given a data set $D = \left\{\left(x^{(i)}, y^{(i)}\right)\right\}_{i=1}^{n}$, in which:

- $x^{(i)} \in R^{w \times h}$ is the i image containing a face of size $w \times h$.
- $y^{(i)} \in \{$angry, disgust, fear, happy, sad, surprise, neutral$\}$ is the label that indicates the image's expression $x^{(i)}$.

Requirement: Develop a model f to determine the label $\hat{y}$ of the image x: $\hat{y} = f(x)$.

4.2 Develop the Data Generation Model

In the data set FER2013, the data volume of the disgust class is only 436 images, quite small compared to other classes, so it makes the dataset imbalanced, leading to training the model to be able to recognize the images of disgust class is very difficult.

To solve this problem, we will develop a data model based on the CycleGAN architecture [7] to generate more data for the disgust class by taking data from the neutral class (called class A) and converting it into data of disgust class (called class B).

Architecture of Data Generation Model. The overview architecture of CycleGAN model is depicted as shown in Fig. 11.

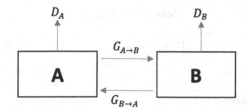

Fig. 11. General architecture of the data generation model.

We use two network generators $G_{A \to B}$ và $G_{B \to A}$, in which the duty of $G_{A \to B}$ is the conversion of image from class A to class B, $G_{A \to B} : A \to B$, duty of $G_{B \to A}$ is the conversion of image from class B to class A, $G_{B \to A} : B \to A$.

Call two networks discriminators are D_A, D_B, in which D_A is used for distinguishing the real image in A from the generated image $G_{B \to A}(B)$ and D_B is used for distinguishing the real image in B from the image generated by $G_{A \to B}(A)$. Our goal is not only to ensure images $G_{A \to B}(A)$ generated similar to the images in class B, but also ensure data integrity means ensure $G_{B \to A}(G_{A \to B}(A)) \approx A$.

Network Architecture for the Generator. Generator Network $G_{A \to B}$ and $G_{B \to A}$ used to generate data from class A to B and from class B to class A consists of 26 floors, details of the classes are described in Table 5.

Table 5. Generator details

No	Floor	Dimensional			Kernel	Stride
		Width	Height	Depth		
0	Input	48	48	1	–	–
1	Convolution	24	24	32	3 × 3	2
2	Instance Normalization	–	–	–	–	–
3	LeakyReLU	–	–	–	–	–
4	Convolution	12	12	64	3 × 3	2
5	Instance Normalization	–	–	–	–	–
6	LeakyReLU	–	–	–	–	–
7	Convolution	6	6	128	3 × 3	2
8	Instance Normalization	–	–	–	–	–
9	LeakyReLU	–	–	–	–	–
10	Convolution	3	3	256	3 × 3	2
11	Instance Normalization			–	–	–
12	LeakyReLU			–	–	–

(continued)

Table 5. (*continued*)

No	Floor	Dimensional			Kernel	Stride
		Width	Height	Depth		
13	Deconvolution	6	6	128	3×3	1/2
14	Instance Normalization			–	–	–
15	LeakyReLU			–	–	–
16	Deconvolution	12	12	64	3×3	1/2
17	Instance Normalization			–	–	–
18	LeakyReLU			–	–	–
19	Deconvolution	24	24	32	3×3	1/2
20	Instance Normalization			–	–	–
21	LeakyReLU			–	–	–
22	Deconvolution	48	48	32	3×3	1/2
23	Instance Normalization			–	–	–
24	LeakyReLU			–	–	–
25	Convolution	48	48	1	3×3	1
26	tanh			–	–	–
	Output	48	48	1	–	–

Network Architecture for Discriminator. Two discriminator networks D_A and D_B is used to evaluate data generated from class A to class B and from class B to class A including ten floors, detailing the floors described in Table 6.

Loss Function of Generation Model. The loss function on generator networks.
Loss function on the network $G_{A \to B}$:

$$\mathcal{L}_{\text{forward\_cyc}} = E_{a \sim p_{\text{data}}(a)} \| G_{B \to A}(G_{A \to B}(a)) - a \|_1 \tag{1}$$

Loss function on the network $G_{B \to A}$:

$$\mathcal{L}_{\text{backward\_cyc}} = E_{b \sim p_{\text{data}}(b)} \| G_{A \to B}(G_{B \to A}(b)) - b \|_1 \tag{2}$$

The loss function of two generator networks:

$$\mathcal{L}_{\text{cyc}} = \mathcal{L}_{\text{forward\_cyc}} + \mathcal{L}_{\text{backward\_cyc}} \tag{3}$$

Loss functions between the generator network and the discriminator network. Loss function between the network $G_{A \to B}$ and the discriminator networks:

$$\mathcal{L}^{(D)}_{\text{forwardGAN}} = E_{b \sim p_{\text{data}}(b)}(D_B(b) - 1)^2 + E_{a \sim p_{\text{data}}(a)}(D_B(G_{A \to B}(a)))^2 \tag{4}$$

$$\mathcal{L}^{(G)}_{\text{forward\_GAN}} = E_{a \sim p_{\text{data}}(a)}(D_B(G_{A \to B}(a)) - 1)^2 \tag{5}$$

Table 6. Dicriminator details

No	Floor	Dimensional			Kernel	Stride
		Width	Height	Depth		
0	Input	48	48	1	–	–
1	Convolution	24	24	32	3×3	2
2	ReLU	–	–	–	–	–
3	Convolution	12	12	64	3×3	2
4	Instance Normalization	–	–	–	–	–
5	ReLU	–	–	–	–	–
6	Convolution	6	6	128	3×3	2
7	Instance Normalization	–	–	–	–	–
8	ReLU	–	–	–	–	–
9	Convolution	3	3	256	3×3	2
10	Instance Normalization			–	–	–
	Output	3	3	1	–	–

Loss function between generator network $G_{B \to A}$ and the discriminator networks:

$$\mathcal{L}_{\text{backward\_GAN}}^{(D)} = E_{a \sim p_{\text{data}}(a)}(D_A(a) - 1)^2 + E_{b \sim p_{\text{data}}(b)}(D_A(G_{B \to A}(b)))^2 \quad (6)$$

$$\mathcal{L}_{\text{backward\_GAN}}^{(G)} = E_{b \sim p_{\text{data}}(b)}(D_A(G_{B \to A}(b)) - 1)^2 \quad (7)$$

The GAN loss function on the discriminator network:

$$\mathcal{L}_{\text{GAN}}^{(D)} = \mathcal{L}_{\text{forward\_GAN}}^{(D)} + \mathcal{L}_{\text{backward\_GAN}}^{(D)} \quad (8)$$

The GAN loss function on the generator network:

$$\mathcal{L}_{\text{GAN}}^{(G)} = \mathcal{L}_{\text{forward\_GAN}}^{(G)} + \mathcal{L}_{\text{backward\_GAN}}^{(G)} \quad (9)$$

$$\mathcal{L}_{\text{GAN}} = \mathcal{L}_{\text{GAN}}^{(D)} + \mathcal{L}_{\text{GAN}}^{(G)} \quad (10)$$

Loss function of the entire model:

$$\mathcal{L} = \lambda_1 \mathcal{L}_{\text{GAN}} + \lambda_2 \mathcal{L}_{\text{cyc}} \quad (11)$$

where λ_1, λ_2 are the weights of the loss function.

4.3 Design Identity Models

Architecture. The general architecture of the model used to learn the features of the expressions is shown in Fig. 12.

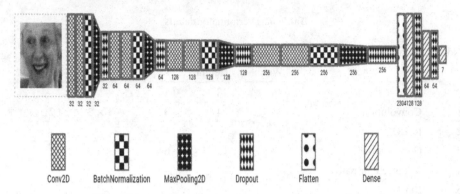

Fig. 12. Architecture of expression recognition model.

Details of the architecture are described in detail in Table 7.

The Loss Function of the Identity Model.

$$\hat{y}^{(i)} = f\left(x^{(i)}\right) \tag{12}$$

$$(y, \hat{y}) = -\sum_{i=1}^{m} y^{(i)} \log\left(\hat{y}^{(i)}\right) \tag{13}$$

where:
$x^{(i)}$: image data i th.
$\hat{y}^{(i)}$: predictive label of data $x^{(i)}$.
$y^{(i)}$: real label of data $x^{(i)}$.

5   Experimental Results

5.1   Model Training Process

Train the Data Generation Model. Optimal parameters used to train the model:
   Weight of loss function: $\lambda_1 = 1$, $\lambda_2 = 10$.
   Coefficients: $2e - 4$.
   Study method: RMSProp.
   Batch size= 32.

Train Expression Recognition Models. Optimal parameters used to train the model.
   Input: image normalized on the value domain [0, 1].
   Weights initialization: Weights are randomly generated according to He [8] method.
   Batch size = 32.
   Loss function optimization method: Using algorithm Adaptive Moment Estimation (Adam) [9] to optimize the loss function by calculating the adaptive arithmetic coefficient for each parameter of the loss function.
   Coefficients: Initially, the coefficient is 1e-3, then the coefficient is adjusted according to Table 8.

Table 7. Network architecture

No	Floor	Dimensional			Kernel	Stride
		Width	Height	Depth		
0	Input	48	48	1	–	–
1	Convolution	48	48	32	3×3	1
2	Convolution	48	48	32	3×3	1
3	Batch Normalization					
4	ReLU					
5	Pooling	24	24	32	2×2	2
6	Convolution	24	24	64	3×3	1
7	Convolution	24	24	64	3×3	1
8	Batch Normalization					
9	ReLU					
10	Pooling	12	12	64	2×2	2
11	Convolution	12	12	128	3×3	1
12	Convolution	12	12	128	3×3	1
13	Batch Normalization					
14	ReLU					
15	Pooling	6	6	128	2×2	2
16	Convolution	6	6	256	3×3	1
17	Convolution	6	6	256	3×3	1
18	Batch Normalization					
19	ReLU					
20	Pooling	3	3	256	2×2	2
21	FC	1	1	128	–	–
22	FC	1	1	64	–	–
23	FC	1	1	7	–	–

Table 8. Coefficients

Epoch number	Coefficients
$1 \to 40$	$1e - 3$
$41 \to 80$	$1e - 4$
$81 \to 120$	$1e - 5$

5.2 Accuracy Evaluation

In FER2013, the test data set is called the private test set. The total number of images in the test set is 3,589 images (accounting for 10% of the database size). The number of images distributed for each class is illustrated by the diagram in Fig. 13.

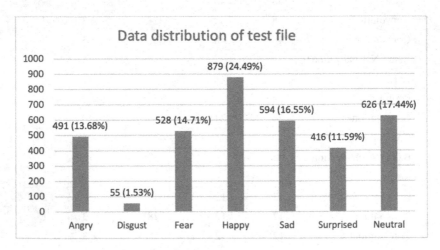

Fig. 13. Statistics on the test set

Figure 13 shows that the data distribution of the test set is similar to the data distribution of the validation set.

Results of accuracy: the model gave the results with an accuracy of 72.24%, with this result, the model built was more accurate than the results of the leader in the FER2013 competition.

6 Conclusion

In this study, we have developed a CycleGAN-based data generation architecture to solve the problem of imbalanced data. After the data is balanced, we develop an architecture that learns facial features to classify facial expressions. The results of accuracy for this solution show that this is a suitable solution to the facial expression recognition.

However, there are some limitations in this work, such as we only use one dataset FER2013 without yet referring to other datasets. Besides, we have used only CycleGAN architecture meanwhile there are still other GANs architectures. The next plan of our future work is to learn more about some other data sets and learn more architectures to be able to offer a more powerful solution to solve the facial expression recognition problem.

References

1. Lapakko, D.: Communication is 93% nonverbal: an urban legend proliferates. Commun. Theater Assoc. Minn. J. **34**, 7–19 (2007)

2. Ekman, P., Friesen, W.V.: Facial Action Coding System: A Technique for the Measurement of Facial Movement. Consulting Psychologists Press, Palo Alto, California, CA (1978)
3. Goodfellow, I.J., et al.: Challenges in representation learning: a report on three machine learning contests. In: Lee, M., Hirose, A., Hou, Z.-G., Kil, R.M. (eds.) ICONIP 2013. LNCS, vol. 8228, pp. 117–124. Springer, Heidelberg (2013). https://doi.org/10.1007/978-3-642-42051-1_16
4. Tang, Y.: Deep Learning using Linear Support Vector Machines. arXiv:1306.0239v4 (2015)
5. Chen, L., Sum, Y., Zhang, F.: Attention Map Comparison between Support Vector Machines and Convolutional Neural Networks (2018)
6. Goodfellow, I., et al.: Generative adversarial nets. NIPS, pp. 2672–2680 (2014)
7. Zhu, J.Y., Park, T., Isola, P., Efros, A.A.: Unpaired image-to-image translation using cycle-consistent adversarial networks. arXiv preprint arXiv:1703.10593 (2017)
8. He, K., Zhang, X., Ren, S., Sun, J.: Delving Deep into Rectifiers: Surpassing Human-Level Performance on ImageNet Classification. arXiv:1502.01852v1 (2015)
9. Kingma, D.P., Ba, J.: Adam: A method for stochastic optimization. arXiv:1412.6980v9 (2014)
10. Minaee, S., Abdolrashidi, A.: Deep-Emotion: Facial Expression Recognition Using Attentional Convolutional Network. arXiv:1902.01019v1 (2019)
11. Wu, C.D., Chen, L.H.: Facial emotion recognition using deep learning. arXiv:1910.11113v1 (2019)
12. Verma, G., Verma, H.: Hybrid-deep learning model for emotion recognition using facial expressions. Rev. Soc. Strateg. 14(2), 171–180 (2020). https://doi.org/10.1007/s12626-020-00061-6

Industry 4.0 and Smart City: Data Analytics and Security

Authorization Strategies and Classification of Access Control Models

Aya Mohamed[1,2]([✉]), Dagmar Auer[1,2]([✉]), Daniel Hofer[1,2], and Josef Küng[1,2]

[1] Institute for Application-oriented Knowledge Processing (FAW), Linz, Austria
[2] LIT Secure and Correct Systems Lab, Linz Institute of Technology (LIT),
Johannes Kepler University (JKU) Linz, Linz, Austria
{aya.mohamed,dagmar.auer,daniel.hofer,josef.kueng}@jku.at

Abstract. Access control enforces authorization policies in order to prohibit unauthorized users from performing actions that could trigger a
security violation. There exist numerous access control models and even
more have recently evolved to conform with the challenging requirements
of resource protection. That makes it hard to classify the models and
choose an appropriate one satisfying security needs. This paper provides
an overview of authorization strategies and proposes a rough classification of access control models providing examples for each category. In
comparison with other comparative studies, we discuss more access control models including the conventional state-of-the-art models and novel
ones. We also summarize each of the literature works after selecting the
relevant ones focusing on database systems domain or providing a survey, a taxonomy/classification, or evaluation criteria of access control
models. Additionally, the introduced categories of models are analyzed
with respect to various criteria that are partly selected from the standard access control system evaluation metrics by the National Institute
of Standards and Technology (NIST). Further studies for extending the
list of access control models as well as analysis criteria are planned.

Keywords: Authorization strategy · Access control model ·
Classification · Criteria

1 Introduction

Access control ensures data security by protecting assets and private information
against unauthorized access by defined subjects. It helps to avoid information
leaks or improper modification by potentially malicious parties. Besides traditional well-known access control models, there are many others that have been
recently evolved to match advanced security requirements. Due to the increase
of access control models, they became hard to classify and the idea of selecting
an appropriate one to fulfill the requirements of the overall system is challenging.
Thus, there is a necessity to clarify the access control concepts (e.g., definitions,

© Springer Nature Switzerland AG 2021
T. K. Dang et al. (Eds.): FDSE 2021, LNCS 13076, pp. 155–174, 2021.
https://doi.org/10.1007/978-3-030-91387-8_11

strategies, and models) along with the commonly used, partly ambiguous, synonyms.

In this paper, we overcome this opaque accumulation of terms and their meaning by guiding researchers and practitioners through the vast amount of available access control models. We further provide support in selecting an appropriate access control model with respect to security requirements. The contributions of our work are the following:

- Distinction between authorization strategies and access control models.
- Rough classification schema of access control models.
- Illustration of classification schema by providing state of the art as well as not commonly discussed models for each class of access control models.
- Review of a selected list of comparative studies on access control that are in the context of databases, include a survey of models, provide evaluation criteria, and/or introduce a taxonomy of models.
- Analysis of the classification schema based on selected criteria of access control models.

We started our research with an extensive literature study on access control models and identified authorization strategies. Then, we derived categories for classifying all these models and analyzed the main features of each category. The remainder of this paper is organized as follows. Section 2 defines authorization strategy and illustrates existing discretionary, mandatory, and hybrid strategies. We introduce a classification of access control models along with examples in Sect. 3. In Sect. 4, we provide a summary of survey works comparing the included access control models. We analyze the proposed categories with respect to selected criteria in Sect. 5. Finally, the paper concludes with a summary and an outlook on future work in Sect. 6.

2   Authorization Strategies

An authorization strategy defines the view point of describing the authorization policies, i.e., owner-centric, administration-centric, or hybrid. Different terms are used as alternatives to authorization strategy. Eckert [24] refers to it as *access control strategy*, Samarati and De Capitani di Vimercati [51] as well as Benantar [9] called it *access control policies class* while others regard it as an *access control model* [16]. Even though we use the term authorization strategy, we keep the well-established DAC (i.e., Discretionary Access Control) and MAC (i.e., Mandatory Access Control) abbreviations.

2.1   Discretionary Strategy (DAC)

The Discretionary Strategy (DAC) is owner-centric, i.e., ownership of each resource is assigned to one or more entities. The subject, who is allowed to access a resource, is either the object creator (i.e., the default owner) or a principal with delegated ownership rights. The resource can be only destroyed by the owner and its ownership may optionally be shared with other subjects as well [9].

DAC systems provide more flexibility to the user, but less administration control. Moreover, they do not scale well and are hard to manage in large environments. Since the propagation and usage of information cannot be controlled after giving access to the legitimate subjects, they are insecure and vulnerable to *Trojan Horse* attacks. A trojan horse program executes more actions, unknown to users, than it seems and should do [14,30].

2.2 Mandatory Strategy (MAC)

The Mandatory Strategy (MAC) is non-discretionary since access decisions are not made at the discretion of the user. A MAC policy is obligatory as the access rights are regulated by a central authority. The owner and subject users can neither control the defined access nor override the policy. This strategy often is based on the security labels concept where the subjects are associated to security clearance and objects to sensitivity classifications [9,34].

Although MAC systems provide stronger security than the DAC ones and overcome the trojan horse problem, they are vulnerable to *Covert Channels* (i.e., tunnels created for transferring information in an unauthorized manner). Furthermore, the required administrative overhead makes it more costly.

2.3 Hybrid Strategy

The advanced access control models are typically based on a middle ground strategy mixing DAC and MAC because the pure mandatory and discretionary strategies often are not sufficient anymore. For instance, the Originator-controlled Strategy (ORCON or ORGCON) [4,46] combines DAC and MAC such that only the originator (i.e., original owner) can alter the privileges on a subject/object basis [42] (cp. DAC). On the other hand, access restrictions on original resources are automatically copied to derived objects without owner control (cp. MAC).

3   Access Control Models

An access control model defines the enforcement of the authorization policy to decide whether to allow or deny access for a subject to a protected resource. We grouped the access control models into five main classes based on their characteristics. In the following subsections, we explain these categories and provide an overview of a selected subset of the access control models including some recent models that are not previously discussed in other surveys (see Sect. 4).

3.1 Access Control by Explicit Object-Subject Assignment (OSA)

The oldest and simplest access control model is the *Access Matrix (ACM)* proposed by Lampson in 1971. It is built upon the strategy of DAC (i.e., identity-based) where the subjects' privileges are described over the objects in a matrix data structure. A single entry in the matrix $A[s,o]$ represents the access rights (i.e., actions) a subject s can take upon an object o [9]. The access rights representation is straightforward and was commonly used in practice, but typically the matrix becomes sparse and oversized due to lots of empty cells. In the following, we give an overview of available access matrix model variants.

Authorization Table. Typically used in database management systems. The non-empty matrix authorization entries are stored as a tuple in a table with three columns for the subject, object, and action [47].

Access Control List (ACL). The most common and basic form of access control for limiting access to data on shared systems. It represents the access matrix in a column perspective (i.e., resource view) where the objects to be accessed are associated with a list of subjects along with the operations allowed to be executed on these objects [47].

Capability List. The conceptual approach is similar to ACL, but with the access matrix stored by row (i.e., subject view). Each subject holds a list of capability certificates containing the access rights to be performed by this principal over a set of resources [47].

3.2   Access Control by Model-Specific Rules (MsR)

Traditionally, this class of models has been used in MAC systems enforcing the concept of rules. A set of predefined rules must be met in order to grant/deny the subject access to a particular resource. The models in this category have fixed rules that apply all the time for all users regardless of their identity. The rules are an implicit part of the access control model specifying detailed situations, i.e., whether a given subject can or cannot access an object and what that subject can do once access is granted. For example, the subject's security level determines the classes of objects to be accessed in the Bell-LaPadula (BLP) and Biba models. Administrators can only manage the basic parameters (e.g., security level) whereas users have no control at all on the rules. In the following, we give an overview of some model-specific rules examples.

Bell-LaPadula (BLP). This model was formulated by Bell and La Padula in 1976 for government and military purposes [8]. The access rights are specified according to subjects and objects associated to different security levels, i.e., *Top Secret (TS)* as the highest sensitivity label, *Secret (S)*, *Confidential (C)*, and *Unclassified (U)* as a public category with the least security clearance [9]. The subject's security level determines the classes of objects to be accessed. The BLP model has the following properties [14]:

- *Simple security*: also known as the *no-read-up* (i.e., *read-down*) property such that a subject is not allowed to read objects with higher sensitivity. The subject security clearance must dominate the object security classification.
- *Star property*: also known as the *no-write-down* (i.e., *write-up*) policy where it is not possible for a subject with some security level to write any object with lower sensitivity. To avoid the leakage of confidential information, the object security classification has to dominate the subject security clearance.
- *Strong star property (optional)*: read and write operations are performed at a single security level such that the subject and object sensitivity are equal.

Biba. The Biba integrity model ensures data security through preventing information flow in an unauthorized direction using a set of access control rules. The integrity labels given to the system's subjects and objects indicate the degree of confidence. A subject is not able to corrupt higher sensitive data and will also not be corrupted by lower security levels. It is quite similar to the BLP in the state transition system architecture and levels classification, but opposite in the characteristics as follows [9]:

- *Simple integrity*: *read-up* rule controls a subject's access from reading lower integrity level data, so that bad information will not flow upwards from lower clearance levels.
- *Star integrity*: also known as *write-down* such that subjects are not allowed to write data or pass information to higher classified levels than theirs.
- *Invocation property*: a service can only be invoked by subjects at a lower integrity level.

Lipner. The purpose of the Lipner model is to preserve confidentiality in addition to addressing commercial integrity concerns. It is a matrix-based model with security levels and functional compartments (i.e., categories) associated with subjects and objects. Each subject/object is assigned to only one of the confidentiality and/or integrity levels, but could be classified to zero or more compartments. Objects are split into data and programs for the first time using Lipner's method. There are two ways of implementing integrity: [42]

1. Based on the (BLP): subjects and objects are assigned to one of the two confidentiality levels. In this case, five defined compartments are responsible for integrity and access control.
2. Full Model: it is a hybrid combination of the BLP and Biba integrity models. Three integrity levels and two categories are added to Lipner's first mechanism, after collapsing some confidentiality compartments, to be assigned to subjects and objects. This is to prevent low-integrity data or programs from impacting those with higher integrity. The purpose of integrity levels is to avoid unauthorized modification of system programs whereas the categories are used to separate domains according to the functional areas.

Clark-Wilson. Clark and Wilson [21] introduced a model for commercial purpose to prevent fraud by requiring subjects to access objects via programs. The model consists of five certification rules (CR) and four enforcement rules (ER) to ensure the external and internal integrity of the data items. The basic components along with the corresponding rules are [42]:

- *Subject*: (ER3): is an authenticated user who attempts to initiate a transformation procedure (TP). (ER4): Only the certifier of a TP can change the list of entities associated with that TP in order to prevent violating the integrity constraints by changing the qualifications of a TP.
- *Object*: is either classified as a *constrained data item (CDI)* with high protection level or *unconstrained data item (UDI)* representing untrusted information entered to the system. (ER2): The system must associate a user with each TP and set of CDIs. (CR1): The validity of CDIs is ensured by *integrity verification procedures (IVPs)*.
- *Transformation procedure (TP)*: a set of operations performed on data items. (CR2): It transforms CDIs in the system from one valid state to another. (CR5): The TP can also take an UDI as input and either produce a CDI or reject the UDI. (ER1): Only TPs certified to run on a CDI can manipulate it so that the certified relations are maintained. (CR3): The relations allowed by the system must enforce the separation of duty principle. (CR4): Transactions are logged using a CDI and the TP only appends to it.

Chinese-Wall (CW). The Chinese-Wall model [18], also known as the Brewer and Nash model according to the inventors names, is designed to avoid conflict of interest (COI) problems. The name is inspired by the Great Wall of China so that no subject can access any object from the wrong side of that wall using a set of rules. The data objects in this model are hierarchically structured to the individual objects, company datasets grouping objects that belong to the same corporation, and conflict of interest classes containing the company datasets of competing corporations. There are two properties for the access definition based on the object organization: [52]

- *Simple security*: the object can be accessed by a specific subject if it belongs to either the same company dataset of the previously accessed objects or a different conflict of interest class.
- *Star property*: for the write access, the simple security rule must be satisfied besides the permission to read the objects which are sanitized (i.e., filtered from sensitive data) and belong to the same company dataset as the one for which write access is requested.

3.3   Access Control by Roles

In this category, system activities and resource permissions are associated to some defined role(s) rather than assigned directly to users. The *Role-based Access Control (RBAC)* model has emerged to adapt with the dynamic organizational needs since individuals change unlike the business functions. The core RBAC components are users, objects, roles, permissions and sessions. In this model, the roles act as an intermediate layer between the subjects and the access rights. The user-role assignment changes over time whereas the role-permission assignment is relatively stable. The access decision depends on the user being a member to the applicable role(s). The session represents the active roles for a user. The RBAC model has several forms: [9, 39]

- *Flat RBAC*: applies basic RBAC, but considers many-to-many relations between users and roles such that a user can have many roles and vice versa. The same applies to the permission-role assignment.
- *Hierarchical RBAC*: organizational and administrative roles are defined in a general or limited hierarchy (i.e., tree) for structuring authorities and responsibilities within the organization. The hierarchies are reflexive and transitive, but anti-symmetric.
- *Constrained RBAC*: adds constraints associated with the user-role assignment relations and/or role activation within user sessions to the hierarchical RBAC. The separation of duty (SoD) concept is applied to prevent the users from being over-authorized and enforce conflict of interest policies. The SoD can be static (SSoD) or dynamic (DSoD). In SSoD, the user cannot be a member of roles having shared principles. However, this is allowed in the DSoD without activating these exclusive roles at the same time even across multiple simultaneous sessions initiated by the same subject.

RBAC Extensions. Several RBAC models are proposed to consider context information for access control decisions. For instance, *Temporal RBAC (T-RBAC)* model extends RBAC such that users are limited to only use the role permissions in specific temporal periods. Depending on the specified time interval(s), the roles are in an either active or inactive state. Furthermore, role triggers are supported for controlling the time of action execution. The priority resolves the conflicts between triggers and periodic activation/deactivation [14]. The language is formally defined and checked for inconsistencies or ambiguities in [13].

Another extension is *GEO-RBAC* [22] that is evolved due to the increasing need for securing mobile applications and location-based services. Spatial capabilities are added to the conventional RBAC model to support location-specific constraints in which a given role can be accessed by a user. The location can be physically or logically expressed in terms of absolute coordinates or relative to spatial objects respectively. In this model, the role is only enabled if the user is located within the spatial boundary of that role [14].

Tie-RBAC [56] extends RBAC to be applied in social networks. It gives full control to the resource owner by allowing users to define their social circle (i.e., contacts) and establish in-between relations to grant access. Thus, the users control which requestor has access to their resources. The access control policies for all users are stored and enforced by a central server.

To sum up, the models in the role-based access control category have many forms and extensions, but all of them are based on the concept of roles which are associated to access permissions and assigned to users.

3.4   Access Control by Content

This category applies the concept of comprehensive data protection where access control decisions are based on data content (e.g., attribute values) [60]. Besides the flexible policy definition, authorizations are dynamically granted and revoked [14]. In content-based models, the policies are only applicable to the users satisfying specific criteria according to the rules defined by users or administrators. On the other hand, the model-specific rules category has static rules that can neither be modified nor controlled by administrators. Selected content-based models are explained in the rest of this section.

Attribute-Based Access Control (ABAC). ABAC overcomes the limitations of other models concerning long-term maintenance as well as representing complex access control requirements. In ABAC, a given subject can have access to a wide range of objects without specifying individual relationships to each resource. Authorization policies are specified in terms of the subject, resource, and environment condition (e.g., time and location) attributes. The access decision is determined by evaluating the attribute values of the applicable policy (or policies). The ACL and RBAC models are even considered as a special case of ABAC using an attribute for the identity and role respectively [33].

Although there are several proposed ABAC policy models, the *eXtensible Access Control Markup Language (XACML)* [1] has become the defacto standard not only in specifying ABAC policies, but also enforcing them in a multi-step authorization process using XACML's reference architecture [25,34]. The second commonly used approach is based on *Next Generation Access Control (NGAC)* [2,3] and its functional architecture.

View-Based Access Control (VBAC). This model is specific to databases. Theoretically, a view is a static typed language construct while from the technical perspective, it is a virtual table having rows and columns defined by a query based on the database tables, but without physical storage. The access control policy is based on a set of predefined interfaces (i.e., views).

Firstly, views are created to handle fine-grained access control requirements. Then, particular principals are allocated to one or more views so that users interact with the resources only via these interfaces. The access decision is denied if (1) the user is not permitted to access the view or (2) the operation to be performed on the object is explicitly denied within one of the views to which the principal or the relevant role is associated [14,50].

New data that satisfy a given policy will be automatically included in the view result. However, new views are created upon modifications to access control policies and their number further increases because users have different permissions. The Oracle VPD mechanism [19] addresses some of these problems where the queries are initially written against the base tables and then, automatically rewritten by the system against the view available to the subject user.

Relation-Based Access Control (ReBAC). The ReBAC model [26,28] addresses the limitations of ABAC to deal with the interpersonal relationships between users in expressing authorization policies. The access control decision is based on the type, depth, and trust level of the relationship between the owner and access requester of the resource. This model has been typically applied in social networks focusing on the privacy of end users [32]. A policy language based on modal logic and an access control model formulated as a state transition system are introduced in [26] for specifying and enforcing complex relations (e.g., friends-of-friends). However, ReBAC supports neither fine-grained access control at attribute level nor entities other than subjects and objects.

Entity-Based Access Control (EBAC). This model takes into account both attributes and relationships in policy evaluation using the concept of entities. The EBAC model addresses the expressiveness limitations of ABAC and ReBAC such that the relationships between entities can be navigated reasoning about these entities along paths of arbitrary length by comparing their corresponding properties. In EBAC, an entity-relationship (ER) model and logical expressions, including logical operators (e.g., *or, and, not*) and quantifiers (i.e., $\forall$ and $\exists$), are introduced into the policy expressions as a generalization to ABAC. The ER-model describes the entities along with their properties and relationships for a particular application which is then represented in an entity graph. This is a directed multi-labelled graph mapping the entities and relationship types to vertices and edges respectively. Authorization policies are specified in terms of the entity model which is then instantiated for evaluating attribute values of the relevant entities (i.e., subject, object, action and environment). An authorization system called Auctoritas provides a policy language and an evaluation engine for EBAC [17]. However, this model is neither popular nor commonly used like the other conventional access control models.

3.5   Access Control by Context

The access decision is not only relying on the policy in terms of subject and resource, but also contextual parameters, such as the sequence of events preceding the access attempt (i.e., history), location, time, and sequence of responses, are taken into account. The permission to access resources is dependent on these contextual information, unlike the content-based access control, which makes access decisions according to the data values. The final decision is based on the result of reviewing the situation [30]. The models in this class, as in the following, are often used as a complement to conventional access control models.

Emotion-Based Access Control. A system could be in danger when an angry user is granted access despite being an authorized subject. The opposite scenario is also valid as there could be unauthorized individuals who need access urgently to save the system from risky incidents. Hence, this model introduces the concept of sensibility to access control systems instead of relying on the authorization component only.

The emotion factor (i.e., feelings of the person trying to access the protected resources) can be used as a complement to the existing access control mechanisms. Firstly, the spontaneous brain signals are recorded from the scalp of the requesting user in the sensing layer. This is primarily a hardware component called Emotiv EPOC headset which collects the EEG signals and transmits them to a listener module. The received data is then analyzed in a signal processing module where the emotions are classified into positive or negative. According to the emotion level, the decision maker determines whether to allow access to the requested resource or not [5]. Although the emotion detection technology is a novel method in access control, it is still an ongoing research and not commonly used in practice.

Risk-Based Access Control. This model, also referred to as *Risk-Adaptive Access Control (RAdAC)*, originated from the need of the enterprise to assess the current situation and possible risks in real-time even when the subjects lack proper permissions. A possible strategy is to deny the access in this case, however, emergency data access is crucial in some domains (e.g., healthcare and military). Hence, this model introduces risk levels into the process of access decision such that the access is determined by computing the security risk and operational need (e.g., subject trustworthiness, information sensitivity, and history events) instead of only using the rigid policies which provide the same decision in different circumstances. After the risky access event, the system will take some mitigating actions for minimizing possible information disclosure in the future. Several methods for estimating access risks are proposed by various works including machine learning (e.g., Molloy et al. [44]), probability theory (e.g., Rajbhandari and Snekkenes [49]), and fuzzy logic (e.g., Chen et al. [20] and Ni et al. [45]). The work of Atlam et al. [7] provides a survey of the state-of-the-art risk-based access control model along with the existing risk estimation techniques (see Sect. 4).

4   Comparative Studies

In this section, we review a selected list of access control model literature studies. The related works are sorted ascendingly by their publishing year. The older surveys focus on data security whereas the newer ones deal with access control in specific domains, e.g., cloud computing, social networks, and internet of things (IoT). In the following, we summarize each of the survey works and map the presented access control models into our classification categories.

Access Control: Principle and Practice (Sandhu and Samarati 1994). This work [52] is one of the earliest works in the area of access control. It provided a concrete explanation for authentication, administration (i.e., centralized, hierarchical, cooperative, ownership, and decentralized), access control, and auditing in addition to how they are related to each other. The difference between policy and mechanism is also illustrated. The DAC and MAC strategies are explained along with the access matrix including its implementation approaches (i.e., ACL, Capabilities, and Authorization Relations). They had a different perspective regarding RBAC due to being relatively recent at the time of publishing this work. They considered the role-based approach as an alternative to traditional DAC and MAC policies with several advantages, e.g., authorization management, hierarchical roles, least privilege, separation of duties, and object classes.

Data Security (Bertino 1998). In this paper [10], Bertino surveys the state of the art in access control for database systems and outlines the main research issues. The *System R* [6] access control is discussed as the basic DAC model for protecting tables and views with specific access modes (i.e., select, insert, update, and delete) in addition to the existing extensions for supporting negative authorizations [15], non-cascading revoke, and temporal duration of authorizations [11,12]. Moreover, RBAC as an extension to access control models is described as well as how MAC strategy is applied in databases enforcing the BLP principles and multilevel relational model using views. Finally, the research directions of access control for database systems are addressed with respect to data protection against intrusions (e.g., trojan horses and covert channels) besides developing authorization and access control models for advanced data management systems.

Access Control: Policies, Models, and Mechanisms (Samarati and de Capitani di Vimercati 2001). One of the earliest literature reviews providing definitions for security policy, model and mechanism [51]. They clarified the basic concepts and explained the access control models along with the current implementations in the context of MAC, DAC, and RBAC categories. Some models belong to only one category while others are hybrid. For instance, the *Access Matrix* is a DAC model whereas the *Chinese Wall* combines DAC with MAC policies. We partly relied on this classification, especially for the DAC and MAC strategies, and the basic state-of-the-art access control models.

Database Security - Concepts, Approaches, and Challenges (Bertino and Sandhu 2005). In 2005, Bertino and Sandhu [16] discuss database security with focus on confidentiality and integrity. They give an overview of the System R model along with its extensions in the context of DAC, BLP principles as well as the multilevel secure database management system (DBMS) model for applying MAC policies, RBAC models, and content-based access control using views to enforce fine-grained authorization policies. Further, requirements and features for XML and object-based database systems are presented.

Survey on Access Control Models (Sahafizadeh and Parsa 2010). A short survey and comparison of access control models are provided in [50]. They briefly explained only five access control models, i.e., MAC, BLP, multilevel secure (MLS) DBMS, RBAC, and VBAC, with a basic description of the main concepts. However, their definition of MAC is not consistent. They considered MAC as a model relying on the comparison of the subjects and objects sensitivity labels. Then, they referred to the MLS model as an implementation of the MAC idea. Thus, they did not clarify whether MAC is a model or a type. Finally, they evaluated all the models with respect to access control criteria: (1) support for fine/coarse grained specification, (2) evaluation using conditions, (3) least privilege, (4) support for single/multiple policy types, (5) information used for making authorization decisions, (6) use of application-specific information only while processing the client request, (7) enterprise-wide consistent access control policy enforcement, and (8) support for changes.

Access Control for Databases: Concepts and Systems (Bertino et al. 2011). This monograph [14] presents a comprehensive state of the art about the access control approaches and models in the DBMS domain. They illustrated the current access control models in relational databases, i.e., the *System R* and *multilevel relational model* which enforces the MAC strategy by classifying tuples and even cells at different access classes. Moreover, they explained how fine-grained access control is applied in databases through query rewriting, SQL extension, and authorization views as techniques for content-based access control. Three case studies implemented in popular commercial DBMSs are presented: (1) Microsoft's SQL Server 2008 DBMS with its base authorization model and access control administration features, (2) the Oracle Virtual Private Database (VPD) technology [19] for controlling database access at the level of columns and rows, and (3) Oracle Label security mechanism implementing the strategy of MAC. Last but not least, they gave an overview of access control models for object databases, XML data, Geographical data, and digital libraries.

Database Security & Access Control Models: A Brief Overview (Kriti 2013). The work of Kriti [39] discussed the access control models in the context of databases presenting security threats and policy requirements as a motivation. An overview of the security models basic terms (i.e., subjects, objects, access modes, policies, authorizations, administrative rights, and axioms) as well as the access control principles of administration (i.e., centralized vs. decentralized), system (i.e., open vs. closed), and privilege (i.e., minimum vs. maximum) are provided. The DAC, MAC, and RBAC are explained in addition to how they are applied in databases along with their vulnerabilities. For instance, the DAC authorization is applied in databases using *System R* model and its extensions, but vulnerable to trojan horse attacks.

Taxonomy and Classification of Access Control Models for Cloud Environments (Majumder et al. 2014). The authors in [41] classified various existing access control models according to a proposed taxonomy of access control schemes for cloud environments. They discussed the access control challenges in cloud computing regarding cost, granularity, data loss, taking the data sensitivity into account, data theft by malicious users, and accessing data from an outside server. Furthermore, they explained 11 models (see Table 1) and analyzed them based on (1) identity vs. non-identity in terms of whether the model is tree-structured or not, and (2) centralized (i.e., per user, group users, and all users) vs. collaborative.

Different Access Control Mechanisms (Sifou et al. 2017). This work [53] analyzes and compares different access control models in the context of cloud computing. Based on the National Institute of Standards and Technology's (NIST) view in [43], the authors illustrated the main features of cloud computing service and deployment models. They demonstrated DAC, MAC, RBAC, ABAC, and organization-based access control (OrBAC) along with the advantages and disadvantages of each access control model. According to the current cloud computing requirements, they defined 9 criteria to evaluate the current access control models: dynamicity, flexibility, reliability, ease of administration, security policy implementation, global management, support scalability, computational costs, and fine-grained access.

Survey on Access Control Mechanisms in Cloud Computing (Karatas and Abkulut 2018). The work of Karatas and Abkulut [37] provided a survey of access control approaches and works related to cloud computing. They reviewed 109 research papers in that domain throughout the past decade. They provide not only a comparative explanation for the existing access control models, but also a unique evaluation using NIST access control metrics [35]. For each access control model, an overview followed by an analysis with respect to the applicable criteria is given. The models are reviewed according to the satisfaction degree for each metric (i.e., *low, medium, high, optional, not applicable,* and *not mentioned*). Additionally, their study is compared versus seven other survey works in terms of the presented approaches, graphical definitions, advantages/disadvantages, the use of NIST metrics, number of reviewed articles, and queried databases (e.g., IEEE, ACM, Springer, etc.).

A Survey on Access Control in the Age of Internet of Things (Qui et al. 2020). The article [48] presents a survey on the access control characteristics, technologies, a taxonomy of access control models requirements, and future development direction in the IoT research field. In the IoT environment, the data are dynamic, massive, need strong privacy, and continuously exchanged between different cooperation organizations. This work is compared with other literature reviews with similar focus in terms of access control policy description method,

combination, conflict resolution and authoring (i.e., attribute discovery mechanism, policy mining, and authorization model) explaining each requirement in
detail. They described the authorization by categories based on the following:
ABAC, RBAC, capability-based access control (CapBAC), usage control-based
access model (UCON), OrBAC, blockchain, and open authorization (OAuth).

**Risk-Based Access Control Model: A Systematic Literature Review
(Atlam et al. 2020).** A systematic review of the risk-based access control
model is provided in [7]. According to their search strategy, they chose 44 recent
studies to summarize their contributions, analyze the various risk factors, and
investigate the used risk estimation techniques. First, they briefly illustrate (1)
the aim of access control, (2) the difference between authentication, authorization and access control, (3) the five core elements of access control models (i.e.,
subjects, objects, actions, privileges, and access policies) as well as (4) the access
control process flow. Furthermore, they compared the static and dynamic access
control models with respect to features, decision, pros/cons, examples, and applications. For the traditional access control approaches, they just mentioned ACL,
DAC, MAC and RBAC with a basic description. Then, an overview of the risk-
based access control model and its elements is provided. Finally, they addressed
the research methodology phases and analysed the results providing answers to
the research questions through comparing all the selected works.

In summary, Table 1 presents the models addressed in the previously discussed surveys with respect to our classification. We indicate whether all the
models (✓) listed for each category in Sect. 3, some of them (**O**), or none
(✗) are addressed in a given citation. We also include the models that are
not stated in our work under the column *Other*. Only two access control
models related to databases (i.e., System R [29] and multilevel secure (MLS)
DBMSs [38]) are not mentioned. The rest belongs to cloud computing and
IoT domains, e.g., gateway-based access control (GBAC) [57], novel data access
control (NDAC) [27], usage control-based access model (UCON) [23], purpose-
based usage access control (PBAC) [54], towards temporal-based access control (TTAC) [61], organization-based access control (OrBAC) [36], fine-grained
access control (FGAC) [40], capability-based access control (CapBAC) [31], hierarchical attribute-based access control (HABE) [58], attribute-based encryption
fine-grained access control (ABE-FGAC) [55], and privacy-preserving ABAC (P-
ABAC) [59].

5   Analysis

Based on the results of our literature study in Sects. 3 and 4, we define ten
criteria to study the differences between the five classes of access control models
presented in Sect. 3. We partly relied on the standard metrics for evaluating
access control systems from NIST [35] and added other criteria to make it feasible
in selecting the appropriate class and access control model according to the

Table 1. The models included for each category in the survey works list

Citation	OSA	MsR	Roles	Content	Context	Other
Sandhu and Samarati [52] (1994)	✓	$O^{b,c}$	✓	✗	✗	
Bertino [10] (1998)	✗	O^b	✓	O^f	✗	System R
Samarati and De Capitani di Vimercati [51] (2001)	✓	✓^{e}	✓	✗	✗	
Bertino and Sandhu [16] (2005)	✗	O^b	✓	O^f	✗	
Sahafizadeh and Parsa [50] (2010)	✗	O^b	✓	O^f	✗	MLS
Bertino et al. [14] (2011)	✓	O^b	✓	$O^{f,g}$	✗	System R
Kriti [39] (2013)	✗	$O^{b,c}$	✓	✗	✗	System R
Majumder et al. [41] (2014)	O^a	✗	✓	O^g	✗	GBAC, NDAC, UCON, PBAC, TTAC, OrBAC, CapBAC
Sifou et al. [53] (2017)	✗	✗	✓	O^g	✗	OrBAC
Karatas and Abkulut [37] (2018)	✗	✗	✓	O^g	✗	FGAC, HABE, ABE-FGAC
Qiu et al. [48] (2020)	✗	$O^{b,c,d}$	✓	$O^{g,h}$	✗	OrBAC, UCON, P-ABAC, CapBAC
Atlam et al. [7] (2020)	O^a	✗	✓	✗	O^i	

^{a} ACL, ^{b} BLP, ^{c} Biba, ^{d} Clark-Wilson, ^{e} Except Lipner, ^{f} VBAC, ^{g} ABAC, ^{h} ReBAC, ^{i} RAdAC

application security requirements. The criteria along with their description are listed below:

- *Authorization strategy*: whether MAC, DAC, or hybrid (recall Sect. 2).
- *Dynamic authorization*: represents the dynamic definition of access rights in terms of rules and policies evaluating their attributes in real-time.
- *Granularity of control*: indicates the objects' levels of granularity, i.e., fine and/or coarse grained.
- *Least privilege principle support*: the minimum access rights required for performing a task.
- *Separation of duty*: ensures that access is only granted to subjects that are duty-related to the objects to limit power and avoid conflict of interests.
- *Vulnerable to attacks*: this is for ensuring the security of the model to avoid the leakage of permissions to an unauthorized principal, e.g., trojan horse and covert channel attacks.
- *Bypass*: it is about whether policy rules are allowed to be bypassed for critical access decisions in emergency situations or not and how tolerant the risk is.

– *Conflict resolution or prevention*: deals with preventing or resolving deadlocks and conflicting rules from the same or different policies.
– *Operational/situational awareness*: considers operational/situational factors (e.g., some environment variables) in access rules specification and enforcement (i.e., decision-making).
– *Privileges/capabilities discovery*: considers the discovery of capabilities/objects (or object groups) of a given subject (or subject group) and vice versa.

After defining the selected criteria, we summarize them against our access control models categories in Table 2. For each criterion, we indicate whether it is satisfied by all the models within a given classification group (✓), partially supported either with further considerations (i.e., based on access control requirements and model implementation) or by specific access control models within that category (O), or not at all (✗). If applicable, the the level of satisfaction is indicated, i.e., low (L), medium (M), or high (H).

Table 2. Access control model categories analysis with respect to the defined criteria

Criteria	OSA	MsR	Roles	Content	Context
Authorization strategy	DAC	MAC	Hybrid	Hybrid	Hybrid
Granularity of control	L	L	M	H	H
Least privilege principle support	L	M	M	H	O[a]
Dynamic authorization	✗	✗	✓	✓	✓
Separation of duty	✗	O[b]	✓	✓	✓
Vulnerable to attacks	✓	✓	✗	✗	✗
Bypass	✗	✗	✗	✗	✓
Conflict resolution or prevention	✗	✗	✗	O[c]	O[a]
Operational/situational awareness	✗	✗	✓	✓	✓
Privileges/capabilities discovery	✓	✗	✓	O[d]	✗

[a]Depends on the underlying access control model
[b]Supported by the *Chinese-Wall* model
[c]Supported by *ABAC* and *EBAC* models
[d]Supported by the VBAC model

6 Conclusion

Access control mitigates the risks of unauthorized access attempts to data, resources, and systems. There are a lot of existing access control models; some of them are commonly known and used in practice while others have evolved recently and are not yet popular like the conventional models. The already available access control survey works are either including the state-of-the-art models at the publishing time or focusing on the taxonomy and classification of models for a particular domain.

In this paper, we first explained authorization strategies. We proposed a general classification for access control models without being restricted to a specific field (e.g., cloud computing and IoT). Moreover, we provided some examples of access control models along with the current implementations and extensions for the five categories, i.e., explicit object-subject assignment, model-specific rules, roles, content, and context. We selected a list of comparative studies about survey, taxonomy, and evaluation of access control models. Then, we summarized each work and compared the included models according to our classification. Finally, we analyzed the categories with respect to several criteria; some of them are selected from the NIST standard access control system evaluation metrics, according to the level of support and considerations (if any).

The comparison result (see Table 1) shows that we discussed more models than other works for the OSA, MsR, content, and context categories. In order to focus on general access control models with a view to databases, we did not include advanced domain-specific models. In our future work, we will extend the access control models list according to our proposed classification. We plan to explain the models in more details with a formal representation besides adding more metrics to our analysis.

Acknowledgement. The research reported in this paper has been partly supported by the LIT Secure and Correct Systems Lab funded by the State of Upper Austria. The work was also funded within the FFG BRIDGE project KnoP-2D (grant no. 871299).

References

1. Extensible access control markup language (xacml) version 3.0 - oasis standard (2013). http://docs.oasis-open.org/xacml/3.0/xacml-3.0-core-spec-os-en.html
2. Information Technology - Next Generation Access Control - Generic Operations And Data Structures (NGAC-GOADS). American National Standard for Information Technology INCITS 526–2016 (2016)
3. Information technology - Next Generation Access Control - Functional Architecture (NGAC-FA). American National Standard for Information Technology INCITS 499–2013 (March 2013)
4. Abrams, M.D.: Renewed understanding of access control policies. In: Proceedings of the 16th National Computer Security Conference-Information System Security: User Choices, pp. 87–96 (1995)
5. Almehmadi, A., El-Khatib, K.: Authorized! access denied, unauthorized! access granted. In: Proceedings of the 6th International Conference on Security of Information and Networks, pp. 363–367 (2013)
6. Astrahan, M.M., et al.: System R: relational approach to database management. ACM Trans. Database Syst. (TODS) **1**(2), 97–137 (1976)
7. Atlam, H.F., Azad, M.A., Alassafi, M.O., Alshdadi, A.A., Alenezi, A.: Risk-based access control model: a systematic literature review. Future Internet **12**(6), 103 (2020). https://doi.org/10.3390/fi12060103
8. Bell, D.E., La Padula, L.J.: Secure computer system: Unified exposition and multics interpretation. Technical report, MITRE CORP BEDFORD MA (1976)
9. Benantar, M.: Access Control Systems: Security, Identity Management and Trust Models. Springer Science & Business Media, Heidelberg (2005)

10. Bertino, E.: Data security. Data Knowl. Eng. **25**(1–2), 199–216 (1998)
11. Bertino, E., Bettini, C., Ferrari, E., Samarati, P.: Supporting periodic authorizations and temporal reasoning in database access control. In: VLDB, pp. 472–483. Citeseer (1996)
12. Bertino, E., Bettini, C., Ferrari, E., Samarati, P.: A temporal access control mechanism for database systems. IEEE Trans. Knowl. Data Eng. **8**(1), 67–80 (1996)
13. Bertino, E., Bonatti, P.A., Ferrari, E.: TRBAC: a temporal role-based access control model. In: Proceedings of the fifth ACM Workshop on Role-based Access Control, pp. 21–30 (2000)
14. Bertino, E., Ghinita, G., Kamra, A.: Access Control for Databases: Concepts and Systems. Now Publishers Inc., Norwell (2011)
15. Bertino, E., Samarati, P., Jajodia, S.: An extended authorization model for relational databases. IEEE Trans. Knowl. Data Eng. **9**(1), 85–101 (1997)
16. Bertino, E., Sandhu, R.: Database security-concepts, approaches, and challenges. IEEE Trans. Dependable Secur. Comput. **2**(1), 2–19 (2005)
17. Bogaerts, J., Decat, M., Lagaisse, B., Joosen, W.: Entity-based access control: supporting more expressive access control policies. In: Proceedings of the 31st Annual Computer Security Applications Conference, pp. 291–300 (2015)
18. Brewer, D.F., Nash, M.J.: The Chinese wall security policy. In: IEEE Symposium on Security and Privacy, vol. 1989, p. 206. Oakland (1989)
19. Browder, K., Davidson, M.A.: The virtual private database in oracle9ir2. Oracle Tech. White Pap. Oracle Corporation **500**(280) (2002)
20. Cheng, P.C., Rohatgi, P., Keser, C., Karger, P.A., Wagner, G.M., Reninger, A.S.: Fuzzy multi-level security: an experiment on quantified risk-adaptive access control. In: 2007 IEEE Symposium on Security and Privacy (SP 2007), pp. 222–230. IEEE (2007)
21. Clark, D.D., Wilson, D.R.: A comparison of commercial and military computer security policies. In: 1987 IEEE Symposium on Security and Privacy, pp. 184–184. IEEE (1987)
22. Damiani, M.L., Bertino, E., Catania, B., Perlasca, P.: Geo-RBAC: a spatially aware RBAC. ACM Trans. Inf. Syst. Secur. (TISSEC) **10**(1), 2-es (2007)
23. Danwei, C., Xiuli, H., Xunyi, R.: Access control of cloud service based on UCON. In: Jaatun, M.G., Zhao, G., Rong, C. (eds.) CloudCom 2009. LNCS, vol. 5931, pp. 559–564. Springer, Heidelberg (2009). https://doi.org/10.1007/978-3-642-10665-1_52
24. Eckert, C.: IT-Sicherheit, 9th edn. De Gruyter Oldenbourg, Munich (2014)
25. Ferraiolo, D., Chandramouli, R., Kuhn, R., Hu, V.: Extensible access control markup language (xacml) and next generation access control (ngac). In: Proceedings of the 2016 ACM International Workshop on Attribute Based Access Control, pp. 13–24 (2016)
26. Fong, P.W.: Relationship-based access control: protection model and policy language. In: Proceedings of the first ACM Conference on Data and Application Security and Privacy, pp. 191–202 (2011)
27. Gao, X.W., Jiang, Z.M., Jiang, R.: A novel data access scheme in cloud computing. In: Advanced Materials Research, vol. 756, pp. 2649–2654. Trans Tech Publ (2013)
28. Gates, C.: Access control requirements for web 2.0 security and privacy. IEEE Web **2**, 12–15 (2007)
29. Griffiths, P.P., Wade, B.W.: An authorization mechanism for a relational database system. ACM Trans. Database Syst. (TODS) **1**(3), 242–255 (1976)
30. Harris, S., Maymi, F.: CISSP All-in-One Exam Guide. McGraw-Hill, New York (2010)

31. Hota, C., Sanka, S., Rajarajan, M., Nair, S.K.: Capability-based cryptographic data access control in cloud computing. Int. J. Adv. Netw. Appl. **3**(3), 1152–1161 (2011)
32. Hu, H., Ahn, G.J., Jorgensen, J.: Multiparty access control for online social networks: model and mechanisms. IEEE Trans. Knowl. Data Eng. **25**(7), 1614–1627 (2012)
33. Hu, V.C., et al.: Guide to attribute based access control (abac) definition and considerations (2014). https://doi.org/10.6028/NIST.SP.800-162, https://nvlpubs. nist.gov/nistpubs/specialpublications/NIST.SP.800-162.pdf
34. Hu, V.C., Ferraiolo, D.F., Chandramouli, R., Kuhn, D.R.: Attribute-Based Access Control. Artech House, London (2017)
35. Hu, V.C., Scarfone, K.: Guidelines for Access Control System Evaluation Metrics. National Institute of Standards and Technology, Gaithersburg, MD (2012). https://doi.org/10.6028/NIST.IR.7874
36. Kalam, A.A.E., et al.: Organization based access control. In: Proceedings POLICY 2003. IEEE 4th International Workshop on Policies for Distributed Systems and Networks, pp. 120–131. IEEE (2003)
37. Karatas, G., Akbulut, A.: Survey on access control mechanisms in cloud computing. J. Cyber Secur. Mobil. (2018). https://doi.org/10.13052/2245-1439.731
38. Keefe, T.F., Tsai, W.T., Srivastava, J.: Database concurrency control in multilevel secure database management systems. IEEE Trans. Knowl. Data Eng. **5**(6), 1039–1055 (1993)
39. Kriti, I.K.: Database security & access control models: a brief overview. Int. J. Eng. Res. Technol. (IJERT) **2**(5) (2013)
40. Li, J., et al.: Fine-grained data access control systems with user accountability in cloud computing. In: 2010 IEEE Second International Conference on Cloud Computing Technology and Science, pp. 89–96. IEEE (2010)
41. Majumder, A., Namasudra, S., Nath, S.: Taxonomy and classification of access control models for cloud environments. In: Mahmood, Z. (ed.) Continued Rise of the Cloud. CCN, pp. 23–53. Springer, London (2014). https://doi.org/10.1007/978-1-4471-6452-4_2
42. Matt, B.: Computer Security: Art and Science. Addison-Wesley Professional, Boston (2018)
43. Mell, P., Grance, T., et al.: The nist definition of cloud computing (2011)
44. Molloy, I., Dickens, L., Morisset, C., Cheng, P.C., Lobo, J., Russo, A.: Risk-based security decisions under uncertainty. In: Proceedings of the second ACM Conference on Data and Application Security and Privacy, pp. 157–168 (2012)
45. Ni, Q., Bertino, E., Lobo, J.: Risk-based access control systems built on fuzzy inferences. In: Proceedings of the 5th ACM Symposium on Information, Computer and Communications Security, pp. 250–260 (2010)
46. Park, J., Sandhu, R.: Originator control in usage control. In: Proceedings Third International Workshop on Policies for Distributed Systems and Networks, pp. 60–66. IEEE (2002)
47. Petkovic, M., Jonker, W.: Security, Privacy, and Trust in Modern Data Management. Springer, Heidelberg (2007)
48. Qiu, J., Tian, Z., Du, C., Zuo, Q., Su, S., Fang, B.: A survey on access control in the age of internet of things. IEEE Internet Things J. **7**(6), 4682–4696 (2020). https://doi.org/10.1109/JIOT.2020.2969326

49. Rajbhandari, L., Snekkenes, E.A.: Using game theory to analyze risk to privacy: an initial insight. In: Fischer-Hübner, S., Duquenoy, P., Hansen, M., Leenes, R., Zhang, G. (eds.) Privacy and Identity 2010. IAICT, vol. 352, pp. 41–51. Springer, Heidelberg (2011). https://doi.org/10.1007/978-3-642-20769-3_4

50. Sahafizadeh, E., Parsa, S.: Survey on access control models. In: 2010 2nd International Conference on Future Computer and Communication, vol. 1. IEEE (2010)

51. Samarati, P., de Vimercati, S.C.: Access control: policies, models, and mechanisms. In: Focardi, R., Gorrieri, R. (eds.) FOSAD 2000. LNCS, vol. 2171, pp. 137–196. Springer, Heidelberg (2001). https://doi.org/10.1007/3-540-45608-2_3

52. Sandhu, R.S., Samarati, P.: Access control: principle and practice. IEEE Commun. Mag. **32**(9), 40–48 (1994). https://ieeexplore.ieee.org/document/312842

53. Sifou, F., Kartit, A., Hammouch, A.: Different access control mechanisms for data security in cloud computing. In: Proceedings of the 2017 International Conference on Cloud and Big Data Computing, pp. 40–44. ACM, New York NY (2017). https://doi.org/10.1145/3141128.3141133

54. Sun, L., Wang, H.: A purpose based usage access control model. Int. J. Comput. Inf. Eng. **4**(1), 44–51 (2010)

55. Tamizharasi, G., Balamurugan, B., Manjula, R.: Attribute based encryption with fine-grained access provision in cloud computing. In: Proceedings of the International Conference on Informatics and Analytics, pp. 1–4 (2016)

56. Tapiador, A., Carrera, D., Salvachúa, J.: Tie-RBAC: an application of RBAC to social networks. arXiv preprint arXiv:1205.5720 (2012)

57. Wu, Y., Suhendra, V., Guo, H.: A gateway-based access control scheme for collaborative clouds. In: Proceedings of the 7th International Conference on Internet Monitoring and Protection, pp. 54–60 (2012)

58. Xie, Y., Wen, H., Wu, B., Jiang, Y., Meng, J.: A modified hierarchical attribute-based encryption access control method for mobile cloud computing. IEEE Trans. Cloud Comput. **7**(2), 383–391 (2015)

59. Xu, Y., Zeng, Q., Wang, G., Zhang, C., Ren, J., Zhang, Y.: A privacy-preserving attribute-based access control scheme. In: Wang, G., Chen, J., Yang, L.T. (eds.) SpaCCS 2018. LNCS, vol. 11342, pp. 361–370. Springer, Cham (2018). https://doi.org/10.1007/978-3-030-05345-1_31

60. Zeng, W., Yang, Y., Luo, B.: Content-based access control: use data content to assist access control for large-scale content-centric databases. In: 2014 IEEE International Conference on Big Data (Big Data), pp. 701–710. IEEE (2014)

61. Zhu, Y., Hu, H., Ahn, G.J., Huang, D., Wang, S.: Towards temporal access control in cloud computing. In: 2012 Proceedings IEEE INFOCOM, pp. 2576–2580. IEEE (2012)

Motorbike Counting in Heavily Crowded Scenes

Chi Kien Huynh[1], Tran Khanh Dang[2(✉)], and Cong An Nguyen[3]

[1] Department of Computer Science, Stony Brook University, New York, USA
kien.huynh@stonybrook.edu
[2] Ho Chi Minh City University of Technology, VNU-HCM,
Ho Chi Minh City, Vietnam
khanh@hcmut.edu.vn
[3] Social Insurance, Dong Nai Province, Ho Chi Minh City, Vietnam

Abstract. Vehicle density estimation has an important role in intelligent traffic systems. As of now, most established studies only focused on areas where people mainly travel by four-wheeled vehicles rather than motorbikes. However, in some countries such as Vietnam where motorbikes are the majority, traffic scenarios will pose different issues. Motorbikes are intrinsically more flexible so they can cause cluttered and chaotic visual. As a result, traffic video data captured in such environment is more challenging to existing systems. In this work, we performed an empirical survey on a set of vision-based counting methods covering a wide range of models and techniques. To our knowledge, there has not been many works dedicated to tackle this problem. Based on our experimental results, some of the top performers is ready to be used in real systems due to their robustness.

Keywords: Motorbike counting · Density estimation · Bag-of-visual words · Least square-suppor vector machine · Convolutional neural network

1 Introduction

In densely populated areas, among many problems we have to face in order to build a sustainable development city such as energy efficiency [14], e-commerce [15], security and privacy [16], easing traffic congestions is one of the most important targets [40,49]. At peak hours, it can be extremely counter-productive to anyone participating in transportation. This is especially true in cities such as Ho Chi Minh or Jakarta where motorbike counts can be up to dozens of millions [21]. Traffic in these cities is not only disoriented but it is also extremely difficult to manage (cf. Fig. 1). A straightforward fix is to improve relevant infrastructures and urban planning, but it might take a long period to execute. Instead, we can take advantage of intelligent traffic applications.

One of the most popular traffic applications is finding the fastest route for drivers. In the past, this was done solely by solving static shortest path problem,

© Springer Nature Switzerland AG 2021
T. K. Dang et al. (Eds.): FDSE 2021, LNCS 13076, pp. 175–194, 2021.
https://doi.org/10.1007/978-3-030-91387-8_12

i.e. we assumed that every subroutines of the final solution are always available. Therefore, these solutions do not work as expected when there are traffic jams or sudden road maintenances. However, modern systems such as [40, 49] and [43] have taken those problems into account. By using appropriate data mining and machine learning methods, they can eliminate trapped routes and recommend better alternatives for GPS users.

To decide whether some road segments are available, those systems either use prior knowledges mined from a large traffic database of GPS users or directly measure the density of vehicles in a certain area. In the latter approach, devices such as accelerometer [30], wireless magnetic censor [1] or cameras [8] can be used to detect vehicles. In the scope of this work, we will only exploit the vision-based approach. This is because most existing traffic data currently available are videos. Cameras are also more common and simpler to setup than other sensors so it would be easier to integrating them into an ITS.

Fig. 1. A scene taken from one of Ho Chi Minh city's 6-way intersections at rush hour.

For the counting problem, all of the methods can be categorized into two approaches: direct and indirect. In direct approaches, the objects' locations will be pinpointed exactly and their boundaries might be estimated. After finishing detecting every objects on any image, we will obviously achieve the number of objects in that image. Some examples of the direct approach are [18, 37, 48] and [7]. In contrast, indirect methods do not need to detect the locations or boundaries of objects in image. These methods work by extracting multiple local feature vectors or a single global vector over the entire image. Then, a mapping function will be learned to compute the number of objects using those vectors. The learning process is done by regression analysis, machine learning or data mining methods. A few examples for the indirect approaches are [6, 25, 26] and [38].

Both of the direct and indirect approaches have their own advantages and disadvantages. The direct methods are very useful for problems requiring the locations of the objects such as tracking or crime detection. Still, most of them are very sensitive to occlusion so their performances would be reduced in chaotic environments. Additionally, robust detectors usually require high-end, dedicated hardware to reach practical speed. On the other hand, indirect methods can still be very strong even if they are to be applied to difficult scenes. But as we already know, they only provide us with the number of objects without telling us their locations. This is a serious restriction because in some applications, density alone is not enough.

In this paper, we explore crowd counting methods that do not rely on object detection. This is the first known work that provides benchmarks for the problem of motorbike counting. We implemented thorough experiments on the motorbike counting problem in two difficult traffic scenarios, unique to some urban areas. Additionally, the methods we selected are distinct in term of features, learning, or classification process [11,26,28,45]. Some of them can work well in extremely difficult scenarios and can be readily applied in practice.

The remaining of this paper is organized as follows. The next section shows a summary of different approaches in the indirect counting literature. The third section explains in detail all methods that will be used in our experiments of counting motorbikes. In the fourth section, experimental methodology and results are shown along with our discussions. Lastly, concluding remarks and future work are presented in the fifth section.

2 Related Work

In the indirect objects counting literature, there are rarely any method designed specifically for motorbike density estimation. Most of the existing works revolve around counting people in human crowds. Regardless, those methods can still be used for our problem since their routines are generic enough to be applied to other subjects other than human.

There are a lot of variants for indirect counting approaches; however, most of them always have a feature extraction step and a regression step. One of the most recurring feature extraction method we have found throughout many studies is based on background subtraction. In [25,39] and [10], background subtraction was used to extract the foreground blobs of the people. Then, simple features such as blob sizes, perimeter, area, histogram of blob sizes, etc. are computed and used as the feature vector. In [25], the authors also add weighting factors to each pixel based on how they are located relatively to the camera plane and the region of interest (ROI). For the mapping function, [25] and [10] do not only use simple linear fitting like [39] does but also use neural networks.

Manara et al. proposed two robust feature extraction methods in [32] and [33]. In the first method [32], the authors used gray level dependence matrix (GLDM), a type of two dimensional histogram. Texture analysis and feature extraction are then performed based on four indicators: contrast, homogeneity, energy and entropy. In their later method [33], Manara et al. first calculate the edge images then use Minkowski's approach to compute the fractal dimension which will be used as the feature vector. In the first work, they used self organizing map neural network to estimate the crowd density while in the second one they used linear regression. It should be noted that their frameworks do not produce the estimated number of people but it only tried to classify images into one of the five categories based on crowd density: very low, low, moderate, high, very high.

Inspired by the works of Manara et al., [38] and [47] also applied GLDM in their frameworks. In their work, Rahmalan et al. [38] do not only use GLDM but also Minkowski fractal dimension and their own proposed method, translation invariant orthonormal Chebyshev moments, to extract the features from images. The method they used to estimate the density is also self organizing map just like Manara et al.'s work. Meanwhile in [47], Xinyu Wu et al. employed cell size normalization based on the camera calibration information. Images are divided into cells with different weights and then the GLDM features will be adjusted accordingly. For regression, the authors used support vector machine [12] with the 2nd degree polynomial kernel.

Unlike the methods mentioned above, Antoni et al. also take the motion information of the people into account in [6] and [5]. The motion segmentation step is performed firsthand on images. Crowds with similar motion will be grouped and extracted from the image. This is similar to background subtraction but it would give us more than one foreground layers depending on how many major motion flows are there in the image. Then, simple features such as area, perimeter, edge pixels, etc. are calculated and used as the global feature vector. In [5], the authors only used Gaussian process to estimate a combined function of the linear the RBF kernels while in [6] they used both Bayesian process and Bayesian Poisson regression.

Other methods for extracting the feature vectors can be found in [11, 22, 26, 45] and [20]. In [26], bag-of-visual words or random forest features are extracted on every single pixel then fed to the mapping function to produce the crowd density map. To estimate the people count, they sum up all of the values on that map. In a similar approach, the method in [22] uses bag-of-visual words as global feature, combined with least square support vector machine for counting [42]. Differently, the authors of [45] apply multiple Gabor filters on the images. Then, primitive features such as mean and variance of gray values are computed for each Gabor channel and used as the feature vector. In [11], Donatello Conte et al. first perform SURF detector to locate the keypoints on the image. The points will then be grouped into clusters using a graph-based clustering algorithm. The main part of the feature vector is the density of the SURF keypoints in a cluster. Unlike all of the listed works, Homa et al. [20] proposed their own method for computing feature vectors using compressed sensing theory. They also employed their own method for estimating the number of people based on K-nearest neighbors approach.

Notice that most of the aforementioned methods work with predefined or handcrafted visual features. A trending approach in recent years is to learn the features automatically instead. One of the most notable models is convolutional neural network (CNN). It is suitable for vision task thanks to its ability to extract features tailored to each specific problem through learning. In [34], the

authors proposed a CNN with architecture similar to the iconic VGG [41] with
fewer layers and larger filters. In the last layer, it will produce a heatmap that
represents the density of the crowd, which can be integrated and return the
predicted count of the image. Two other similar methods that also train the
network to output heatmaps are [28] and [27]. In the first one, the authors
employ dilated CNN layers so that their network can extract features of different
scales. In the latter, a set of contextual features called contrast features were
introduced. The calculation of these features was done by taking the difference
between features of coarser scales and the original scale. The resulted contrast
features reflect the saliency of objects in extremely crowded scenes.

3   Crowd Counting Methods

In each following subsection, we present the details of each counting method
that will be used in our experiments. Although the methods of choice employ
widely different techniques, the general workflow for all of them stays the same
and always comprises of a training phase and a testing phase. The same feature
extractor is shared by both phases. The outcome of the training phase is a
mapping function which will provide us the number of vehicles when it is fed
with a feature vector. We will reuse this learned function to estimate vehicle
density in the testing phase.

3.1   Bag-of-Visual-Words

For the model to work, we first need to extract local feature vectors on the image
and then quantize them into visual words. The quantizing process is essentially
clustering the vectors into groups. Vectors in the same group normally have
similar values and they will be labelled as the same word; we treat them the
same after the quantizing process. At the end, we will count the occurrences of
each word and use the counting results as our global feature vector. As of when
this work was conducted, there were many concrete methods for computing local
feature vectors. We chose to use the well-known features such as [29], SURF [2]
or HOG [13] as the basis for the Bag-of-words model.

The entire set of every local feature vectors computed over all training images
will be used to build the visual vocabulary. We used k-means [31] to group those
vectors into K clusters and then label the vectors in the same cluster under the
same word. The centroids of the clusters will be kept for the testing phase. We
will use them later for the word classification step without repeating k-means
clustering again.

After attaining the visual vocabulary, we can start computing the global
feature vector for each image I_p as follow:

$$h_p = [c_1, c_2, ..., c_K]^T \tag{1}$$

Here, c_j is the number of times the j-th word appears in image I_p. To count
the number of motorbikes in an image, we need to find the following mapping
function:

$$y(x) = \phi(x)^T w + b \tag{2}$$

where x is the feature vector, $y(x)$ is the desired number of vehicles, $\phi(x)$ is a mapping kernel that maps the vectors to a higher feature space if needed. LS-SVM is used as our regression model here and we experimented with two common kernels, namely linear and radial basis function (RBF).

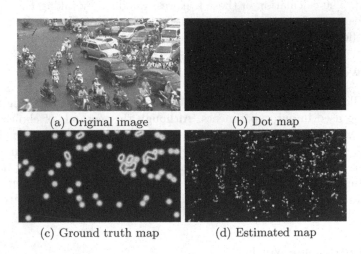

(a) Original image   (b) Dot map

(c) Ground truth map   (d) Estimated map

Fig. 2. A traffic image and its respective dot map, ground truth map and density estimation map computed using Lempitsky-Zisserman method. The feature which is being used here to produce this map is bag-of-visual words built on top of dense SIFT. The bottom maps are shown in jet color scheme for visualization purpose.

3.2   Lempitsky-Zisserman's Method

In the original paper [26], the authors employed feature extraction on every pixel. Each of the feature vectors is then used for the dot product with a weighting term w to compute the object density map (Fig. 2-d). Finally, we can easily attain the number of vehicles as it is the sum of all values making up that map.

To achieve the desired w factor, the authors proposed a specific training routine. Initially, a few images are chosen from the training dataset for manual labeling. On each of these image, a dot of the size 1 pixel is placed at the center of every object, producing a dot map (Fig. 2-b). Each dot on this map equals 1 while every other pixels amount to 0. The sum of this map is essentially the object count. Then, a Gaussian kernel is applied on those dot images to spread the value of the dots across its neighborhood (Fig. 2-c). These filtered images now act as ground truth maps for training.

In the learning process, the estimated density map is produced by computing dot product between the extracted feature vectors and w. The aim of the learning process is to find w so that we could produce an estimated map which has as short the distance to the original map as possible. A specific distance between two maps F_1 and F_2 called MESA (Maximum Excess over Subarrays) was used:

$$D_M(F_1, F_2) = \max_{B \in \mathbf{B}} |\sum_{p \in B} F_1(p) - \sum_{p \in B} F_2(p)| \tag{3}$$

where $\mathbf{B}$ is the set of all box subarrays of the map, p is the pixel index of any pixel belong to $B \in \mathbf{B}$, $\sum_{p \in B} F_1(p)$ is the sum of all values over the box subarray B of F_1 while $\sum_{p \in B} F_2(p)$ is the same sum of F_2. Intuitively, the MESA distance equals to the biggest absolute difference computed over all box subarrays between the ground truth and the estimated maps.

For feature extraction, the authors used bag-of-visual words built from dense SIFT [3] and random forest [4] in their experiments. It is worth noting that any feature extraction method could work here as long as it can be used on every single pixel.

3.3 Qing Wen et al.'s Method

As in [45], the images are first convolved with a set of different Gabor filters:

$$g(x, y, \theta, \sigma) = e^{-\frac{x^2+y^2}{2\sigma^2}} e^{2\pi \omega i (x \cos \theta + y \sin \theta)} \tag{4}$$

Note that the Gaussian component of this version doesn't have any bias toward x or y, normally it should have been $e^{-\frac{x^2+\gamma y^2}{2\sigma^2}}$. Each change in θ or ω will produce a different channel of the original image. The values of σ is chosen to be $\frac{1}{\omega}$, same as [45]. The average gray value and variance of every channel will be concatenated and used as our feature vector:

$$h_g = [E_1, var_1, E_2, var_2, ..., E_{N_c}, var_{N_c}] \tag{5}$$

E_i and var_i are respectively the mean and variance of gray value of the i-th channel, N_c is the total number of channels. Although every Gabor filter has a real part and an imaginary part, we don't simultaneously use both of them on a single image. We performed experiments using the real part and the imaginary part separately. To estimate the number of people in their original work, Qing Wen et al. used LS-SVM with the RBF kernel. However, in this work we also used the linear kernel for more variety in the results.

3.4 Donatello Conte et al.'s Method

In the feature extraction step of this method, SURF keypoints are first detected over an image and then grouped into clusters using a graph-based algorithm [19], as shown in Fig. 3. This algorithm starts by constructing a minimum spanning tree using distances between keypoints as edge weights. Then, edges with weights larger than the threshold λ is removed from the tree to create a forest:

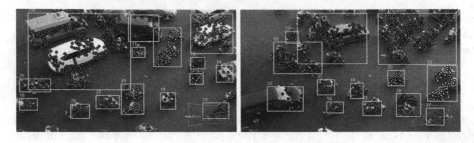

Fig. 3. Clustering results of two different images produced by the method of Donatello Conte et al. with a small change in parameters. Vehicles locating closely are likely to be grouped into the same cluster by the method.

$$\lambda = \gamma \frac{1}{N_e} \sum_{i=1}^{N_e} e_i \tag{6}$$

with N_e being the number of edges, e_i being the weight of the i-th edge and γ is a hyperparameter. Each new tree in the forest is now a keypoint cluster. Next, bounding boxes are constructed for every tree. Each box will provide its own feature vector which will be used to estimate the number of objects inside of it.

The feature vector has three main components: keypoints count, density of the keypoints in the bounding box and the distance from the lower bound of the box to the camera. After performing SURF detection, the authors also eliminate unmoving SURF keypoints between frames which belong to the background. Because of the feature choice, there is no distinction between the SURF keypoints belong to motorbikes and those belong to cars. As a consequent, keypoints of cars and motorbikes are sometimes grouped together if they stay close enough.

It should also be noted that keypoints of cars also affect the estimated motorbikes count because there is no distinction between keypoints of both. Therefore, apart from using the original features proposed by the author, we tried adding the bag-of-visual words features to maintain this distinction. Although the authors originally proposed to use ϵ-support vector regression [9] for estimation purpose, we still used LS-SVM in this step instead since it would be easier to compare this method to bag-of-visual words and Qing Wen et al.'s.

3.5   Deep Convolutional Neural Network for Object Counting

The architecture being used here is inspired by VGG in [41]. In our work, we use 6 stacks of convolutional layers, each stack has 4 convolutional layers. The depth of all layers in each stack is 64, 128, 256, 512, 512, 512 respectively. All convolutional layers have a 3×3 kernel. Between two stacks, we also add 2×2 max-pooling to reduce the width and height of the feature map. By the end of the network, 3 fully-connected layers are used, with the number of output nodes being 4096, 4096 and 1. Thus, the network we use has 27 layers in total. The aim of the last output node is to produce the number of vehicles in the image.

3.6   Context-Aware Crowd Counting (CAN) [28]

In this method, the end goal is to produce a heatmap of objects like Lempitsky-Zisserman's. The neural network model they use comprises of three components. The first is one, called front-end network, consists of the first 10 layers of the VGG model. Its primary purpose is to extracting deep local features (f_v) across the image. The second component, which is also the main contribution of this works, consists of steps that allow the model to compute contextual features based on the output of the front-end network. Finally, the output of the second component will be combined with that of the first one and fed through a back-end decoder. The back-end decoder is essentially a network consisting of several dilated convolutional layers. Its output is the heatmap of the scene, representing the density of crowd. This heatmap can be integrated to produce the count of objects.

To extract contextual features with the second component, they first need to apply four different average pooling operators over f_v, producing four feature maps of size (width × height) 1×1, 2×2, 3×3 and 6×6. Each of them represents the information of the crowd in multiple different scales, where 1×1 is the global level. These are then fed through a scaling convolutional layer with kernel size 1×1 and upsampled using bilinear upsampling to match the width and height of f_v. By the end, there are four feature maps s_j belongs to four scales. Notice that the number of channels does not change in this entire process. The contrast features that encodes contextual information for each scale will be computed by taking the difference between s_j and f_v:

$$c_j = s_j - f_v \tag{7}$$

Each of these c_j contains the contextual features of its corresponding scale, which is essentially the saliency areas resulted from the subtraction. These will be fed again through another scaling layer, before being concatenated with the original f_v and used as input for the back-end decoder.

The job of the decoder is to compute the heatmap of the crowd. In this component, multiple dilated convolutional layers were used. They are believed to provide larger effective receptive fields with fewer parameters. In the final layer, they will output a heatmap and this will be used for learning purpose with the following loss function:

$$\frac{1}{2B} \sum_{i=1}^{B} \left|\left| D_i^{gt} - D_i^{est} \right|\right|_2^2 \tag{8}$$

4   Experiments and Discussion

This section provides the results of our experiments and explanations of them. The details of our datasets are shown in Subsect. 4.1. Then, the evaluation method and implementation details are described in the next Subsections. Finally, the results are presented in Subsect. 4.4.

4.1   The Dataset

In this work, we performed experiments on two different datasets labeled A and B (Fig. 4). Each of the datasets has 1000 images, sampled from two 4 min videos with a step of 5 frames per second. The resolution of both videos is 1280×720 pixels. The number of vehicles was manually counted for all 1000 frames of both datasets. No interpolation was used due to many motorbikes only appears in a few frames and they could be lost in the interpolating process. Although there are other kinds of vehicle in the video, we only focused on motorbikes in our experiments since the others don't appear frequently enough to provide us reliable train and test results.

Dataset A

Dataset B

Fig. 4. A few examples taken from dataset A and B. The plots below each row of photos show the motorbike count over 1000 frames for each respective dataset.

Dataset A is taken from a video recorded by [46] from a top-down viewpoint directly above the street. The number of motorbikes ranges from 11 to 56 with an average of 36. Dataset B is extracted from a video we captured from a moderately high viewpoint with mild perspective distortion. The motorbike count in this dataset is about 24 to 98 with an average of 63. With higher perspective distortion and motion blurs, the second dataset tends to be more difficult than the first one.

4.2   Evaluation Method

To measure the accuracy of each of the mentioned methods in the testing phase, we used the mean absolute error (MAE), mean relative error (MRE) and mean squared error (MSE):

$$MAE = \frac{1}{N_t} \sum_{i=1}^{N_t} |y_i - y(x_i)| \tag{9}$$

$$MRE = \frac{1}{N_t}\sum_{i=1}^{N_t}|\frac{y_i - y(x_i)}{y_i}| \tag{10}$$

$$MSE = \frac{1}{N_t}\sum_{i=1}^{N_t}(y_i - y(x_i))^2 \tag{11}$$

Additionally, the following standard deviations are also being calculated:

$$\sigma_a = \sqrt{\frac{1}{N_t}\sum_{i=1}^{N_t}(|y_i - y(x_i)| - MAE)^2} \tag{12}$$

$$\sigma_r = \sqrt{\frac{1}{N_t}\sum_{i=1}^{N_t}\left(\left|\frac{y_i - y(x_i)}{y_i}\right| - MRE\right)^2} \tag{13}$$

Here, N_t is the number of images used for testing, y_i is the number of motorbikes in ground truth, $y(x_i)$ is the number of motorbikes predicted using one of the experimental methods. The mean errors indicate how close the predictions are to ground truth values while the standard deviations represent the stability of the predictions. If a sample has $y_i = 0$, its relative error is set to 0 when $y(x_i) = 0$, otherwise it is set to 1.

In all experiments, we divided the dataset into F subsets and perform cross-testing [24]. For each iteration, one subset is used for testing and the rest of $F-1$ subsets are used for training. For methods which have specific training processes like [26] and [11], a smaller subset of fewer images is used for training instead of the entire set. At the end of the iteration, we calculate the four performance values mentioned above. This process is repeated until every subset has been the test set exactly once. Finally, we will compute the average scores over F iterations and use them to evaluate. Notice that this is different from cross-validation in that we did not perform testing on the same subset multiple time to find the best set of parameters for each method. Each test subset is treated as a black box in its fold.

We chose our F to be 5, which means we divided dataset A and B into 5 subsets of 200 images each. The division was done according to the video time instead of random subsampling, the reason is to correctly measure the generalization ability of each method. When there are many consecutive, roughly identical frames in the video which happened to be included in both of the test set and the train set due to random subsampling, the error rates are likely to be very low. Consequently, the results won't reflect the performance of the method in a fair manner.

4.3   Implementation Details

Hardware and Programming Language. For all experiments except CNN, the device we used to test the processing speed of each method was a personal computer with core i7, 4th generation, 8 GB RAM and a dedicated GTX-850M GPU. Those were done using MATLAB with the help of the following 3-rd party libraries: VLFeat [44], LS-SVM [36].

For CNN, the experiments were run on a dedicated computer with a P40 GPU 128 GB RAM. Programming was done in Python with the help of PyTorch [35].

Bag-of-Visual Words. Each image will be downscaled to 70% of the original size before the very first step of this method. After then, they will be processed by a 5 × 5 pixels median filter for noise reduction.

In the local feature extraction step, we used the standard configuration of SIFT and SURF so each keypoint descriptor is 128-dimensional. For HOG, the original version in [13] was used to compute 36-dimensional vectors. The cell size of each HOG descriptor was 16 × 16 pixels. For dense SIFT, the spatial bin size was 8 pixels, the distance between the keypoints was 12 pixels and every descriptors have the same orientation unlike sparse SIFT.

For K-means clustering, the choice of K in the experiments was 300 taken from the validation process in each train subset of each test fold. At the beginning of the clustering stage, the initial centroids were randomly generated. In the experiments using LS-SVM, we chose either linear and radial basis function (RBF) with $\sigma = 0.8$.

Lempitsky-Zisserman's Method. Since the approach here is to perform feature extraction on every pixel, it would be computationally more effective if we downscale the images to a small size. The resolution we chose in our experiments is 320 × 180 pixels. Even so, each image will still provide us numerous feature vectors hence training all 800 images per cross-validation fold is highly redundant. We uniformly picked 100 images from $F - 1$ subsets instead of using all 800 images for training here. The Gaussian kernel we used to create the ground truth map had a size of 24 × 24 pixels and a sigma of 4.

The parameters for the bag-of-visual words feature extraction here were very similar to the previous method except for the step of dense SIFT being 1 pixel and the number of clusters $K = 512$. With these settings, each pixel provided a feature vector of with a length of 512. For the random forest variation, we extracted the features from six channels: gradient in the x and y directions, subtraction and absolute difference between the gray image and its filtered result produced by a 3 × 3 median filter, difference between the images filtered by two 3 × 3 Gaussian kernel with a sigma of 0.5 and 1. The number of decision trees used for bagging was 6 and the minimum number of observations per tree leaf is 10000.

Qing Wen et al.'s Method. As mentioned before, we used the sine and cosine versions of Gabor filter separately for different experiments and then compare them. The other parameters are the same for both versions. The set of θ and ω values chosen in our experiments were $\Theta = \{0, 45, 90, 135\}$ and

$\Omega = \{2, 4, 8, 16, 32, 64\}$, which led to a total of 24 possible combinations. Since we computed both of the mean and variance of gray value for each channel, the length of our feature vectors will be 48. Similar to the experiments for Lemptisky-Zisserman's method, we resized the images to 320×180 pixels so that the feature extraction step would be faster. When performing training LS-SVM with the RBF-kernel, we also chose $\sigma = 0.8$.

Donatello Conte et al.'s Method. The image size we chose when testing this method was 640×360, i.e. half of the original size. To perform training for this specific method, we first extracted patches of motorbike clusters from the original images. For each fold of 200 images, about 145 to 190 patches were sampled. Every patch was symmetrically padded with 16 pixels on all four edges so that some SURF keypoints near the boundary won't be lost during detection stage.

Regarding the calculating of the distance to camera and the area of the bounding box, the authors originally used real world distance inferred from camera calibration information. But dataset A and B do not have such information so we used pixel distance instead. For our own variation that integrates bag-of-visual words feature into the original method, we replaced the number of SURF keypoints by the bag-of-visual words feature vector and kept the density and distance to camera. The number of clusters here was also 300. Additionally, we chose γ to be 1.5 instead of 2 as in the original paper since it resulted in better performance. For the regression, the setting of LS-SVM kernels here was the same as the one used in bag-of-visual words.

Deep Convolutional Neural Network. The training of CNN faced the following problem: there were not enough training examples for the network to generalize, which leads to overfitting. To tackle this, we added dropout layers with $p = 0.5$ in front of each stack except the first one. We also added a dropout layer before the first and the second fully-connected layer.

The learning algorithm that was used is Adam [23] with batch size 2. The whole training process has 120 epochs. To add more varieties to the train examples, each time 2 images were chosen, we randomly perform a set of image transformation to further avoid overfitting. These include vertical flip, horizontal flip and random rotation with an angel between $[-10°, 10°]$.

Context-Aware Crowd Counting. The network model configuration of this method is taken directly from the original paper [28], which is available at the corresponding code repository. In the training process, we set the batch size to 4 and train each fold in 35 epochs. Apart from that, we keep the training process the same as in the paper. To avoid overfitting, we split each train dataset into one smaller train set and one validation set, they respectively consist of 700 and 100 images. Among all epochs, the model with the best validation score will be used for testing.

Table 1. The results of our best variant and the baseline methods on two datasets

Method			[26]			[45]	[11]		CNN	[28]
Feature		BoVW on HOG	BoVW on dense SIFT	RF		Even Gabor channels	Original	Our variant		
Estimation method		LSL	Original	Original		LSL	LSL	LSL		
Dataset A	MAE	**2.26**	2.87	3.19		3.06	7.63	6.71	2.36	3.06
	σ_a	1.86	2.04	2.26		2.37	5.86	5.10	**1.81**	2.97
	MRE	**0.07**	0.08	0.09		0.09	0.26	0.23	**0.07**	0.08
	σ_r	0.07	**0.06**	0.07		0.09	0.27	0.24	**0.06**	0.08
	MSE	**8.58**	13.2	18.7		15.0	165	71.1	8.66	18.2
Dataset B	MAE	5.99	7.08	7.29		7.70	13.15	9.64	6.11	**3.74**
	σ_a	4.18	5.14	4.61		5.43	9.29	6.99	4.13	**2.86**
	MRE	0.10	0.11	0.12		0.13	0.23	0.16	0.10	**0.06**
	σ_r	0.08	0.07	0.08		0.11	0.20	0.14	0.08	**0.05**
	MSE	53.3	76.6	84.4		88.7	371	146	57.0	**22.2**
Time (second per frame)		**0.08**	0.86	0.21		0.09	1.66	1.68	0.96	0.6

4.4   Experimental Results

Table 1 shows the average results of all methods over both datasets. BoVW, CNN and CAN have the highest accuracy overall. The model with the strongest performance in Dataset B is CAN, with the smallest MSE. For Dataset A, BoVW and CNN share similar results with the latter has better standard deviations. The main reason is the random data augmentation used in training which leads to a more reliable estimation. CAN outperforms the rest in Dataset B with its contextual features. Although it tends to overestimate the number of motorbikes in images of higher count, the saliency features of multi-scales work well even with small objects.

Lempitsky-Zisserman's method is behind the above three. Even though their methods also used bag-of-visual words features, our variant has lower error values. This is due to their method resizes the images to a small scale before extracting features and some important visual information might be lost in the resizing process. However, the method extracts feature on every pixel, so it is still stable enough and has a low error standard deviation. The random forest feature of Lempitsky-Zisserman's method produces lower results than the bag-of-visual words feature. It is because we used only 6 basic channels to calculate the random forest feature hence it is not as robust as in the original paper.

Qing Wen et al.'s method has lower accuracy mainly because the chosen features, mean and variance of grey values, are too primitive. The results come from using odd channels are not shown here since they are worse than the results from even counterparts, in all metrics. These features are also very sensitive to moderate changes in illumination. Moreover, those values are computed over the entire image hence they might lack important local information. The loss of information after downscaling is also responsible for the lower results. Nevertheless, the bank of 24 Gabor filtered images still provide a strong basis for the features to work out. Combining with LS-SVM, this method produced average results in the end.

The results of Donatello Conte et al.'s method are the lowest among all of the presented. The major reason for this is that it is very hard to group keypoints in a cluttered environment into clean clusters. For most of the test images in dataset B, the clustering process could not perform correctly due to vehicles being too close. This led to bounding boxes with bloated size and wrong distance to camera. The original 3-dimensional feature vectors have lower accuracy than our bag-of-visual words variant since it is not able to differentiate between keypoints of cars and of motorbikes.

Although all results are not shown in Table 1 for brevity, a noticeable trend is that linear regression methods always provided better results than the RBF-kernels. It is also worth mentioning that the estimation method of Lempitsky-Zisserman is also linear, i.e. we can achieve the number of motorbikes by calculating the $w^T x$ dot product. Although we believe that with further tuning, an RBF-kernel will provide similar or even better performance. However, we deem it unnecessary to employ an RBF-kernel when there is already a strong linear relationship between the features and the count. Moreover, RBF-kernel is generally slower to compute than the linear counterpart.

The detailed estimation results of the best variants of six methods can be seen in Fig. 5. Like Table 1, BoVW, CNN and CAN stay closer to the ground truth than the others. For dataset A, which is not strongly affected by perspective, most methods produced good results. However, all methods have lower accuracy in dataset B when the number of motorbikes is too low or too high. This is mainly due to the lack of data for those situations. Another reason for the worsen performance in high-density cases is occlusion. As we can see in Fig. 5, a lot of motorbikes arc only visible for in small parts. Currently, except for CAN, the features being used in the observed methods are not robust enough to express small key elements. Lempitsky-Zisserman's method might be able to work better if we applied it to larger images; however, it would be much slower since the method need to compute features on every pixel.

The execution time for each method is shown in the last row of Table 1. Note that some of them ran on different hardware and some require high-end GPU to work while the other don't. BoVW and Qing Wen et al.'s methods are faster than the rest. Although taking most of the total time, HOG descriptor extraction is still fast enough to be less than 0.1 s. Lempitsky-Zisserman's method has solid accuracy and standard deviation, but the total time of their bag-of-visual words variant is almost one second since it must compute SIFT descriptor on all pixels. The random forest feature extraction is faster, but it has lower accuracy. Donatello Conte et al.'s method is the slowest due to the cost of computing the minimum spanning tree. The computation time of the CNN is the second longest and behind BoVW by more than an order of 10, even with the help of a powerful GPU. The speed of CAN is around the middle with a score of 0.6 s per frame.

Since the k-nearest neighbours process mostly takes under 0.01s, we should only focus on the HOG computation if we want to tweak our system for real-

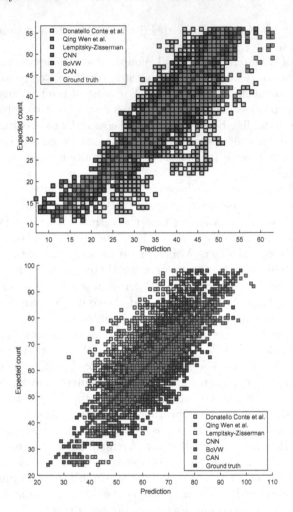

Fig. 5. The number of motorbikes estimated over 1000 frames of dataset A (top) and B (bottom) produced from the best settings of all four methods. They are aggregated from 5 different test folds, each with a test set of 200 frames. The closer the predictions are to the diagonal red line, the better they are.

time purpose. A proposal here is that we can replace the original HOG with a fast-HOG method proposed in [17] which is able to run four times faster. It was also proven to achieve better performance than the original.

5   Conclusion and Future Work

In this paper, we have done experiments on six methods for the motorbike counting problem. Some of them provide good results and are practical in real applications. Among them, the bag-of-visual words model is robust and fast enough to tackle the motorbike counting problem. It was able to work on two different datasets, one with top-down viewpoint and one with mild perspective distortion and extreme occlusions. The other method, CAN, is also powerful in handling difficult scenarios. The only downside is that it requires more hardware resources and longer training time.

The main drawback from all the approaches included in this paper is that they do not generalize well to other scenes with vastly different lighting or viewpoint. These can be solved by collecting more data from different scenes. In the future, we also want to address the problem of extreme occlusions and weather conditions where it is difficult even for humans to count objects correctly. For the first category, the estimated number started to deviate from the actual count when the motorbikes are mostly hidden by other vehicles leaving a few small parts visible. Moreover, some specific issues with motorbikes also need to deal with such as improving feature extraction techniques so that they are suitable for deciding if more than two people are on the same motorbike or if they are too close to each other. With extreme weather conditions, our main concern are videos captured under the rain. Usually, this specific kind of data are hard to find, and the lighting of each scene is also very varied. Finding out a way to make the models work with bad weather conditions is extremely valuable.

With the CNN and CAN, we suspect that with more training data, it should be able to achieve an even better performance. However, using them requires high hardware demands therefore it is not always practical, especially in small and low-cost traffic systems. As mentioned before, although the result was not as good as other methods, Donatello Conte et al.'s idea of grouping keypoints into clusters is still valuable. Instead of completely detecting every single motorbike on the image, which becomes very difficult as the environment is too chaotic likes dataset B, we can try detecting groups of motorbike. If conducted correctly, it should be informative enough to predict the flows of motorbike groups presented in the scene. Among all baseline methods, Lempitsky-Zisserman's approach has a high potential to be improved. Although the random forest feature only has average accuracy comparing to the others, there are multiple proposed improvements for it that we would like to experiment. With Qing Wen et al.'s method, although their features are too primitive (even though Gabor filter was very robust), it is possible to replace mean and variance of grey values with more novel features to obtain a better accuracy of the model. Overall, each of these methods have its own strong and weak points, and finding a way to take advantage of those strengths and combine them with our method to create a better model will be also of our great interest in the future.

Acknowledgements. Tran Khanh Dang is supported by a project with the Department of Science and Technology, Ho Chi Minh City, Vietnam (contract with HCMUT No. 42/2019/HD-QPTKHCN, dated 11/7/2019). We also thank all members of AC Lab and D-STAR Lab for their great supports and comments during the preparation of this paper.

References

1. Amine Haoui, R.K., Varaiya, P.: Wireless magnetic sensors for traffic surveillance. Transp. Res. Part C: Emerg. Technol. **16**(3), 294–306 (2008)
2. Bay, H., Tuytelaars, T., Van Gool, L.: SURF: speeded up robust features. In: Leonardis, A., Bischof, H., Pinz, A. (eds.) ECCV 2006. LNCS, vol. 3951, pp. 404–417. Springer, Heidelberg (2006). https://doi.org/10.1007/11744023_32
3. Bosch, A., Zisserman, A., Muñoz, X.: Scene classification via pLSA. In: Leonardis, A., Bischof, H., Pinz, A. (eds.) ECCV 2006. LNCS, vol. 3954, pp. 517–530. Springer, Heidelberg (2006). https://doi.org/10.1007/11744085_40
4. Breiman, L.: Random forests. Mach. Learn. **45**(1), 5–32 (2001)
5. Chan, A.B., Liang, Z.S., Vasconcelos, N.: Privacy preserving crowd monitoring: counting people without people models or tracking. In: IEEE Conference on Computer Vision and Pattern Recognition, CVPR 2008, pp. 1–7. IEEE (2008)
6. Chan, A.B., Vasconcelos, N.: Counting people with low-level features and Bayesian regression. IEEE Trans. Image Process. **21**(4), 2160–2177 (2012)
7. Chang, W.C., Cho, C.W.: Online boosting for vehicle detection. IEEE Trans. Syst. Man Cybern. B Cybern. **40**(3), 892–902 (2010)
8. Chen, B.H., Huang, S.C.: Probabilistic neural networks based moving vehicles extraction algorithm for intelligent traffic surveillance systems. Inf. Sci. **299**, 283–295 (2015)
9. Cherkassky, V., Ma, Y.: Selecting of the loss function for robust linear regression. Neural Comput. (2002)
10. Cho, S.Y., Chow, T.W., Leung, C.T.: A neural-based crowd estimation by hybrid global learning algorithm. IEEE Trans. Syst. Man Cybern. B Cybern. **29**(4), 535–541 (1999)
11. Conte, D., Foggia, P., Percannella, G., Tufano, F., Vento, M.: A method for counting moving people in video surveillance videos. EURASIP J. Adv. Signal Process. **2010**, 5 (2010)
12. Cortes, C., Vapnik, V.: Support-vector networks. Mach. Learn. **20**(3), 273–297 (1995)
13. Dalal, N., Triggs, B.: Histograms of oriented gradients for human detection. In: IEEE Computer Society Conference on Computer Vision and Pattern Recognition, CVPR 2005, vol. 1, pp. 886–893. IEEE (2005)
14. Dang, T.K., Pham, C.D., Nguyen, T.L.: A pragmatic elliptic curve cryptography-based extension for energy-efficient device-to-device communications in smart cities. Sustain. Cities Soc. **56**, 102097 (2020)
15. Dang, T.K., Pham, D.M.C., Ho, D.D.: On verifying the authenticity of e-commercial crawling data by a semi-crosschecking method. Int. J. Web Inf. Syst. (2019)
16. Dang, T.K., Tran, K.T.: The meeting of acquaintances: a cost-efficient authentication scheme for light-weight objects with transient trust level and plurality approach. Secur. Commun. Netw. **2019** (2019)
17. Dollár, P.: Piotr's Computer Vision Matlab Toolbox (PMT) (2016). http://vision.ucsd.edu/~pdollar/toolbox/doc/index.html
18. Felzenszwalb, P., McAllester, D., Ramanan, D.: A discriminatively trained, multiscale, deformable part model. In: IEEE Conference on Computer Vision and Pattern Recognition, CVPR 2008, pp. 1–8. IEEE (2008)
19. Foggia, P., Percannella, G., Sansone, C., Vento, M.: A graph-based algorithm for cluster detection. Int. J. Pattern Recogn. Artif. Intell. **22**(05), 843–860 (2008)

20. Foroughi, H., Ray, N., Zhang, H.: People counting with image retrieval using compressed sensing. In: 2014 IEEE International Conference on Acoustics, Speech and Signal Processing (ICASSP), pp. 4354–4358. IEEE (2014)
21. Huynh, C.K., Dang, T.K., Le, T.S.: Motorbike detection in urban environment. In: Dang, T.K., Küng, J., Wagner, R., Thoai, N., Takizawa, M. (eds.) FDSE 2018. LNCS, vol. 11251, pp. 286–295. Springer, Cham (2018). https://doi.org/10.1007/978-3-030-03192-3_22
22. Huynh, K.C., Thai, D.N., Le, S.T., Thoai, N., Hamamoto, K.: A robust method for estimating motorbike count based on visual information learning. In: Sixth International Conference on Graphic and Image Processing (ICGIP 2014), vol. 9443, p. 94431T. International Society for Optics and Photonics (2015)
23. Kingma, D.P., Ba, J.: Adam: a method for stochastic optimization. arXiv preprint arXiv:1412.6980 (2014)
24. Kohavi, R., et al.: A study of cross-validation and bootstrap for accuracy estimation and model selection. In: IJCAI, vol. 14, pp. 1137–1145 (1995)
25. Kong, D., Gray, D., Tao, H.: Counting pedestrians in crowds using viewpoint invariant training. In: BMVC. Citeseer (2005)
26. Lempitsky, V., Zisserman, A.: Learning to count objects in images. In: Advances in Neural Information Processing Systems, pp. 1324–1332 (2010)
27. Li, Y., Zhang, X., Chen, D.: CSRNet: dilated convolutional neural networks for understanding the highly congested scenes. In: Proceedings of the IEEE Conference on Computer Vision and Pattern Recognition, pp. 1091–1100 (2018)
28. Liu, W., Salzmann, M., Fua, P.: Context-aware crowd counting. In: Proceedings of the IEEE/CVF Conference on Computer Vision and Pattern Recognition, pp. 5099–5108 (2019)
29. Lowe, D.G.: Object recognition from local scale-invariant features. In: The Proceedings of the Seventh IEEE International Conference on Computer Vision, vol. 2, pp. 1150–1157. IEEE (1999)
30. Ma, W., et al.: A wireless accelerometer-based automatic vehicle classification prototype system. IEEE Trans. Intell. Transp. Syst. 15(1), 104–111 (2014)
31. MacQueen, J., et al.: Some methods for classification and analysis of multivariate observations. In: Proceedings of the Fifth Berkeley Symposium on Mathematical Statistics and Probability, Oakland, CA, USA, vol. 1, pp. 281–297 (1967)
32. Marana, A., Velastin, S., Costa, L., Lotufo, R.: Estimation of crowd density using image processing. In: IEE Colloquium on Image Processing for Security Applications (Digest No.: 1997/074), pp. 11–1. IET (1997)
33. Marana, A.N., da Fontoura Costa, L., Lotufo, R., Velastin, S.A.: Estimating crowd density with Minkowski fractal dimension. In: Proceedings of the 1999 IEEE International Conference on Acoustics, Speech, and Signal Processing, vol. 6, pp. 3521–3524. IEEE (1999)
34. Marsden, M., McGuinness, K., Little, S., O'Connor, N.E.: Fully convolutional crowd counting on highly congested scenes. arXiv preprint arXiv:1612.00220 (2016)
35. Paszke, A., et al.: Pytorch: an imperative style, high-performance deep learning library. In: Advances in Neural Information Processing Systems 32, pp. 8024–8035. Curran Associates, Inc. (2019). http://papers.neurips.cc/paper/9015-pytorch-an-imperative-style-high-performance-deep-learning-library.pdf
36. Pelckmans, K., et al.: LS-SVMlab: a MATLAB/C toolbox for least squares support vector machines. Tutorial. KULeuven-ESAT. Leuven, Belgium (2002)
37. Rabaud, V., Belongie, S.: Counting crowded moving objects. In: 2006 IEEE Computer Society Conference on Computer Vision and Pattern Recognition, vol. 1, pp. 705–711. IEEE (2006)

38. Rahmalan, H., Nixon, M.S., Carter, J.N.: On crowd density estimation for surveillance (2006)
39. Ryan, D., Denman, S., Fookes, C., Sridharan, S.: Crowd counting using multiple local features. In: Digital Image Computing: Techniques and Applications, DICTA 2009, pp. 81–88. IEEE (2009)
40. Sanaullah, I., Quddus, M., Enoch, M.: Developing travel time estimation methods using sparse GPS data. J. Intell. Transp. Syst. 20(6), 532–544 (2016)
41. Simonyan, K., Zisserman, A.: Very deep convolutional networks for large-scale image recognition. arXiv preprint arXiv:1409.1556 (2014)
42. Suykens, J.A., et al.: Least Squares Support Vector Machines, vol. 4. World Scientific (2002)
43. Tang, Y., Zhang, C., Gu, R., Li, P., Yang, B.: Vehicle detection and recognition for intelligent traffic surveillance system. Multimed. Tools Appl. 76(4), 5817–5832 (2015). https://doi.org/10.1007/s11042-015-2520-x
44. Vedaldi, A., Fulkerson, B.: VLFeat: an open and portable library of computer vision algorithms (2008). http://www.vlfeat.org/
45. Wen, Q., Jia, C., Yu, Y., Chen, G., Yu, Z., Zhou, C.: People number estimation in the crowded scenes using texture analysis based on gabor filter. J. Comput. Inf. Syst. 7(11), 3754–3763 (2011)
46. Whitworth, R.: Ho Chi Minh City (Saigon), Vietnam Rush Hour Traffic in Real Time (2013). http://www.robwhitworth.co.uk/. Accessed 6 Jan 2016
47. Wu, X., Liang, G., Lee, K.K., Xu, Y.: Crowd density estimation using texture analysis and learning. In: IEEE International Conference on Robotics and Biomimetics, ROBIO 2006. pp. 214–219. IEEE (2006)
48. Yaghoobi Ershadi, N., Menéndez, J.M.: Vehicle tracking and counting system in dusty weather with vibrating camera conditions. J. Sens. 2017 (2017)
49. Zheng, Y., Zhang, L., Xie, X., Ma, W.Y.: Mining interesting locations and travel sequences from GPS trajectories. In: Proceedings of the 18th International Conference on World Wide Web, pp. 791–800. ACM (2009)

Pesticide Label Detection Using Bounding Prediction-Based Deep Convolutional Networks

An C. Tran[1(✉)], Hung Thanh Nguyen[1], Van Long Nguyen Huu[1], and Nghia Duong-Trung[2]

[1] Can Tho University, Can Tho city, Vietnam
tcan@ctu.edu.vn , nhvlong@cit.ctu.edu.vn
[2] Technische Universität Berlin, Berlin, Germany
nghia.duong-trung@tu-berlin.de

Abstract. The paper addresses the un-explored scenario in intelligent agriculture and computer science, e.g., pesticide label detection. The problem opens to an exciting challenge in image recognition where the deployed system heavily depends on the performance of machine learning models despite unconstrained environments. To build up the system, the authors collect a real-world dataset to evaluate several state-of-the-art object detection algorithms. The authors select a dataset of 1221 photos containing 32 common pesticides on mango trees. Then we evaluate off-the-shelf deep convolutional networks to detect pesticide labels and take into account the detection accuracy. Finally, we integrate the best model into our self-developed mobile application that (i) correctly detects pesticide labels online and offline and (ii) provides essential pesticide information to facilitate further integrated treatment and services.

Keywords: Pesticide · Image detection · Convolutional neural networks · YOLO · SSD

1 Introduction and Motivation

The GDP contribution of agriculture in Vietnam has reached the proportion of 14.85% in 2020 [11]. Fruit trees are a group of plants with strengths in exporting to foreign countries and domestic consumption. Mango is one of the strongest and potential crops with high economic value. By 2020, the total mango production of Vietnam reached near 900 thousand tons, which has made Vietnam the 13th biggest mango producer in the world [1]. The international integration with VietGAP and GlobalGAP standards brings farmers many benefits. However, there are also observably complex challenges in producing clean and safe products. One of them is the rational and proper use of pesticides [13,17]. When we examine a wide range of drugs on the market today, it is difficult to identify suitable pesticides for mango and its diseases [12].

© Springer Nature Switzerland AG 2021
T. K. Dang et al. (Eds.): FDSE 2021, LNCS 13076, pp. 195–210, 2021.
https://doi.org/10.1007/978-3-030-91387-8_13

With the rapid development of digital technology, we can see the positive influence of artificial intelligence (AI) in many sections. Significantly, the development of computer vision (CV) with the application of deep learning methodologies in the field of image recognition and processing [7,16,20]. Therefore, the exciting topic of proposing an automatic system to identify pesticides on mango trees with deep learning techniques has been explored in this paper. This research has not been investigated in the literature. The authors aim at building a deep convolutional neural networks model to develop a system that can identify pesticides on mango trees. Generally, it shows the ability to build an effective practical application related to object recognition through images.

2   Technical Background

2.1   Deep Learning and Convolutional Neural Networks (CNNs)

In Machine learning perspective, convolutional neural networks (CNNs) [8] is a general term to refer to methods built on a multi-layer neural network architecture to solve complex problems by automatically extracting the abstraction level of input data through each layer. They have recently reached remarkable achievements in most problems related to image data processing, including image recognition and detection. The architecture of a CNN is usually a composition of different layers that can be grouped according to their function. The high performance of CNNs is achieved by (i) its ability to capture rich-level image representations and (ii) being trained on large amounts of data. Several CNN models can consist of tens of millions of estimated parameters to characterize the network. The deeply complex network has achieved outstanding performance in the majority from image classification [14], transfer learning [3–5] and object detection [2,18,19]. Compared to other image-related problems, object detection issue is complex. The fundamental process of dejection is as follows. First, some candidate object regions must be processed. Second, these suggested regions must be fine-tuned to achieve the correct localization. Fortunately, many researchers have worked hard to achieve modern solutions that combine speed, accuracy, and simplicity.

2.2   You Only Look Once (YOLO)

YOLO [9,15] is a type of CNN that predicts multiple bounding rectangles and the probability of classes for those rectangles. Object detection is viewed as a single regression problem, from image pixels to bounding rectangle coordinates and their probability of classes. Models are designed in an end-to-end fashion should be trained entirely using gradient descent. The YOLO core is presented in Fig. 1. In this study, the authors apply YOLO version 3, which is the most used model to address object detection. For ease of expression, the authors used the abbreviation of YOLO as version 3 of the architecture hereafter.

YOLO, by default configuration, splits the image into a 13 × 13 grid. Then, it predicts in each cell whether there is an object whose center point falls on

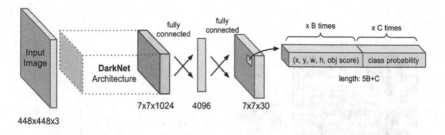

Fig. 1. The illustration of YOLO core.

that cell. If an object is detected, it produces the center point, the object's size, and the probability. Each of these cells is responsible for predicting five bounding boxes. Each rectangular descriptor box surrounding an object consists of 5 values b_x, b_y, b_w, b_h and the confidence score where b_x and b_y are the coordinates of the central point of the rectangle, b_h and b_w are its height and weight respectively. The confidence score indicates how likely the object is to appear. If the confidence score value is lower than a given threshold, then the remaining values are no longer matter. In addition to the above values, each box also includes C values where C is the number of classes representing the probability of each class. The most considerable C_i value means that the identified object has an index of i. An example of these values is presented in Fig. 2.

To find the bounding box for the object, YOLO requires the anchor boxes, see Fig. 3, as the basis of the estimation. These anchor boxes are predefined and surround the object in a relatively precise manner. Then, the regression bounding box algorithm will refine the anchor box to create a predicted bounding box for the object. Each object in the training image is distributed to an anchor box. In the case of two or more anchor boxes surrounding the object, YOLO determines the anchor box with the highest Intersection over Union (IoU) with the ground truth bounding box. Each object in the training image is distributed to a cell on the feature map that contains the object's midpoint.

The loss function of YOLO is described belows:

$$
L = \lambda_{coord} \sum_{i=0}^{S^2} \sum_{j=0}^{B} \mathcal{O}_{ij}^{obj} [(x_i - \hat{x}_i)^2 + (y_i - \hat{y}_i)^2 + (\sqrt{w_i} - \sqrt{\hat{w}_i})^2 + (\sqrt{h_i} - \sqrt{\hat{h}_i})^2]
$$

$$
+ \sum_{i=0}^{S^2} \sum_{j=0}^{B} (\mathcal{O}_{ij}^{obj} + \lambda_{noobj}(1 - \mathcal{O}_{ij}^{obj}))(C_{ij} - \hat{C}_{ij})^2 + + \sum_{i=0}^{S^2} \sum_{c \in C} \mathcal{O}_i^{obj} (p_i(c) - \hat{p}_i(c))^2
$$

$$(1)$$

where $\lambda_{coord} = 5$ and $\lambda_{noobj} = 0.5$ are hyperparameters that control the loss from bounding box coordinate predictions and the loss from confidence predictions. $\mathcal{O}_i^{obj}$ denotes if an object exists in cell i. $\mathcal{O}_{ij}^{obj}$ denotes if the j bounding box predictor in cell i. $p_i(c)$ is the probability that cell i contains object c while $\hat{p}_i(c)$ is the estimation.

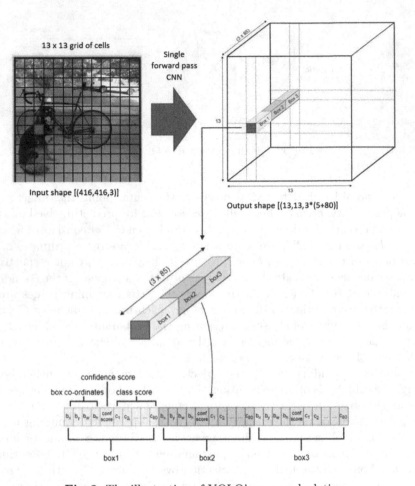

Fig. 2. The illustration of YOLO's score calculation.

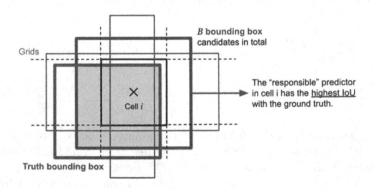

Fig. 3. Anchor box in YOLO.

2.3   Single Shot Detection (SSD)

The SSD model [10,21] contains a convolutional-based network, e.g., VGG-16 or any standard architecture used for high-quality image classification auxiliary structure to produce object detection. SSD's input is an image and ground-truth rectangles of each object during training. The model evaluates a small set of default rectangles of different aspect ratios of each location in several feature maps with different shapes, e.g., 8×8 and 4×4. The main idea of SSD comes from using bounding boxes by pre-initializing boxes at each location on the image. SSD computes and evaluates information at each location to see if that location has objects or not. If any, and based on the proximity results, the model calculates a best-fit box that covers the object. The SSD model has two main phases: feature map extraction and object detection. Regarding extracting feature maps, the SSD approach uses VGG16 or any off-the-shelf CNNs models, see Fig. 4. Then it detects the objects using the Conv4_3 class. For each cell it makes 4 object predictions, see Fig. 5. Each prediction consists of a bounding box and a $C + 1$ score, where C is the number of classes and selects the class with the highest score as the feature class. Conv4_3 generates a total of $38 \times 38 \times 4$ predictions. For non-object predictions, the SSD reserves a class of 0 to indicate that it has no objects. SSD uses multi-scale feature maps to detect objects independently. As CNN decreases in spatial size, the resolution of feature maps also decreases. Therefore, SSD uses lower resolution layers to detect larger objects. For example, a 4×4 feature map is used for larger objects.

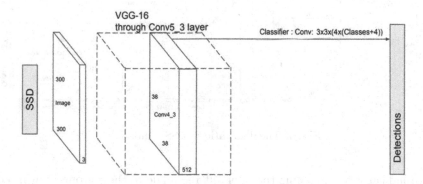

Fig. 4. The illustration of SSD core.

Regarding object detection, SSD calculates both position and layer points using small convolutional filters. After extracting the feature maps, the SSD applies a 3×3 convolution filter to each cell to make predictions. Each filter yields $C + 5$ channels where $C + 1$ points for each class plus a boundary. For each default rectangle, the model predict both the offsets and the confidences for all object classes $(c_1, c_2, \ldots, c_p)$. The default rectangles are matched to the ground-truth squares in training. The loss is calculated by weighting the summation

Fig. 5. The illustration of SSD core prediction.

between the confidence loss and the localization loss. The key features of the added auxiliary structure at the end of the network are (i) multi-scale feature maps for detection, (ii) convolutional predictors for detection, and (iii) default boxes and aspect ratios. The model of SSD is presented in Fig. 6.

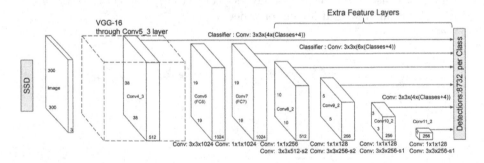

Fig. 6. The illustration of SSD model.

An indicator for matching the i default rectangle to the j ground-truth rectangle of class p is denoted as $x_{i,j}^p = \{1, 0\}$. N is the number of matched default rectangles. If $N = 0$, the loss is set to 0. The objective loss function is a weighted sum of the confidence loss L_{conf} and the localization loss L_{loc}:

$$L(x, c, l, g) = \frac{1}{N}(L_{conf}(x, c) + \alpha L_{loc}(x, l, g)) \tag{2}$$

The confidence loss is the softmax loss over multiple classes' confidence (c).

$$L_{conf}(x, c) = -\sum_{i \in Pos}^N x_{i,j}^p \log\left(\frac{\exp(c_i^p)}{\sum_p \exp(c_i^p)}\right) - \sum_{i \in Neg} \log(\hat{c}_i^0) \tag{3}$$

The localization loss is a Smooth L1 loss [6] between the ground-truth rectangle g and the predicted rectangle l parameters. We denote cx, cy as the offsets center of the default bounding rectangle d and its height h and width w.

$$L_{loc}(x, l, g) = \sum_{i \in Pos}^{N} \sum_{m \in \{cx, cy, w, h\}} x_{i,j}^{k} \text{smooth}_{L1}(l_i^m - \hat{g}_i^m) \tag{4}$$

where for each $m \in \{cx, cy, w, h\}$, we calculate

$$\hat{g}_j^{cx} = \frac{g_j^{cx} - d_i^{cx}}{d_i^w} \tag{5}$$

$$\hat{g}_j^{cy} = \frac{g_j^{cy} - d_i^{cy}}{d_i^h} \tag{6}$$

$$\hat{g}_j^w = \log\left(\frac{g_j^w}{d_i^w}\right) \tag{7}$$

$$\hat{g}_j^h = \log\left(\frac{g_j^h}{d_i^h}\right) \tag{8}$$

3   Experiments

3.1   Datasets

According to the Plant Protection Department of Vietnam, the authors collected a dataset of 1221 photos containing 32 common pesticides on mango trees. All photos are taken manually with high resolution, clear images of pesticide labels without blur. Then we resized all images to the exact size of 416 × 416 pixels. The reduction in size helps to load images faster without sacrificing the model's prediction accuracy. We divided the data set into two parts: train set 80% and test set 20%. Image labeling is done by labelImg software[1]. We summarized the experimented datasets in Table 1.

Table 1. Dataset of collected pesticide labels.

#	Label	# Train	# Test
1	AminoQuelant05	19	5
2	AmistarTop_325SC	12	3
3	Antracol_70WP	24	6

(*continued*)

---

[1] https://github.com/tzutalin/labelImg.

Table 1. (*continued*)

#	Label	# Train	# Test
4	Anvil_5SC	27	7
5	Chess_50WG	32	8
6	Cyrux_25EC	20	5
7	Daconil_75WP	28	7
8	Dipomate_430SC	18	4
9	Eagle_50WG	26	7
10	FlintPro_648WG	22	5
11	Flower95_0.3SL	44	12
12	Kamsu_2SL	28	7
13	MapRota_50WP	33	8
14	RholamSuper_50SG	27	7
15	RidomilGold_68WG	26	7
16	Tasieu_1.9EC	32	8
17	TiltSuper_300EC	14	3
18	TungMectin_5.0EC	32	8
19	Ortiva_600SC	34	9
20	RevusOpti_440SC	26	7
21	ActiNoVate_1SP	40	10
22	Avalon_8WP	35	9
23	CabrioTop_600WG	38	9
24	Daconil_500SC	38	10
25	DithaneM45_80WP	41	10
26	Kozuma_3SL	37	9
27	MapWinner_5WG	34	9
28	Newtracol_70WP	44	11
29	Oshin_20WP	38	9
30	Radiant_60SC	35	9
31	Reasgant_1.8EC	41	10
32	TPThanDien_78SL	34	8

3.2 Models' Architecture Configuration

YOLO. We implement the YOLO architecture with five types of layers:

- 75 convolutional layers.
- 23 skip connection layers connecting between 2 convolution layers, adding feature maps to the later layer, and making the model convergence better.
- 2 route layers navigating the input feature vectors for the next layer.

– 2 upsampling layers increasing the size of the feature vectors for YOLO layers.
– 3 YOLO layers performing the identification through 3 grid ratios 19×19, 38×38, and 76×76.

The model was trained on a Google Colab server with two cores, 12.7 GB of RAM, and a Tesla K80 GPU for 60 h with a batch size of 32 images. The model converges after more than 66000 iterations.

SSD. We implement the SSD with 54 convolution layers:

– 1 3×3 convolutional layer.
– 16 expansion layers to increase the number of input channels.
– 17 depthwise layers to extract features.
– 17 projection layers to reduce the number of input channels.
– 1 1×1 convolutional layer.
– 1 average pooling layer.
– 1 classification layer.

The model was trained on a Google Colab server with two cores, 12.7 GB of RAM, and a Tesla K80 GPU within 50 h with a batch size of 24 images. The model converges after more than 88000 iterations.

3.3 Evaluation Metrics

The criteria for evaluating the models are as follows. True Positive (TP): The model correctly predicts the object's class. The object bounding box predicted by the model and the grounding-truth box has a considerable IoU value of more than or equal to 0.5. False Negative (FN): The object is present in the image, but the model cannot correctly recognize it. False Positive (FP): The model incorrectly predicts the object's class. Or the model still rightly forecasts the object's class, but the object bounding box predicted by the model and the grounding-truth box has a smaller IoU less than 0.5. Then, we can define Precision, Recall and F1-score to judge models' performance per class.

$$\text{Precision} = \frac{\text{TP}}{\text{TP} + \text{FP}} \tag{9}$$

$$\text{Recall} = \frac{\text{TP}}{\text{TP} + \text{FN}} \tag{10}$$

$$\text{F1-score} = 2\frac{\text{Precision} \times \text{Recall}}{\text{Precision} + \text{Recall}} \tag{11}$$

We also evaluate the overall performance of models through Micro-average scores of Precision, Recall and F1-score as follows.

$$\text{Micro-average precision} = \frac{\sum^{C} TP_C}{\sum^{C}(TP_C + FP_C)} \tag{12}$$

$$\text{Micro-average recall} = \frac{\sum^C TP_C}{\sum^C (TP_C + FN_C)} \tag{13}$$

$$\text{Micro-average F1 score} = 2\frac{\text{Micro-average precision} \times \text{Micro-average Recall}}{\text{Micro-average Precision} + \text{Micro-average Recall}} \tag{14}$$

3.4 Experimental Results

Two CNN-based models used for pesticide label detection on mango trees (YOLO and SSD) have achieved several experimental goals, ensuring prediction accuracy. All experiment models have reached a noticeably high performance, of which SSD produces the highest detection accuracy of 98.76% (micro average). In terms of recognition speed, the YOLO gives the fastest processing speed. Table 2 presents the performance of YOLO while Table 3 shows that of SSD. A comparison between two models is briefly summarized in Table 4.

Table 2. Performance evaluation of YOLO.

#	Label	Precision	Recall	F1-score	Support
1	AminoQuelant05	1	1	1	5
2	AmistarTop_325SC	1	1	1	3
3	Antracol_70WP	1	0.666	0.800	6
4	Anvil_5SC	0.857	1	0.923	7
5	Chess_50WG	1	0.750	0.857	8
6	Cyrux_25EC	1	1	1	5
7	Daconil_75WP	0.857	1	0.923	7
8	Dipomate_430SC	1	1	1	4
9	Eagle_50WG	1	1	1	6
10	FlintPro_648WG	1	1	1	5
11	Flower95_0.3SL	1	1	1	11
12	Kamsu_2SL	1	1	1	7
13	MapRota_50WP	1	1	1	8
14	RholamSuper_50SG	1	1	1	7
15	RidomilGold_68WG	1	1	1	7
16	Tasieu_1.9EC	1	1	1	8
17	TiltSuper_300EC	1	1	1	3
18	TungMectin_5.0EC	1	1	1	8
19	Ortiva_600SC	1	1	1	9
20	RevusOpti_440SC	1	1	1	7
21	ActiNoVate_1SP	1	1	1	10

(*continued*)

Table 2. (*continued*)

#	Label	Precision	Recall	F1-score	Support
22	Avalon_8WP	1	1	1	9
23	CabrioTop_600WG	1	1	1	9
24	Daconil_500SC	1	1	1	10
25	DithaneM45_80WP	1	1	1	10
26	Kozuma_3SL	1	1	1	9
27	MapWinner_5WG	1	1	1	9
28	Newtracol_70WP	1	1	1	11
29	Oshin_20WP	1	0.888	0.941	9
30	Radiant_60SC	1	1	1	9
31	Reasgant_1.8EC	1	1	1	10
32	TPThanDien_78SL	1	1	1	8
Micro average		0.991	0.979	0.985	248
Macro average		0.991	0.978	0.982	248

Table 3. Performance evaluation of SSD.

#	Label	Precision	Recall	F1-score	Support
1	AminoQuelant05	1	1	1	5
2	AmistarTop_325SC	1	1	1	3
3	Antracol_70WP	1	1	1	6
4	Anvil_5SC	1	1	1	7
5	Chess_50WG	1	1	1	8
6	Cyrux_25EC	1	1	1	5
7	Daconil_75WP	0.857	1	0.923	7
8	Dipomate_430SC	1	1	1	4
9	Eagle_50WG	1	1	1	6
10	FlintPro_648WG	1	1	1	5
11	Flower95_0.3SL	1	0.818	0.900	11
12	Kamsu_2SL	1	1	1	7
13	MapRota_50WP	1	1	1	8
14	RholamSuper_50SG	1	1	1	7
15	RidomilGold_68WG	0.714	1	0.833	7
16	Tasieu_1.9EC	1	1	1	8
17	TiltSuper_300EC	1	1	1	3
18	TungMectin_5.0EC	1	1	1	8
19	Ortiva_600SC	1	1	1	9

(*continued*)

Table 3. (*continued*)

#	Label	Precision	Recall	F1-score	Support
20	RevusOpti_440SC	1	1	1	7
21	ActiNoVate_1SP	1	1	1	10
22	Avalon_8WP	1	1	1	9
23	CabrioTop_600WG	1	1	1	9
24	Daconil_500SC	1	1	1	10
25	DithaneM45_80WP	1	1	1	10
26	Kozuma_3SL	1	1	1	9
27	MapWinner_5WG	1	1	1	9
28	Newtracol_70WP	1	1	1	11
29	Oshin_20WP	1	1	1	9
30	Radiant_60SC	1	1	1	9
31	Reasgant_1.8EC	1	1	1	10
32	TPThanDien_78SL	0.875	1	0.933	8
Micro average		0.983	0.991	0.987	248
Macro average		0.982	0.994	0.987	248

Table 4. Comparison of YOLO and SSD.

Criteria	YOLO	SSD
Model size (MB)	235	20.7
# correct prediction	241	242
# incorrect prediction	2	4
# in-recognized image	5	2
# total image	248	248
Accuracy (%) (micro average)	98.54	98.76
Processing time (s) (at resolution 416×416)	[2,7]	[3,5]

4   Mobile-App Deployment

Our software is a mobile application that (i) recognizes the pesticides through captured images and real-time camera posture, and (ii) inquiry pesticides' infor-

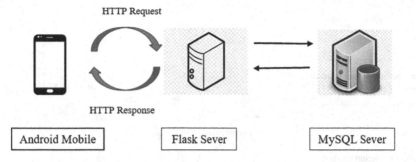

Fig. 7. Client - Server architecture used in our solution.

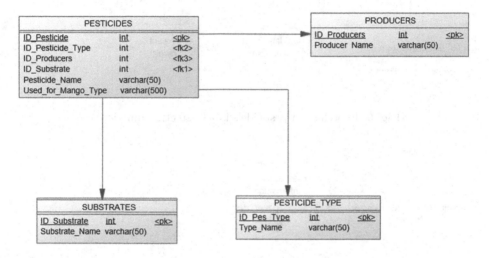

Fig. 8. Database design.

mation. The authors visualize the mobile-app's flowcharts of two functions and selected several screenshots in Figs. 9, 10, and 11. The mobile app is written for Android-based devices that communicate with a server through the Flask framework's support. Client-Server is a well-known architecture in computer networks and is widely applied. The idea of this architecture, see Fig. 7, is that a client sends a request to a server to process and return the results to the client. The software uses the MySQL database management system to store data, connecting via MySQL Server. Since this is just a relatively simple application, it only

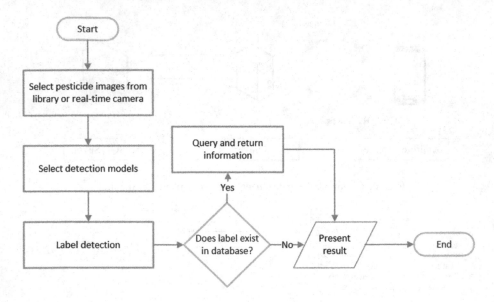

Fig. 9. Flowchart of pesticide's label detection function.

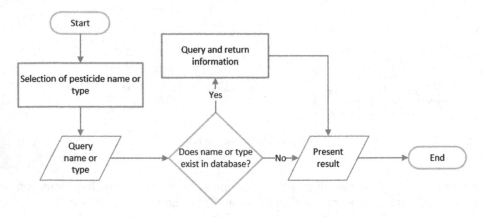

Fig. 10. Flowchart of querying pesticide's information function.

needs a small database with four tables to store data and required information, see Fig. 8. A regular camera on mobile devices takes images as the input of the system.

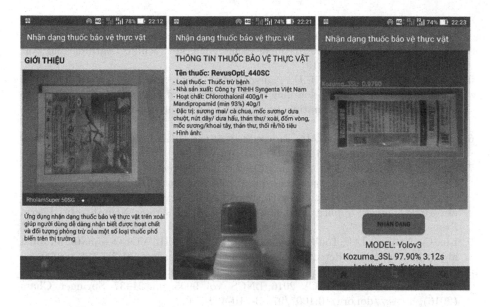

Fig. 11. Selected several screenshots of the mobile application.

5   Conclusion

In this research, we have proposed integrating CNNs for the task of pesticide label detection and mobile-app development. An un-explored scenario has been successfully addressed via utilizing state-of-the-art deep networks upon our collected data. First, we collect a pesticide image dataset. Then, the extensive experiments show that the detection accuracy of 98% has been achieved within the acceptable online training facilities. The best approach is selected based on detection performance. Third, we develop a mobile application integrating the best model. Consequently, we can conclude that the exciting scenario of intelligent agriculture and computer science, e.g., proposing an automatic system to identify pesticides on mango trees with deep learning techniques, has been explored.

References

1. Vietnam is the 13th biggest mango producer in the world. https://www.mard.gov.vn/en/Pages/vietnam-is-the-13th-biggest-mango-producer-in-the-world.aspx
2. Agarwal, S., Terrail, J.O.D., Jurie, F.: Recent advances in object detection in the age of deep convolutional neural networks. arXiv preprint arXiv:1809.03193 (2018)
3. Duong-Trung, N., Quach, L.D., Nguyen, C.N.: Learning deep transferability for several agricultural classification problems. Int. J. Adv. Comput. Sci. Appl. **10**(1) (2019)

4. Duong-Trung, N., Quach, L.D., Nguyen, M.H., Nguyen, C.N.: Classification of grain discoloration via transfer learning and convolutional neural networks. In: Proceedings of the 3rd International Conference on Machine Learning and Soft Computing, pp. 27–32 (2019)
5. Duong-Trung, N., Quach, L.D., Nguyen, M.H., Nguyen, C.N.: A combination of transfer learning and deep learning for medicinal plant classification. In: Proceedings of the 2019 4th International Conference on Intelligent Information Technology, pp. 83–90 (2019)
6. Girshick, R.: Fast R-CNN. In: Proceedings of the IEEE International Conference on Computer Vision, pp. 1440–1448 (2015)
7. Kumar, S., Mankame, D.P.: Optimization driven deep convolution neural network for brain tumor classification. Biocybern. Biomed. Eng. **40**(3), 1190–1204 (2020)
8. Li, Z., Liu, F., Yang, W., Peng, S., Zhou, J.: A survey of convolutional neural networks: analysis, applications, and prospects. IEEE Trans. Neural Netw. Learn. Syst. (2021)
9. Liu, G., Nouaze, J.C., Touko Mbouembe, P.L., Kim, J.H.: YOLO-tomato: a robust algorithm for tomato detection based on YOLOv3. Sensors **20**(7), 2145 (2020)
10. Liu, W., et al.: SSD: single shot multibox detector. In: Leibe, B., Matas, J., Sebe, N., Welling, M. (eds.) ECCV 2016. LNCS, vol. 9905, pp. 21–37. Springer, Cham (2016). https://doi.org/10.1007/978-3-319-46448-0_2
11. Nguyen, M.N.: Topic: agriculture in Vietnam. https://www.statista.com/topics/5653/agriculture-in-vietnam/
12. Ploetz, R.: The major diseases of mango: strategies and potential for sustainable management. In: VII International Mango Symposium, vol. 645, pp. 137–150 (2002)
13. Rajmohan, K., Chandrasekaran, R., Varjani, S.: A review on occurrence of pesticides in environment and current technologies for their remediation and management. Indian J. Microbiol. **60**(2), 125–138 (2020). https://doi.org/10.1007/s12088-019-00841-x
14. Rawat, W., Wang, Z.: Deep convolutional neural networks for image classification: a comprehensive review. Neural Comput. **29**(9), 2352–2449 (2017)
15. Redmon, J., Divvala, S., Girshick, R., Farhadi, A.: You only look once: unified, real-time object detection. In: Proceedings of the IEEE Conference on Computer Vision and Pattern Recognition, pp. 779–788 (2016)
16. Smys, S., Chen, J.I.Z., Shakya, S.: Survey on neural network architectures with deep learning. J. Soft Comput. Paradigm (JSCP) **2**(03), 186–194 (2020)
17. Srisookkum, T., Sapbamrer, R.: Health symptoms and health literacy of pesticides used among Thai cornfield farmers. Iran. J. Public Health **49**(11), 2095 (2020)
18. Tran, A.C., Thoa, P.K., Tran, N.C., Duong-Trung, N., et al.: Real-time recognition of medicinal plant leaves using bounding-box based models. In: 2020 International Conference on Advanced Computing and Applications (ACOMP), pp. 34–41. IEEE (2020)
19. Tran, A.C., Tran, N.C., Duong-Trung, N.: Recognition and quantity estimation of pastry images using pre-training deep convolutional networks. In: Dang, T.K., Küng, J., Takizawa, M., Chung, T.M. (eds.) FDSE 2020. CCIS, vol. 1306, pp. 200–214. Springer, Singapore (2020). https://doi.org/10.1007/978-981-33-4370-2_15
20. Traore, B.B., Kamsu-Foguem, B., Tangara, F.: Deep convolution neural network for image recognition. Eco. Inform. **48**, 257–268 (2018)
21. Zhai, S., Shang, D., Wang, S., Dong, S.: DF-SSD: an improved SSD object detection algorithm based on DenseNet and feature fusion. IEEE Access **8**, 24344–24357 (2020)

Intelligent Urban Transportation System to Control Road Traffic with Air Pollution Orientation

Binh Thanh Nguyen[1]([✉]), Pham Lu Quang Minh[2], Huynh Vu Minh Nguyet[2],
Do Huu Phuoc[2], Pham Dinh Tai[2], and Huy Truong Dinh[2]

[1] International Institute for Applied Systems Analysis (IIASA), Laxenburg, Austria
nguyenb@iiasa.ac.at
[2] Duy Tan University, Da Nang, Vietnam

Abstract. Enhancing transportation services and reducing vehicle emissions at intersections are main challenges for megacities. In this paper, an Intelligent Urban Transportation System is proposed as smart green traffic lights. In this context, vehicles are detected, and categorised. Furthermore, emission factors of each vehicle category are specified and used as an additional factor to optimize the traffic light cycle. IUTAR dashboard has also been developed and presented in this paper.

Keywords: Urban transportation system · Smart traffic light · GAINS · Vehicle · Emission factor · Object detection

1 Introduction

Nowadays, cities around the world are expanding dramatically, with urban increment reaching nearly 2.5 billion people in urban areas and road traffic growth exceeding 1.2 billion cars by 2050 [27]. This issue puts a burden on the overall conveyance infrastructure and creates numerous issues like overcrowding, delayed services, and commuter dissatisfaction [13]. Public administration has adopted information and communication technology to construct new intelligent transportation systems and elegance new risk prevention strategies in transportation management [3, 12, 13]. In this context, Intelligent Transportation Systems (ITS) [14, 16, 22] is considered as a vital enabler for the smart cities paradigm. Currently, such systems generate massive amounts of granular data which can analyze to higher understand people's dynamics [1].

On the other hand, road vehicle emissions contribute significantly to a large range of pollution problems, particularly in urban areas [11, 17]. The author also indicated that vehicle emissions vary by manufacturer, vehicle model, emission standard, engine size and fuel type and plenty of other factors [5, 9, 11]. Furthermore, [20] shows a case study in India, which road vehicles are considered as uncontrolled ones and therefore the proposed technology layer specified vehicle category, fuel type, operation (e.g. taxi and personal use for passenger cars), engine type (two and 4 stroke) and emission control technologies.

© Springer Nature Switzerland AG 2021
T. K. Dang et al. (Eds.): FDSE 2021, LNCS 13076, pp. 211–221, 2021.
https://doi.org/10.1007/978-3-030-91387-8_14

In this context, our study goal is to enhance the standard of the transportation services and also to confirm public transportation safety [12], and to reduce vehicle emissions. The research motivation is the unreasonable waiting time for the red lights, leading to the difference in traffic density in each lane. Especially that is the increase in traffic accidents caused by bypassing red lights, vehicles jostling the road, encroaching lanes. For example, the report of the Vietnam Insider newspaper [8] shows that most of the accidents were caused by violations on lanes (23.31%) and speed (8.15%). Besides, the increase in vehicle emissions causes air pollution worse day by day [11, 20].

In this paper, an Intelligent Urban Transportation System (IUTAR) is studied to control road traffic with air pollution orientation, which can be considered as a main contribution of our study in comparison with other ITS approaches and systems. First, YOLOv4 and DeepSORT algorithm [4, 10, 15] have been used to detect and classify vehicles per lane based on IIASA GAINS vehicle category [5, 9] in an intersection. Afterwards, emisssions per vehicle category per lane in the intesection are calculated based on IIASA GAINS vehicle emission factors. Hereafter, an Optimal Traffic Light Cycle formula is specifed as an extension of the Webster method [26] and used to control the traffic lights. To proof of our concepts, an IUTAR dashboard has been developed and illustrated as a visualization tool to display the traffic data and to handle errors in object detection.

The rest of this paper is presented as follows: Sect. 2 introduces typical approaches and projects related to our work; after introducing the IUTAR concepts, i.e. system context overview and its specifications, and the modeling optimal traffic light cycle in Sect. 3, Sect. 4 will present our IUTAR application results. And lastly, Sect. 5 gives a summary of our achieved as well as future works.

2   Related Work

According to [4], real-time and accurate detections of vehicles in images and video data have been considered as very important and challenging work. Especially in traffic situations with complex scenes, different vehicle models, and high density, it is too difficult to accurately locate and classify vehicles during traffic flows. Therefore, YOLOv3-D [10] has been proposed as a single-stage deep neural network, which is used to predict on the Tensorflow framework to improve this problem.

During the previous few decades, significant research efforts are dedicated to using CCTV 3 cameras [24, 28] to see real-time traffic parameters like volume, density, and speed. Furthermore, the authors in [24] also presented these methods which are often broad-ly divided into three categories: (a) detection-based methods, (b) motion-based methods, and (c) holistic approaches [24].

In the context of using deep learning methods for traffic counting tasks, faster recurrent convolution neural networks (RCNNs) are used in [19] to calculate traffic density. Moreover, [19] used two variations of CNNs, namely counting CNN and hydra CNN, to conduct vehicle counting and predict traffic density. Recently, [28] used both deep learning and optimization-based methods to perform vehicle counts from low framerate, high occlusion videos [24]. They mapped the image to a vehicle density map using rank-constrained regression and full convolution networks.

The concepts and implementation approaches of the Intelligent Transportation Systems (ITS) have been introduced in [3, 4, 24]. Such systems have been connected with transport vehicles to communicate with the surrounding infrastructure. In this context, the SURTRAC Intelligent Transportation Systems of Rapid Flow Technologies [23] is an artificial intelligence-activated traffic light system that allows the traffic light to adapt to traffic conditions instead of relying on a pre-set sequence. Table 1 is used to compare our IUTAR with SURTRAC. This comparison can be considered as a main contribution of our research and development work.

Table 1. Features comparison between SURTRAC and IUTAR

		SURTRAC	IUTAR
Functions	**Calculate vehicle emissions**	No	Yes
	Control effective the flow of road traffic	Yes	Yes
	Ensuring time-saving for a traffic light	No	Yes
	Reduce accident	Yes	Yes
	Reduce traffic congestion	Yes	Yes
	Manage the information of traffic violation	No	Yes
	Report the data of traffic flow	No	Yes
Installment cost		High	Low
Performance		Low because of a big system	Faster

3   IUTAR Concepts and Solutions

In this section, IUTAR system context overview is presented as an application framework of our research and development. Its component descriptions show how the data are specified from camera systems to the simulation traffic lights and linked to the GAINS system [5] and visualized in IUTAR dashboard.

3.1   System Context Overview

The camera system uses Real-Time Broadcasting Protocol (RTSP) [15, 16, 24] to import real-time images of the streaming media server from intersections into the IUTAR system. Those real-time images can be used to detect vehicles by using YOLO3 [10]. Furthermore, the just-detected vehicle objects can be classified based on GAINS vehicle categories [12]. As a result, with the input of video from the camera systems, the IUTAR system will detect, track vehicles based on the GAINS vehicle categories. Afterwards, number of vehicles and their emission factors per category per lane in a specific intersection can be calculated. Based on this result, the IUTAR system can analyse and provide a green traffic light cycle. In the IUTAR Dashboard, a traffic map has been developed and used to display road junctions, each of which will simulate a smart traffic light (Fig. 1).

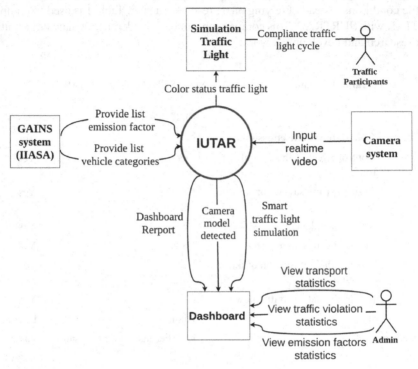

Fig. 1. System context overview

3.2 GAINS Vehicles Categories and Emission Factors

The GAINS model operates with emission factors averaged over the whole lifetime of the vehicle, effects of fuel, age and driving patterns can be seen as be averaged ones [5]. For example, in case of emission factors for light duty vehicles, EURO 5 emission factors have been based on preliminary assessments derived from recent measurements [9]. In particular, measured NOx emissions of EURO 5 diesel cars are higher than those of EURO 4 and EURO 3 cars and exceed the limit value defined for type approval several times [9]. Table 2 shows a subset of emission factors for gasoline vehicles provided by GAINS model.

3.3 Modeling Optimal Traffic Light Cycle

In the IUTAR system, after detecting, tracking, classifying, and counting vehicles in each lane, we will calculate optimal traffic light cycles. We propose a formula for calculating

Table 2. GAINS emission factors for gasoline vehicles.

Categories	Name	Fuel Activity	Kt PM PM_2_5/Act.unit	Technology	Emission factor [kt/act.unit]
TRA_RD_HDB	Bus	GSL	[PJ]	NOC	0.02396
TRA_RD_HDT	Truck	GSL	[PJ]	NOC	0.02396
TRA_RD_LD2	Bike	GSL	[PJ]	NOC	0.372
TRA_RD_LD4C	Car	GSL	[PJ]	NOC	0.03616
TRA_RD_LD4T	Minibus	GSL	[PJ]	NOC	0.03617
TRA_RD_M4	Motorcycle	GSL	[PJ]	NOC	0.00904

the optimal lamp period by extending the Webster's method [26]. The system monitors lanes in the intersection using traffic cameras. The green light cycle corresponding to that route will be given precedence to go first since the lane with the highest traffic volume and with highest emissions. As a result, the following formula is used to calculate the optimal traffic light cycle:

$$Cop = \frac{1.5L + 5}{1 - \sum_{i=1}^{n} \frac{V_A^{CE}}{S_A^E}} \tag{1}$$

where:

L: Lost time. Start-up lost time happens when a traffic signal changes from red (stop) to green (go).

n: The number of lanes per street

V_A^{CE}: number of vehicles per category C having emission factor E in lane A

S_A^E: number of saturated average emissions at lane A lane A.

4 Application Results

The IUTAR system has been developed and can be used to detect and track vehicles in Danang city. The following allocation view shows the IUTAR implementation system. In this section, the preprocessing of training datasets is presented to describe how camera photos can be collected, resized, box-bounded and labelled. Afterwards, we show the how vehicle catergories can be classfied from traning data sets based on YOLOv4. Hereafter, the test runing process can be specified. Moreover, the IUTAR dashboard represents our application results (Fig. 2).

4.1 Training Dataset with Image Pre-processing

An image used for training purpose can be retrieved from camera systems and stored in JPG or JPEG format. There are currently no constraints on the average number of

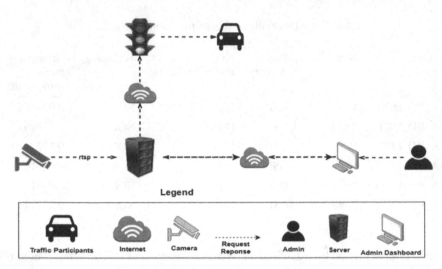

Fig. 2. Allocation view

objects and classes on each image. The raw images can be of any resolution, but it is recommended that its width and height do not go below 50px and above 700px. In case

(a)

```
File  Edit  Format  View  Help
4 0.578077 0.526151 0.146923 0.656904
4 0.290769 0.558577 0.143077 0.654812
4 0.425385 0.424163 0.104615 0.498954
4 0.893077 0.550209 0.184615 0.692469
```

(b)

Fig. 3. a. An example of training images. **b.** And labelled file

the objects' sizes in pixels break these constraints, the image ought to be resized before we proceed to the next step Then a labelling tool can be used to create bounding boxes and label objects on the collected images. Figure 3a shows an example of input images and Fig. 3b show its labelled description.

As a result, each training item will consist of 2 files: the original image and the labelling text file. Figure 4 shows an example of our pre-processed image data sets.

Fig. 4. Example of training data sets

Hereafter, the training process will then invoked and produces constantly output training logs specifying the value of the loss function as well as periodically generating training checkpoint files. The process can be stopped once the loss function consistently outputs values under 0.01 and a training checkpoint has been generated after that point.

Finally, a script and a command-line call are used to generate a frozen inference graph, which can be used to start detecting objects. Figure 5 shows object detection application results in the context of the GAINS vehicle categories.

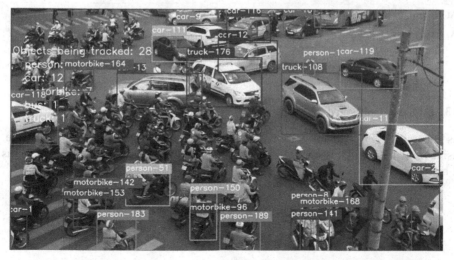

Fig. 5. Object detection application results based on the GAINS vehicle categories.

4.2   Test Run Results

We have developed a python script which can be used along with a command-line for detecting vehicle objects. The detector script will have the following directories declared: label map and frozen inference graph (generated at the end of the training process. Test run results are currently only recorded when performing single capture detection since the detection can vary under certain circumstances during real-time ones. For example, if an object goes undetected when it goes past a size threshold, the improved training dataset is supposed to cover that size to a possible degree. The same goes for special features like artistic fonts, colorful background, minimalistic style, etc. After the dataset has been updated, a new training-testing cycle is undertaken.

The splitting between traning and testing datasets can be described in Table 3.

Table 3. Training and testing results of the experimented model

	Accuracy (%)	Predict (%)	Recall (%)
Training	91,55	98.81	85.53
Testing	75.11	79.43	66.32

The evaluation for training results for each training set corresponding to 117 labels is indicated in Table 4.

Table 4. Comparative training result on different training set size

Case	Data training set (KB)	Total hour of training	Accuracy (%)
At Day	1	117	16.8
At Night	10	1170	55.25
At Peak Hour	35	4095	96.69

4.3   IUTAR Dashboard

As illustrated in Fig. 6, the IUTAR map can be seen as the IUTAR dashboard main page. Aside from that, the image depicts a traffic light icon, as well as locations of the cameras at high-traffic intersections.

Fig. 6. IUTAR dashboard

5 Conclusion and Future Works

This paper introduced the Intelligent Urban Transportation System (IUTAR) to control road traffic with a consideration of vehicle emissions at intersections. Vehicles are detected and classified by using YOLOv4 and DeepSORT algorithm [4, 10, 15]. Furthermore vehicle categories and their emission factors are provided by GAINS model [5, 9]. As a result, vehicle emissions per category per lane at an intesection can be calculated. Hereafter, the Optimal Traffic Light Cycle formula is specified as an extension of the Webster method [25] and used to control the traffic lights. The IUTAR dashboard with its training and testing data sets have been presented to proof of our concepts.

Our future work can then be able to support local government to predicting in related application domains, e.g. vehicle detection by urban traffic roads with additional data collected from measurements. Furthermore, emission factors can be retrieved from detail level of vehicle model, fuel types.

References

1. Abduljabbar, R., Dia, H., Liyanage, S., Bagloee, S.A.: Applications of artificial intelligence in transport: an overview. Sustainability **11**(1), 189 (2019). https://doi.org/10.3390/su1101 0189
2. Advantech's MIC-720AI, NVIDIA JETSON: Intelligent Video Traffic Monitoring for Self-Adaptive Traffic (2019)
3. AI Infrastructure Solutions, 10 March 2021. https://www.ibm.com/it-infrastructure/soluti ons/ai
4. Asha, C.S., Narasimhadhan, A.V.: Vehicle counting for traffic management system using YOLO and correlation filter. In: 2018 IEEE International Conference on Electronics, Computing and Communication Technologies (CONECCT), pp. 1–6 (2018). https://doi.org/10. 1109/CONECCT.2018.8482380

5. Borken-Kleefeld, J., Ntziachristos, L.: The potential for further controls of emissions from mobile sources in Europe. [TSAP Report #4], Version 1.0, DG-Environment of the European Commission, Belgium, June 2012
6. Circular 54/TT-BGTVT: National technical regulation on road signs (2019). https://luatvi etnam.vn/giao-thong/thong-tu-54-2019-tt-bgtvt-quy-chuan-ky-thuat-quoc-gia-ve-bao-hieu-duong-bo-181213-d1.
7. Darwish, T., Bakar, K.A.: Traffic density estimation in vehicular ad hoc networks: a review. Ad Hoc Netw. 24(PA), 337–351 (2015). https://doi.org/10.1016/j.adhoc.2014.09.007
8. Dtinew, Vietnam Inside: Over 600 people killed by traffic accidents a month (2019). https://vietnaminsider.vn/over-600-people-killed-by-traffic-accidents-a-month/
9. Hausberger, S.: Fuel Consumption and Emissions of Modern Passenger Cars. TU Graz, Institute for Internal Combustion and Thermodynamics. 2012. Overview of the Measurement Programs on LDV and HDV Presented at the Annual Plenary Meeting of ERMES, September 27, Brussels, Belgium (2010)
10. Huang, Y.-Q., Zheng, J.-C., Sun, S.-D., Yang, C.-F., Liu, J.: Optimized YOLOv3 algorithm and its application in traffic flow detections. Appl. Sci. 10(9), 3079 (2020). https://doi.org/10.3390/app10093079
11. Davison, J., et al.: Distance-based emission factors from vehicle emission remote sensing measurements. Sci. Total Environ. 739, 139688 (2020). ISSN 0048-9697
12. Kouziokas, G.N.: The application of artificial intelligence in public administration for forecasting high crime risk transportation areas in Urban environment. Transp. Res. Procedia 24, 467–73 (2017). 3rd Conference on Sustainable Urban Mobility, 3rd CSUM 2016, 26–27 May 2016, Volos, Greece, 1 January 2017. https://doi.org/10.1016/j.trpro.2017.05.083
13. Kuberkar, S., Singhal, T.K.: Factors Influencing Adoption Intention of AI Powered Chatbot for Public Transport Services within a Smart City (2020). https://www.semanticscholar.org/paper/Factors-Influencing-Adoption-Intention-of-AI-for-a-Kuberkar-Singhal/e253f96e9345 1f17ae766a4906d5cc76b0f3e55a
14. Mahrez, Z., Sabir, E., Badidi, E., Saad, W., Sadik, M.: Smart Urban Mobility: When Mobility Systems Meet Smart Data. ArXiv:2005.06626 [Cs], 9 May 2020. http://arxiv.org/abs/2005.06626
15. Morera, Á., Sánchez, Á., Moreno, A.B., Sappa, Á.D., Vélez, J.F.: SSD vs. YOLO for detection of outdoor urban advertising panels under multiple variabilities. Sensors 20(16), 4587 (2020). https://doi.org/10.3390/s20164587
16. Nikitas, A., Michalakopoulou, K., Njoya, E.T., Karampatzakis, D.: Artificial intelligence, transport and the smart city: definitions and dimensions of a new mobility era. Sustainability 12(7), 2789 (2020). https://doi.org/10.3390/su12072789
17. Patania, F., Gagliano, A., Nocera, F., Galesi, A., D'Amico, A.: The environmental impact of Urban transport: a case study for a new road in Catania Province. In: Urban Transport XIII: Urban Transport and the Environment in the 21st Century, I, pp. 699–709. WIT Press, Coimbra, Portugal (2007). https://doi.org/10.2495/UT070661
18. Larson, P.: Orijen Elltrom, Toyota Motor Corporation. ITS: Intelligent Transport System (2014)
19. Ren, S., He, K., Girshick, R., Sun, J.: Faster R-CNN: towards real-time object detection with region proposal networks. IEEE Trans. Pattern Anal. Mach. Intell. 39(6), 1137–1149 (2017)
20. Baidya, S., Borken-Kleefeld, J.: Atmospheric emissions from road transportation in India. Energy Policy 37(10), 3812–3822 (2009). ISSN 0301-4215
21. Smith, S.F., Barlow, G.J., Xie, X.F., Rubinstein, Z.B.: SURTRAC: Scalable Urban Traffic Control. 20 (2013)
22. Sobral, T., Galvão, T., Borges, J.: Visualization of urban mobility data from intelligent transportation systems. Sensors (Basel Switz.) 19(2), 332 (2019). https://doi.org/10.3390/s19020332

23. Surtrac - Real-time Adaptive Traffic Signal Control. https://www.rapidflowtech.com/. Rapid Flow

24. Traffic Congestion Detection from Camera Images Using Deep Convolution Neural Networks - Google Search. https://www.google.com/search?client=firefox-b-d&q=Traffic+congestion+detection+from+camera+images+using+deep+convolution+neural+networks. Accessed 24 Aug 2021

25. Velastin, S.A., Fernández, R., Espinosa, J.E., Bay, A.: Detecting, tracking and counting people getting on/off a metropolitan train using a standard video camera. Sensors **20**(21), 6251 (2020). https://doi.org/10.3390/s20216251

26. Webster, F.V.: Traffic Signal Settings, Road Research Technical Paper No. 39.27 (1957)

27. Zhang, F., Li, C., Yang, F.: Vehicle detection in urban traffic surveillance images based on convolutional neural networks with feature concatenation. Sensors **19**(3), 594 (2019). https://doi.org/10.3390/s19030594

28. Zhang, S., Wu, G.: Understanding Traffic Density from Large-Scale Web Camera Data (2015). 15

A Data Union Method Using Hierarchical Clustering and Set Unionability

Manh Huy Ta[1], Tran Khanh Dang[1(✉)], and Nhan Nguyen-Tan[2]

[1] Ho Chi Minh City University of Technology (HCMUT), VNU-HCM,
268 Ly Thuong Kiet Street, District 10, Ho Chi Minh City, Vietnam
{1870399,khanh}@hcmut.edu.vn
[2] H2A Technology Solutions Joint-stock Company,
Bien Hoa, Dong Nai, Vietnam

Abstract. Data nowadays is an extremely valuable resource. The owner of the data could be an organization, a company, the government, or just a normal person. Because of that, the content of the datasets that came from those sources would be varied: the data content could be about primary education, it could be about medical care in the U.S or it could be about agriculture in Vietnam, etc. It is intuitive that some datasets would be about the same topic so that they would have the same structures, or at least, similar structures. It is beneficial that we can union those datasets into a more meaningful dataset. In this paper, we proposed a data union method based on hierarchical clustering and Set Unionability. For simplicity, the method will used JSON data format as input data type.

Keywords: Data integration system · Data union · Hierarchical clustering · Open data

1 Introduction

With the development of technology, data is becoming an extremely valuable resource. Data is being created, analyzed, and used in a massive scale in every modern system. As a result, data analysis and data mining are essential in each aspect of social applications [4,5]. Data will be used much better if we combine the data from different sources into bigger, more informative datasets, making the data even more suitable for analyzing and mining tasks. The integration of data is especially useful for solving current social problems [9,12]. However, in order to make data integration feasible, there are multiple challenges that need to be solved: the challenge of data transformation, data cleansing, data semantic identifying, data unifying, etc.

Data union is an interesting research field. It can be understood as the operation to combine data from multiple sources into a combined dataset (Fig. 1). The data from the sources must be related to each other. As one may notice, data

T. K. Dang et al. (Eds.): FDSE 2021, LNCS 13076, pp. 222–235, 2021.
https://doi.org/10.1007/978-3-030-91387-8_15

union could be the last step of data integration as it is the real step to combine the data into collective datasets. However, data union is challenging not
only because of the volume of the data but also because of different structures
of the data as well as because of the underlying problem to unionize the data:
to unionize the data, we have to find a measurement to measure the similarity
of the data. The datasets with higher similarity will likely be about the same
subjects therefore can be combined.

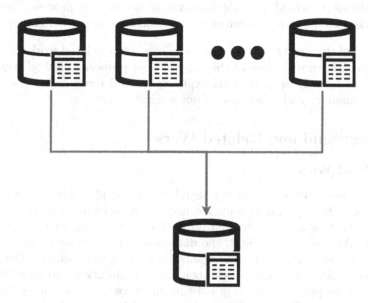

Fig. 1. Data Union - the process of combining data from multiple sources into a combined dataset.

But as one may intuitive, the similarity of the data is very hard to define as
it consisted of multiples problems: one first have to define the way to measure
the similarity between the attributes of the data as this is the cornerstone to
define the similarity between the records and the similarity between datasets.
Measuring the similarity is a classic discipline. Many similarity measurements
have been proposed like the cosine similarity, Jacob similarity, Euclid similarity,
etc. However, these similarity measurements can not be applied or at least, can
not be directly applied in the problem of finding the similarity between datasets.
The reason is that these similarity measurements only concerned with numeric
values. However, the similarity of data consisted of the similarity of the values
of the attributes and semantic similarity as in [6] stated.

As mentioned above, [6] have proposed that the similarity between data
should consist of set similarity, natural language similarity, and semantic similarity. However, despite proposing the similarity measurement, [6] only proposed
the method the find the top-k most similar datasets to a querying datasets rather
than a method to unionize datasets. Because of that, in this papers, we proposed
a data union method using the similarity measurement proposed in [6] and data

clustering technique. However, as we want to know the feasibility of our method, we proposed these following restrictions:

- The similarity measurement would consist of only similarity of the attributes of the records. The remaining similarity measurement will be included in a future paper aimed at the similarity measurement as this paper aimed at the overall method and the similarity measure can be "plugged-in" and improved.
- The datasets would be of the same field.
- The data source would be of the same format to ease the process. The JSON format is chosen here for convenience and could be replaced with other format.

The rest of this paper is organized as follows, some related works and background knowledge are discussed in Sect. 2 while our proposed method is explored in Sect. 3. In Sect. 4, we performed experiments and evaluations. Section 5 is about the summary and conclusion of our work.

2 Background and Related Works

2.1 Related Works

Since 2010, there have been a lot of researches in the field of data conversion and the researchers have proposed some methods for data conversion. In 2013, Ivan et al. proposed a data transformation system based on a community contribution model [11]. As depicted in Fig. 2, the data shared on the *publicdata.eu* portal included data from many different organizations of various formats. The system would then make initial mappings, then the community could contribute by creating new mappings, re-editing existing mappings, transforming the data, and using the data. The accuracy in data conversion would be improved over time with the contribution of the community.

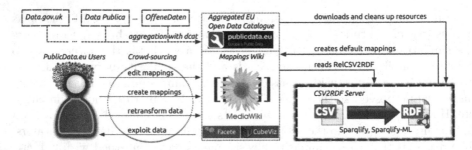

Fig. 2. Data transformation system based on a community contribution model (Ivan et al. 2013)

In 2015, Rocha et al. proposed a method to support the migration of data from relational databases (RDBMS) to NoSQL [16]. This method included 2 main module: data migration and data mapping.

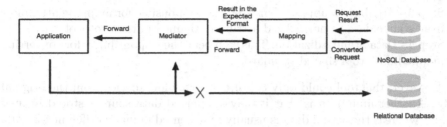

Fig. 3. The migration of data from relational databases (RDBMS) to NoSQL - data migration module (Rocha et al. 2015)

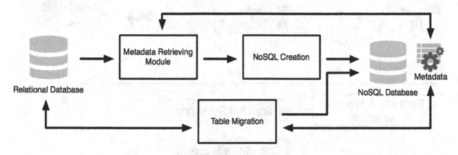

Fig. 4. The migration of data from relational databases (RDBMS) to NoSQL - data mapping module (Rocha et al. 2015)

– The data migration module's (Fig. 3) responsibility included automatically identifying all elements from the original relational databases (e.g. tables, properties, relationships, indexes, etc.), creating equivalent structures using the NoSQL data model and exporting the data to the new model.
– The data mapping module (Fig. 4) consisted of an abstract class, it was designed as an interface between the application and the DBMS. This module oversaw all SQL transactions from the applications, and translated these operations then moved to the NoSQL model that was created in data migration module.

Hyeonjeong et al. developed a semi-automatic tool for converting ecological data in Korea in 2017 [10]. The goal of this tool was to gather data in different formats from various research organizations and institutes specializing in environment in Korea and then convert to a shared standard ecological dataset. To accomplish this goal, the authors proposed 4 transformation steps as described in Fig. 5 including:

– Step 1 - Data File & Protocol Selection: This step provided an interface that allowed users to select data from the source file and the corresponding protocol.
– Step 2 - Species Selection: The user chose which species in the data to be converted.

– Step 3 - Attribute Mapping: This step was responsible for mapping attributes from source data to normalized attributes defined in the protocol.
– Step 4 - Data Standardization: This step bear the responsibility for converting mapped data to a shared standard.

However, this tool could only convert data for few species from the original data. Another limitation is that it only supported data sources stored in **.csv** format whereas the actual data is usually represented in many different formats.

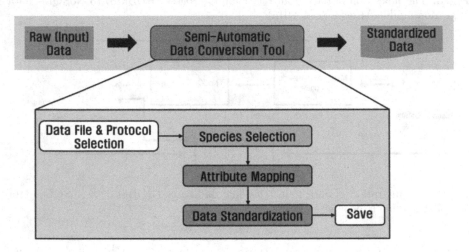

Fig. 5. Semi-automatic tool for converting ecological data in Korea (Hyeonjeong et al. 2017)

Besides, data transformation solutions are also embedded in data integration systems. Dong and Srivastava in [8], based on traditional data integration architectures as depicted in Fig. 6, highlighted three main challenges of three main phrases of big data integration:

– Schema Alignment: This phase solves semantic ambiguity challenges.
– Record Linkage: This phases solves ambiguity challenges of data representation.
– Data Fusion: This phase solves data quality challenges.

Fig. 6. Challenges for big data integration (Dong et al. 2015)

Knoblock and Szekely developed the Karma system, an integrated data system in the cultural heritage domain [13]. This system integrated data with high data heterogeneity from different museums. The process is described through four main stages:

- Data import phase: The data from any different source including databases, spreadsheets, or web services provided in XML or JSON format would be imported into the system.
- Data cleansing and normalization phase: In this phase, unusual data components and normalizing the data according to similar formats of related sources would be identified.
- Modeling phase: Semantic description of each resource would be created.
- Phase integration: The data would be converted into a single format using a description on semantics and data integration in an unified framework.

In 2018, Michael J. Mior et al. (2018) have proposed a methodology to transform a de-normalized schema into a normalized schema [15]. They argued that transforming a de-normalized schema into a normalized schema will provide information that can be used to "guide application and database evolution" [15].

In 2020, Tran Khanh Dang et al. (2020) have proposed a data conversion framework for data integration system [2]. The authors also did some experiments to find the most suitable data format to use in their system [3]. In 2021, the work in [2] had been extended in [1].

The problem of dataset querying is a traditional problem. In 2009, [22] have tried to solved the problem of keyword-based search on the content and context of Web tables. These tables would then be clustered into groups of unionable tables to provided keyword search functionality on metadata of datasets.

Ling et al. (2013) have proposed "table stitching" - the process of unioning tables with identical schemas within a given site into a union table [14]. This work heavily relies on the identical schemas of the tables. Based on [14], Lehmberg and Bizer have had union tables of stitched tables using schema-based and instance-based matching techniques [23].

In 2016, Renee J. Miller et al. (2016) had studied the problem of domain search and proposed LSH Ensemble, a new index structure based on MinHash LSH [18], as a solution to the domain search problem [17]. By using the results of [17], Renee J. Miller et al. (2018) proposed a solution for finding tables that are unionable with a query table within massive repositories [6].

2.2 Dataset Similarity Measurement

Dataset Similarity Measurement - Ensemble Unionability. The core problem of clustering problem is the problem of finding the distance or the similarity between the items in the dataset. Some of the commonly used measures of similarity or measure of distance:

- Cosine similarity.
- Euclidean distance.
- Hamming distance.
- Spearman correlation.
- Jaccard distance.

All of the above-mentioned distances/similarities have contributed greatly to the advancement of science and technology. They have been applied in many research problems and achieved great results. However, to measure the similarity between two datasets, these measures are not sufficient. Because of that, in [6], the authors have proposed a new way to measure the similarity between the datasets based on three criteria: Set Unionability, Semantic Unionability, and Natural Language Unionability. These three criteria can be described as:

- Set Unionability: This criterion represents the similarity of the values of the attributes of the datasets, if the attributes have a lot of common values, this value is higher.
- Semantic Unionability: This criterion represents the similarity of the semantics values of the attributes by mapping attribute values to entities in the ontology. For example, the "City" attribute and "Location" attribute of two different datasets could have high Semantic Unionability because these two attributes values can be mapped into a "city" ontology.
- Natural Language Unionability: This criterion also represents the similarity of the semantics values of the attributes. Natural Language Unionability is calculated by using word embeddings [6]. This criterion comes into existence as not all values can be mapped into a semantic ontology like YAGO[7]. In fact, the author of [6] said that "only 13% of values in attributes can be mapped to YAGO entities" [6].

The Ensemble Unionability is then computed by finding the maximum value of Set Unionability, Semantic Unionability, and Natural Language Unionability. The author of [6] used Ensemble Unionability to measure the similarity between the attributes of the datasets. Then, they computed the similarity between the datasets by using the similarity of the attributes. By using Ensemble Unionability, Renee J. Miller et al. have proposed a table union search on open data that have good performance time [6].

Dataset Similarity Measurement - Set Unionability. In this work, we only used the Set Unionability to measure the similarity between the datasets as this paper's purpose is to proposed a table union method using similarity measurement and clustering methods. The two remaining measurements: Semantic Unionability and Natural Language Unionability would be included in a future work to improve the performance of this method.

A Domain D could be understood as a finite set of discrete values. To evaluate if attributes A and B are both from the same Domain D, we could assume that A is a random sample of D and we want to know if B is also a sample of D. Because D is unknown and we assumed that A is a sample of D, the values of

A are the only known values in D. We can then use the size of intersection of the values in A and B as the test statistic for evaluating the probability that A and B are samples from the same domain D. The size of the intersection of A and B are assumed to be drawn from same domain follows a hypergeometric distribution [19].

Suppose that D contains the values of A as success values and we draw— α—samples from D, without replacement. If a draw would be a success if it is in the intersection of A and B, otherwise it is a failure. In this case, the number of successful draws indicates the likelihood that A and B are from the same Domain D. The number of successful draws is limited by intersection of A and B and has the maximum value equal to size of intersection of A and B. Then, we can use the hypergeometric test to test whether A and B are from the same domain D by using the size of the intersection.

Let $n_a = |A|$ Let $n_b = |B|$ and Let $n_D = |D|$. If A is in the domain of D and B is drawn from D then the distribution of s successful draws ($s \in \{1, 2, ..., A \cap B\}$) is:

$$p(s|n_a, n_b, n_D) = \frac{C(n_a, s)C(n_D - n_a, n_b - s)}{C(n_D, n_b)}$$

The probability p(s) allows the evaluation of the statistical significance of the intersection size i.e. whether A and B are both from the same domain D given an intersection size. By using the intersection of A and B, t = $|A \in B|$, we can define the cumulative distribution of t:

$$F(t|A, B) = \sum_{0 \le s \le t} p(s|n_a, n_b, n_D)$$

The cumulative distribution of a hypergeometric distribution, F, can be used as a hypothesis test meaning we can reject the hypothesis that A and B are both of the same domain if the value of $F(t|A, B)$ is lesser than a threshold θ. Because of that, we can use F as a similarity measurement or unionability of two domains. The Set Unionability $U_{set}(A, B)$ is therefore:

$$U_{set}(A, B) = F(t|A, B)$$

As we can see, to calculate Set Unionability, the size (or the cardinality) of D is required. However, in practice, it is impractical or impossible to know the true domain D. Thus, we have to make a practical assumption that D is approximately the disjoint union of A and B therefore $n_D \approx n_a + n_b$.

3   Proposed Method

In this work, we proposed a data union method including three steps (Fig. 7):

- Step 1 - Schema Step: Getting datasets schema or structure: The datasets schema would be used in the following steps.

– Step 2 - Clustering Step: Using clustering algorithm on the datasets to get the best "mapping" of the datasets.
– Step 3 - Union Step: Union the original datasets into union datasets.

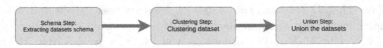

Fig. 7. Data Union Steps consisted of three steps: Schema Step, Clustering Step, and Union Step

3.1 Schema Step

The schema step, as mentioned before, extracted the schema or structure of the datasets. As we mentioned in the Introduction section, the datasets will be of JSON format for simplicity and easiness. However, there are many ways to store datasets in JSON format: the datasets could be stored in binary files, they could be stored as a string file or they could be stored in a database management systems like MongoDB, CouchDB or OrientDB. In our work, to simplify the process and take advantage of already-existing advancements in technology, we used MongoDB as JSON storage to make use of its various features to ease the process. The choosing of MongoDB is just a preference and can be interchange with other database of choice or even files of JSON dataset, provided that the schema step would be much time-consuming as this will limit the support of technology.

The process of schema extracting is simple, for each dataset (or collection in MongoDB):

1. Read the entire dataset to find the schema of the dataset.
2. Save the findings to the database for further use.

The schema of the dataset is only the information about the attribute of the records in the datasets. The schema is a dictionary consisting of the structure or schema of every table in the database - each database's name is the key of its schema. An example of the schema could be viewed in Fig. 8.

3.2 Clustering Step

This step uses hierarchical clustering and Set Unionability (Sect. 2.2) to unionize the data. The main idea is the more similar the datasets are the least "distance" they are from each other. In this works, Set Unionability is used as a similarity measurement for simplicity and will be improved in future work for better results. The main goal of this step is the "mappings" of the datasets and the datasets clusters. The "mappings" is a list of mappings between the datasets.

```
"table_union_test_schema": {
    "t_013a2f8c584d44d7____c1_0____4": {
        "pk": "_id",
        "_id": {
            "dest": "_id",
            "type": "TEXT"
        },
        "Quality Indicator Indicateurs de qualité": {
            "dest": "Quality_Indicator_Indicateurs_de_qualité",
            "type": "TEXT"
        }
    },
    "t_013a2f8c584d44d7____c1_0____1": {
        "pk": "_id",
        "_id": {
            "dest": "_id",
            "type": "TEXT"
        },
        "Quality Indicator Indicateurs de qualité": {
            "dest": "Quality_Indicator_Indicateurs_de_qualité",
            "type": "TEXT"
        }
    },
    "t_013a2f8c584d44d7____c1_1____3": {
        "pk": "_id",
        "_id": {
            "dest": "_id",
            "type": "TEXT"
        },
        "Quality Indicator  Indicateurs de qualité": {
            "dest": "Quality_Indicator__Indicateurs_de_qualité",
            "type": "TEXT"
        }
    },
    "t_013a2f8c584d44d7____c2_0____2": {
        "pk": "_id",
        "_id": {
            "dest": "_id",
            "type": "TEXT"
        },
        "Quality Indicator  Indicateurs de qualité Indicateurs de qualité": {
            "dest": "Quality_Indicator__Indicateurs_de_qualité_Indicateurs_de_qualité",
            "type": "TEXT"
        },
```

Fig. 8. The schema of a sample database.

Each mapping represents a pair of datasets that can be unionized. Each of the above-mentioned mappings consisted of a list of "sub-mappings". Each "sub-mappings" represents a pair of attributes of two datasets that can be unionized. Each "sub-mapping" only represents the relation between exactly two attributes from two different datasets and can not represent the relationship between an attribute from one dataset and multiple attributes from another dataset. Each data cluster represents a combination of datasets that could be unionized into a single dataset.

The process to create the "sub-mapping" between two datasets:

- Calculate the similarity for each attribute in the first dataset to every attribute in the second dataset to create the "sub-mappings" and their corresponding scores - the similarity between the attributes (Set Unionability, in this case).
- Find all valid combinations of "sub-mappings" - a mapping. A combination of "sub-mappings" is valid if there is no exist two "sub-mappings" mapped to the same attribute.
- Calculate the score of all mappings. The score of a mapping is the product of all the scores of its "sub-mapping".

- Find the "best" combination of "sub-mappings", we defined the "best" combination of "sub-mappings" of dataset A and B as the mapping that has the highest score and covers a sufficient amount of attributes of the two datasets. If no such combination existed then "best" mapping is empty and its corresponding score is 0.
- Return the "best" mapping and its score as it will be used in other steps.

The process to create the "mappings" of datasets:

- Step 1: Loading all datasets from database to memory.
- Step 2: Creating the distance matrix:
  - For each collection, find the optimal mapping and its corresponding similarity score of that collection to every other collection in database. The method to find optimal mapping and its corresponding similarity score had been discussed above.
  - Save the results to a distance matrix.
- Step 3: Using hierarchical clustering on the distance matrix to cluster the data.
- Step 4: Save the clustering results.

To summarized, in clustering step, we generated the "mappings" of the datasets and the datasets clusters. These datasets clusters represent the way the datasets in database could be combined to create the union dataset and the "mappings" are the steps to unionize the datasets.

3.3   Union Step

In this step, we union the datasets into the union datasets. The main idea is to use the "mappings" and "clustering results" acquired in the "Clustering Step" to unionize the data. The "clustering results" in the Clustering Step should provide us with the step-by-step on how to merge two datasets in the database into a bigger dataset until only one dataset remaining. By using "clustering results" and "mappings" we could devise a process to unionize the datasets in the database into union datasets. The process of the steps is as followed:

For each "step" in the "clustering results":

- Step 1: Check if the distance between two datasets in this "step" is sufficiently small (i.e. the distance between two datasets is smaller than a threshold). If it is, go to step 2, otherwise, go to step 4.
- Step 2: Merge those two datasets into a "merged" dataset using the "mappings" acquired from Clustering Step.
- Step 4: Get the next "step" from the "clustering results" and go to "Step 1".
- Step 5: Move the remaining datasets in the "original database" into destination database.
- Step 6: Save all "merged datasets" into destination database.

4   Experiment and Evaluation

The test was conducted on several databases that contained the synthesized benchmark [20] that were created data from Canadian and UK Open Data that were used in [6]. Each database contained a different number of datasets from [20]. We applied our method to those databases and recorded the union datasets. The experiments only concerned about the clustering of the datasets, not the result of the union process since the Clustering Step is the step that decided the result of our methods and Union Step is mainly a technical step.

The data of the testing datasets include:

- Date data which includes the days of the weeks, the months of the year, the date in the full format.
- Railroad data from Canada which includes names of the companies, railroad operating data, Canada geography data (provinces, suburb, streets, etc.), the data about the soils.
- Research grant programs data which includes the applications of the researches, area of the researches' applications, the research committee, etc.

The results could be view in Table 1, the testing datasets and the results could be viewed at[21].

Table 1. Experiment results

Experiment no	Number of datasets	Number of original datasets	Number of union dataset
Experiment 1	25	2	8
Experiment 2	19	4	8
Experiment 3	39	4	11
Experiment 4	54	4	13
Experiment 5	15	1	11

In Experiment 1, 2, 3 and 4 the datasets were chosen from the original datasets that have high Set Unionability. However, upon further inspection, we found out that some "mapping" between two datasets didn't cover a sufficient amount of attributes of those two datasets, hence they were not clustered together. Another reason for the high number of union datasets in Experiment 1, 2, 3 and 4 is that the attributes from some datasets from one particular original dataset didn't have a lot of common values as we expected. However, all of the above mention issues were expected since we only used Set Unionability and should be solved by using better unionability measurement in further researches.

In Experiment 5, we chose datasets from only one original dataset. In this experiment, we especially chose the datasets that should have low Set Unionability beside those that should have high Set Unionability to check if those low Set

Unionability datasets would be clustered into any union datasets. The results of Experiment 5 showed that these datasets were not clustered as we expected.

Finally, in all experiments, all datasets in a cluster were from the same original datasets, and no datasets had been clustered with datasets from other original datasets. The results showed that the datasets in all databases had been reasonably clustered.

From the experiments, we concluded that our method is feasible. However, the experiments also show the insufficiency of Set Unionability for measuring the similarity between datasets. A better similarity measurement should be used instead of just Set Unionability. This will be our next work.

5   Conclusion

Data union is an emerging topic closely related to various fields, including open data, data integration. In this papers, we have proposed a method to unionize JSON data in database using hierarchical clustering and Set Unionability. To evaluate this method, we did an experiment using data from Canadian and UK Open Data that used in [6]. The experiments showed that datasets that have high Set Unionability were unionized and other weren't, thereby prove the feasibility of our method.

However, there is still existed some shortcomings in our method: the similarity measurement as we only use Set Unionability and the best "mapping" between two datasets, the process to create the "mapping" between two datasets is not optimal. Another problem is that the performance of our method still needs to be improved to be applied in real-world problems. These shortcomings will be the object of our next future works.

Acknowledgements. This work is supported by a project with the Department of Science and Technology, Ho Chi Minh City, Vietnam (contract with HCMUT No. 42/2019/HD-QPTKHCN, dated 11/7/2019). We also thank all members of AC Lab and D-STAR Lab for their great supports and comments during the preparation of this paper.

References

1. Dang, T.K., Ta, M.H., Dang, L.H., Hoang, N.L.: An elastic data conversion framework - a case study for MySQL and MongoDB (2021)
2. Dang, T.K., Ta, M.H., Dang, L.H., Hoang, N.L.:. An elastic data conversion framework for data integration (2021)
3. Dang, T.K., Huy, T.M., Hoang, N.L.: Intermediate data format for the elastic data conversion framework. In: International Conference on Ubiquitous Information Management and Communication IMCOM 2021 (2021)
4. Dang, T.K., Anh, T.D.: A pragmatic blockchain based solution for managing provenance and characteristics in the open data context. In: Dang, T.K., Küng, J., Takizawa, M., Chung, T.M. (eds.) FDSE 2020. LNCS, vol. 12466, pp. 221–242. Springer, Cham (2020). https://doi.org/10.1007/978-3-030-63924-2_13

5. Ha, T., Dang, T.K.: Investigating local differential privacy and generative adversarial network in collecting data (2020)
6. Nargesian, F., Zhu, E., Pu, K.Q., Miller, R.J.: Table union search on open data (2018)
7. Suchanek, F.M., Kasneci, G., Weikum, G.: YAGO: a core of semantic knowledge. In: WWW, pp. 697–706 (2007)
8. Dong, X.L., Srivastava, D.: Big Data Integration, p. 198. Morgan & Claypool Publishers (2015)
9. McLaren, D., Agyeman, J.: Sharing Cities: A Case for Truly Smart and Sustainable Cities. MIT Press, Cambridge (2015)
10. Lee, H., Jung, H., Shin, M., Kwon, O.: Developing a semi-automatic data conversion tool for Korean ecological data standardization. J. Ecol. Environ. **41**(11), 1–7 (2017)
11. Ermilov, I., Stadler, C., Martin, M., Auer, S.: CSV2RDF: user-driven CSV to RDF mass conversion framework. In: Proceedings of the 9th International Conference on Semantic Systems (2013)
12. Lai, C.S., et al.: A review of technical standards for smart cities. Clean Technol. **2**, 290–310 (2020)
13. Knoblock, C.A., Szekely, P.: Exploiting semantics for big data integration. AI Mag. **36**(1), 25–38 (2015)
14. Ling, X., Halevy, A. Y., Wu, F., Yu, C.: Synthesizing union tables from the web. In: International Joint Conference on Artificial Intelligence, IJCAI 2013 (2013)
15. Mior, M.J., Salem, K., et al.: Renormalization of NoSQL database schemas. In: Trujillo, J.C. (ed.) ER 2018. LNCS, vol. 11157, pp. 479–487. Springer, Cham (2018). https://doi.org/10.1007/978-3-030-00847-5_34
16. Rocha, L., et al.: A Framework for Migrating Relational Datasets to NoSQL1. Proc. Comput. Sci. **51**, 2593–2602 (2015)
17. Zhu, E., Nargesian, F., Pu, K.Q., Miller, R.J.: LSH ensemble: internet-scale domain search. Proc. VLDB Endow. (2016)
18. Broder, A.: On the resemblance and containment of documents. In: Proceedings of the Compression and Complexity of Sequences (1997)
19. Rice, J.A.: Mathematical Statistics and Data Analysis (2006)
20. https://github.com/RJMillerLab/table-union-search-benchmark
21. https://github.com/ligthsworn/table_union_benchmark
22. Cafarella, M.J., Halevy, A.Y., Khoussainova, N.: Data integration for the relational web. In: Proceedings of International Conference on Very Large Data Bases (2009)
23. Lehmberg, O., Bizer, C.: Stitching web tables for improving matching quality. In: Proceedings of International Conference on Very Large Data Bases (2017)

Blockchain and IoT Applications

A Consensus-Based Load-Balancing Algorithm for Sharded Blockchains

M. Toulouse[1(✉)], H. K. Dai[2], and Q. L. Nguyen[1]

[1] School of Information and Communication Technology, Hanoi University of Science and Technology, Hanoi, Vietnam
michel.toulouse@soict.hust.edu.vn, lam.ntq166334@sis.hust.edu.vn
[2] Computer Science Department, Oklahoma State University, Stillwater, OK 74078, USA
dai@cs.okstate.edu

Abstract. Public blockchains are decentralized networks where each participating node executes the same decision-making process. This form of decentralization does not scale well because the same data are stored on each network node, and because all nodes must validate each transaction prior to their confirmation. One solution approach decomposes the nodes of a blockchain network into subsets called "shards", each shard processing and storing disjoint sets of transactions in parallel. To fully benefit from the parallelism of sharded blockchains, the processing load of shards must be evenly distributed. However, the problem of computing balanced workloads is theoretically hard and further complicated in practice as transaction processing times are unknown prior to be assigned to shards. In this paper we introduce a dynamic workload-balancing algorithm where the allocation strategy of transactions to shards is periodically adapted based on the recent workload history of shards. Our algorithm is an adaptation to sharded blockchains of a consensus-based load-balancing algorithm. It is a fully distributed algorithm inline with network based applications such as blockchains. Some preliminary results are reported based on simulations that shard transactions of three well-known blockchain platforms.

Keywords: Sharded blockchains · Dynamic load-balancing · Distributed average consensus · Distributed algorithms

1 Introduction

Replication in computer systems improves fault tolerance to network failures, computer hardware components failures, software bugs, malicious activities or to reduce systems access latencies. Databases have developed a variety of replication techniques as these systems face most of the above issues. Blockchain technologies extend the reach of replication to implement transparent decentralized control.

© Springer Nature Switzerland AG 2021
T. K. Dang et al. (Eds.): FDSE 2021, LNCS 13076, pp. 239–259, 2021.
https://doi.org/10.1007/978-3-030-91387-8_16

The extensive use of replication in blockchains raises a whole set of familiar problematic issues which are often cited as a barrier to the adoption of this new technology. Cryptocurrency platforms such as Bitcoin and Ethereum can only process a limited number of transactions per second given that transaction validation is replicated across all the nodes of the blockchain network. Similarly, storage requirement grows proportionally to the # of transactions × the number of nodes as the blockhain is identically copied on each node. Several solutions are proposed to address these two particular issues. One of them borrows from database systems, it is called sharding.

Sharding is a parallelization strategy for databases. Sharding partitions a large data set into multiple databases, where each database runs on a different server. Access requests to the data set are directed to the database where the requested data are stored. As the databases run on different servers, requests can be served in parallel. Designers of sharded blockhains aim to exploit sharding parallelism to reduce the amount of memory needed to store blockchains and to speedup transactions processing.

Although sharding of blockchains has only been proposed recently, there are already several sharded blockchain protocols or fully implemented sharded blockchain platforms. Elastico [15], 2016, has been historically the first proposed sharding protocol. It has been followed by several others, among them Zillica [23], Omniledger [12], Chainspace [1], RapidChain [32], Ethereum 2.0 [3], Monoxide [27], Ostraka [16], Harmony [22], Logos [33], SSChain [4], Stegos [20], OptChain [17], Sharper [2]. The most widely known in the blockchain community is probably Ethereum 2.0, a sharded version of Ethereum, to be released in 2021. The following three recent surveys [14,26,31] among others are summarizing design strategies related to sharded blockchains.

The full sharding of a blockhain partitions the blockchain network into a set of sub-networks, i.e. shards, in which the nodes of each shard evaluates and stores disjoint subsets of the transactions generated by users. Known issues about this approach are cross-shard transactions and improper workload-balance among shards. Cross-shard transactions connect accounts which are stored in different shards, the evaluation of these transactions requires some form of synchronization, such as atomic commit protocols, among the shards involved in the cross-shard transactions. Load-balancing for sharded blockhains consists to assign transactions to shards such that the sum of the transaction processing times is distributed evenly across shards. Mathematically, load-balancing is an NP-Hard optimization problem, it is unlikely polynomial time algorithms can be found to solve this problem. For sharded blockchains, load-balancing has an extra layer of complication as the processing time of transactions is unknown prior to assigning transactions to shards.

We are only aware of three recent publications addressing the issue of load-balancing in sharded blockchains [13,18,29]. All these papers propose centralized algorithms to periodically balance the workload of each shard. In the present paper, given that control decisions for blockchains are bottom-up, and achieved through consensus, we propose a fully distributed algorithm for sharded blockchain load-balancing problem, inline with the decentralization of blockchain designs. Our solution is based on distributed average consensus, a class of distributed algorithms to compute a global average of initial parameters using only local interactions among processing nodes. Distributed average consensus has been used for problems in diverse fields, control theory [30], multi-agent systems [10], physics [25], distributed optimization [24] including load-balancing, [6], 1989, where it is called *diffusion algorithm*. The diffusion algorithm solves the dynamic load-balancing problem, which is how we formulate the load-balancing problem for sharded blockhains. The work of [6] has been extended to several dynamic load-balancing problems, parallel processing [28], cloud computing [8], many others. Our algorithm is based on one of these extensions [28], adapted to the load-balancing problem for sharded blockchains from dynamic load-balancing techniques in distributed processing.

Lastly, we have adapted two well-known scheduling algorithms for solving the shard load-balancing problem. The first one is the "longest processing time" (LPT) algorithm from [9] which assigns tasks to multiprocessors in decreasing order of their processing time. LPT minimizes the finishing time of the last executed task. Mathematically it is an approximation algorithm, which means it is proved the solutions return by LPT are no worst than some constant factor of the optimal solution. Our implementation of LPT is centralized. The second algorithm is the MULTIFIT algorithm [11], also a scheduling algorithm minimizing the makespan of the schedule. MULTIFIT is also an approximation algorithm, compared to LPT it has a better approximation factor but its asymptotic running time is slightly worst than LPT. Our implementation of MULTIFIT is also centralized. We use the approximation factor of these two algorithms to validate the performance of the consensus-based algorithm.

The paper is organized as follow. Next section briefly provides relevant background about blockchains, introduces a classification of sharding techniques for blockchains and finally describes the workload problem specifically addressed in this paper. Section 3 describes three algorithms for our workload problem. Section 4 reports preliminary results of sharding simulations for three well known blockchain platforms. Finally, last section summarizes the paper and provides avenues for future investigations.

2   Problem Definition

A blockchain can be described as a network of computing nodes running the client part of a blockhain application. The main purpose of a blockchain is to record the transfer of assets among nodes (clients) of the network. A transfer of assets is manifested by the creation of a file called "transaction" which has an id (the digest of a cryptographic hash function), and which holds data describing the transfer itself and potentially some metadata for managing the transaction.

The main processing activity of blockchain network nodes is to validate transactions, as each transaction, prior to its confirmation, must be independently verified by all the nodes in the blockchain. Details of the verification process vary across blockchain platforms. Minimally, each node verifies whether the client node initiating the transaction holds the assets it transfers. The independent transaction verification is based on nodes having access on their local disk to the file of each previously posted transaction across the whole blockchain network. Once a transaction is confirmed, the transaction file is indexed in a data structure called "block". A block can hold a finite number of transactions, once this limit is reached, the block is closed and broadcast to the blockchain network to be stored on local hard disk of each node. Each block holds (in its header) the cryptographic hash of the previously stored block. Because of this cryptographic hash, the blocks form a chain, the "blockchain". The blockchain must have the same sequence of blocks at every node, otherwise nodes will not be able to verify consistently ownership of the assets written for transfer in newly initiated transactions. All the transactions stored in the blockchain are visible to any user having an account on the blockchain.

2.1   Types of Blockchains

There are several criteria along which one can classify blockchain platforms. Two significant ones in regard of sharding protocols is the difference between public versus private blockchains and how the ownership of assets is recorded in the blockchain.

Public vs Private Blockchains. Blockchain platforms can be divided into two categories: public, permissionless versus private, permissioned. In permissionless blockchains, an user only need to install the client side of the blockchain platform to attach itself to the blockchain network, without any form of authentication. Like credit cards, permissionless blockchains support applications targeting the consumer layer of the economy, they usually have a large network of nodes, Bitcoin and Ethereum, the two largest blockchain networks, are permissionless blockchains. Private blockchains support applications at the corporate level, which may demand a great level of trust among users, a high level of access securities, and which may place restrictions on which user may access which data in the blockchain. Private, permissioned, blockchain networks restrict entries, they have far fewer nodes compared to permissionless blockchains. In general, permissioned blockchains do not face the same pressing scalability issues as for public blockchains, they have much less transactions to process and to store. As a consequence, there are few private sharded blockchain proposals. In the list of sharded blockchains in Sect. 1, only Chainspace [1] and Sharper [2] are permissioned platforms.

Transaction Models. A second distinction among blockchains that impacts sharding designs is how assets are recorded. Most blockchain adopts one of two approaches: the Unspent Transaction Output (UTXO) model or the account-based model, the blockchains adopting one of these two models are said to be UTXO-based or account-based blockchains. In UTXO blockchains, assets are recorded only in the transactions. In this model, transactions have an input and an output section. Once a new transaction tx is created to transfer assets, the input section of tx refers to the output section of existing transactions as the source of assets to transfer. The output section of tx list the public keys part of private keys that can unlock the assets transferred by tx. It is like someone write you a check, you give this check to pay for a given service, if your check is larger then the value of the service, you receive a new check for the difference in values. Your assets sums up to the checks you hold in your hand. In terms of UTXO blockhains, your assets are openly display on the blockchain, except it is not possible (in principle) to link the displayed assets to any specific user.

The account-based transaction model is more intuitive. Users have anonymous accounts which are made up of a cryptographic pair of keys: public and private. Like bank accounts, each account has a balance. The account balance represents the value of the assets held by the owner of an account. Each transaction transfers assets from a source account to a destination account, the assets transfer is reflected in the balance of the accounts listed in the transaction.

Account-based blockchains often have two types of accounts: user accounts and smart contract accounts. Smart contract accounts are computer programs stored on the blockchain. User accounts and smart contract accounts are stored on every node of the blockchain network. When a transaction involves as source or destination a smart contract, the verification process of such transaction requires the smart contract be executed by each validating node. Similarly, the balance of user accounts listed in a transaction is verified by each node. The verification of transactions could not be executed independently if the user accounts or the smart contract accounts were not stored on each of the blockchain network node.

In this paper, we address the load-balancing problem for public, permissionless account-based sharded blockchains.

2.2   Sharding Strategies

Sharding strategies can be classified into 3 groups [19]: state sharding, transaction sharding and full sharding. State sharding means that the storage capacity collectively available from all the client nodes in the blockchain network is partitioned into shards, where each shard stores a disjoint subset of confirmed transactions. In state sharding, each transaction is verified by each node of the blockchain network. We are not aware of any state sharding protocol alone, but state sharding is often combined with transaction sharding in full sharding protocols.

In transaction sharding, computing nodes of the blockchain network are partitioned into subsets, called committees, where each committee verifies a disjoint subset of transactions. Transactions that are assigned to different committees are evaluated in parallel. Unlike state sharding, confirmed transactions are stored on all the nodes of the blockchain network, the storage requirement in transition sharding is the same as for un-sharded blockchains. Zilliqa [23] and Elastico [15] implement this form of sharding. Zilliqa is an account-based blockchain. Computing nodes are partitioned into a pre-defined number of committees. Transactions are assigned to committees for processing based on the address of the account that initiates the transaction. For example, if there are l committees from 0 to $l-1$, the last $\lfloor \log_2 l \rfloor + 1$ bits of the initiating account address, a number between 0 and $l-1$, identifies the committee where the transaction is sent for processing. Elastico, an UTXO transaction model, assigns transactions to committees according to the transaction hash id. For example, if there are 8 committees, the decimal conversion of the first 3 bits of a transaction id defines the committee assignment of the transaction.

In full sharding, storage and transaction verifications are sharded. Several recent blockchain platforms/protocols apply this sharding design: RapidChain (UTXO) [32], SSChain [4] (UTXO), Omniledger [12] (UTXO), Harmony [22] (account-based), Monoxide [27] (account-based) and Ethereum 2.0 (account-based). Computing nodes are assigned to committees according to rules that vary across these sharded platforms. In full sharding for account-based blockchains, the storage of account balances as well as smart contract codes is also sharded, that is user accounts and smart contracts are stored only on the computing nodes of the committee where they are assigned. Consequently, the assignment of transactions to shards is implicitly done through accounts, i.e. transactions are verified in the committee where the account initiating the transaction is stored. In full sharding, each shard is responsible for generating blocks from transactions confirmed in the shard, for maintaining its own blockchain and for storing a subset of the account balances. Full sharding divides the storage requirement of a blockchain platform by the number of shards. Note that in order to defend against security vulnerabilities, some or all the nodes forming a committee are re-assigned periodically to other shards.

2.3 The Load-Balancing Problem for Sharded Account-Based Blockchains

The general form of the load-balancing problem is formulated as follow: assign m tasks $t_1, t_2, \ldots, t_m$ with respectively computing times $c_1, c_2, \ldots, c_m$, to n processing units $p_1, p_2, \ldots, p_n$ such to minimize the maximum load on any processing unit. Assume that $\mathcal{T}_i$ is the set of tasks assigned to processing unit p_i, the load on machine p_i is $C_i = \sum_{k \in \mathcal{T}_i} c_k$, the objective function is $\min \max\{C_i \mid i = 1..n\}$.

The load-balancing problem for account-based sharded blockchains can be modeled as follow: given a set of m accounts a_1, a_2, a_m, assign the m accounts to n shards $s_1, s_2, \ldots, s_n$ such to minimize the maximum load on any shard. Let $\mathcal{R}_j$ be the set of transactions initiated by account a_j and p_k^j the processing time of transaction k initiated by a_j. The computing time c_j of account a_j is the sum of the processing times p_k^j of the transactions initiated by account a_j, $c_i = \sum_{k \in \mathcal{R}_j} p_k^j$. Let $\mathcal{T}_i$ be the set of accounts assigned to shard s_i, the workload of shard s_i is $C_i = \sum_{k \in \mathcal{T}_i} c_k$, the objective function is $\min \max\{C_i \mid i = 1..n\}$.

The sharded load-balancing problem is not static, new accounts are constantly created from existing users or new users joining the blockchain network, and transactions are constantly initiated from the account holders. Thus the load of shards is changing, it has to be continuously re-evaluated, which is the definition of a dynamic load-balancing problem. The computing time of shards is divided in periods between which load-balancing is performed. The duration of the periods could be variable based on measurements of the shards load imbalances. In this paper, periods have fixed elapse times, we name these periods "epochs".

The processing time of an account is not known prior to assign an account to a shard, thus load prediction must be applied. Consider two consecutive epochs e_k and e_{k+1}. During epoch e_k, the real processing time of each account a_i is recorded in c_i. The processing time c_i from epoch e_k is used as a prediction of the processing time of account a_i in epoch e_{k+1}. Load-balancing is performed before epoch e_{k+1} starts, the balanced workload for e_{k+1} is computed based on predictions from epoch e_k.

3   Algorithms

This section describes three algorithms for solving the dynamic load-balancing problem formulated in the previous section: the diffusion algorithm, LPT and MULTIFIT.

3.1   Diffusion Algorithm

The diffusion algorithm belong to a class of algorithms called distributed average consensus algorithms which compute global averages in parallel in decentralized networks such as ad-hoc networks, peer-to-peer networks or sensor networks. Mathematically the network is represented as an undirected graph where adjacency among the graph vertices stands for direct communication links among the network computing nodes. Let $G = (V, E)$ be such a graph where $V = \{v_1, v_2, \ldots, v_n\}$ is a set of vertices (modeling the network computing nodes) and E denotes a set of edges pairing vertices (direct communication links). Graphs like G have an adjacency structure represented by some $n \times n$ *adjacency matrix* (denoted by A here) where $a_{ij} = 1$ if and only if $(v_i, v_j) \in E$, $a_{ij} = 0$ otherwise. The adjacency structure of G defines for each node $v_i \in G$ a *neighborhood* $\mathcal{N}_i$ where $\mathcal{N}_i = \{v_j \in V | (v_i, v_j) \in E\}$. The global average computed by distributed consensus is $\frac{1}{n}(\sum_{i=1}^{n} x_i(0))$ where $x_i(0)$ is some initial value of vertex i. The global average is obtained from the execution by each node of the network of the following iterative algorithm:

$$x_i(t+1) = w_{ii}x_i(t) + \sum_{j \in \mathcal{N}_i} w_{ij}x_j(t) \qquad (1)$$

where w_{ij} is a weight associated to edge (i, j). The weight values w_{ij} belong to a $n \times n$ weight matrix W satisfying some properties such as $W = W^T$ (transpose of W) and others algebraic properties that are used to prove the local updates converge asymptotically to the global average $\frac{1}{n}(\sum_{i=1}^{n} x_i(0))$. The Metropolis-Hasting weight matrix below satisfies those conditions [30]:

$$W_{ij} = \begin{cases} \frac{1}{1+\max(d_i,d_j)} & \text{if } i \neq j \text{ and } j \in \mathcal{N}_i \\ 1 - \sum_{k \in \mathcal{N}_i} W_{ik} & \text{if } i = j \\ 0 & \text{if } i \neq j \text{ and } j \notin \mathcal{N}_i \end{cases} \tag{2}$$

where $d_i = |\mathcal{N}_i|$.

In order to solve the shard load-balancing problem using a consensus algorithm, shards must be embedded in some network topology defining the neighborhood of each shard. For simplicity, we assume shards are connected through a ring network, thus a regular graph, where each shard has two neighbors. The weight matrix is Metropolis-Hasting matrix, wherein a ring network, $w_{ij} = 0.33$ if $a_{ij} = 1$, $w_{ij} = 0$ otherwise. Shard processing is synchronous, divided into fix length epochs. The initial value $x_i(0)$ in the consensus model is the sum of the processing times of shard i, its workload. Each iteration of Eq. (1) computes the local workload average of shard i and its neighbors. This procedure converges locally to the global average of the shards workload.

The diffusion algorithm computes local load averages and the amount of load to transfer between neighbor shards once the distributed average consensus algorithm has converged to a workload balance among all shards. We use the same distributed average consensus as in [6], it differs slightly from Eq. (1):

$$Load_i(t+1) = Load_i(t) - \sum_{j \in \mathcal{N}_i} w_{ij}(Load_i(t) - Load_j(t)) \tag{3}$$

where $Load_i(t)$ is the load of shard i at iteration t. The full description of the diffusion algorithm appears in the pseudo-code of Algorithm 1. This algorithm is executed by each shard.

Algorithm 1. Diffusion algorithm for sharded blockchain load-balancing

---

Diffusion_algorithm(i, $\mathcal{N}_i$, n, W, workload of shard i)
  int $Load_i(0) = $ workload of shard i
  int $t = 0$
  float $\Delta_i[n] = 0$
  while no convergence
    for $(j = 0; j < n, j++)$
      if $j \in \mathcal{N}_i$
        $\Delta_i[j](t+1) = \Delta_i[j](t) + w_{ij}(Load_i(t) - Load_j(t))$
    $Load_i(t+1) = Load_i(t) - \sum_{j \in \mathcal{N}_i} w_{ij}(Load_i(t) - Load_j(t))$
    $t = t + 1$

$Load_i(0)$ is the load of shard i at the end of an epoch (equivalent of $x_i(0)$ in the consensus model). Δ_i is a vector of n entries, called *transfer vector* in [28]. At each iteration t of the "while" loop, $\Delta_i[j](t+1)$ stores the load that shard i must transfer to its neighbor j such that the load of shard i at iteration $t+1$ is the local average of its load and its neighbor loads at iteration t, i.e.
$Load_i(t+1) = \frac{Load_i(t)+\sum_{j\in\mathcal{N}_i} Load_j(t)}{|\mathcal{N}_i|+1}$, note $\Delta_i[j] = 0$ if $j \notin \mathcal{N}_i$.

Computationally, instruction $Load_i(t+1) = Load_i(t)-\sum_{j\in\mathcal{N}_i} w_{ij}(Load_i(t)-Load_j(t))$ is not necessary as the load of a shard at iteration $t+1$ can be computed using the transfer vectors and the loads of the previous iteration. However this instruction allows us to claim without proof that Algorithm 1 converges asymptotically to $\frac{\sum_{i=1}^{n} Load_i(0)}{n}$ based on numerous convergence proofs in the literature for iterative procedures such as Eqs. (1) and (3).

Algorithm 1 converges to a state where the local averages are all approximately the same, the same as the global average, and where the loads are balanced among all shards. This is however not yet the case after the first iteration of the algorithm. At iteration 1 of the "while" loop, the load of shard i, $Load_i(1)$, is the average load of shard i and its neighbors at iteration 0, i.e. $Load_i(1) = \frac{Load_i(0)+\sum_{j\in\mathcal{N}_i} Load_j(0)}{|\mathcal{N}_i|+1}$. However the load of shard i is not the same as the average load of shard i and its neighbors at iteration 1: $Load_i(1) \neq \frac{(\sum_{j\in\mathcal{N}_i} load_j(1))+load_i(1)}{|\mathcal{N}_i|+1}$, the loads are not balanced at iteration 1. Average loads is not equivalent to balanced loads initially. It is only once the average loads are all the same across all shards, where they are the same as the global average, that we have the workloads to be balanced.

For all the iterations of Algorithm 1, the transfer vector $\Delta_i[j]$ always contents the load to transfer from shard i to shard j to get local average loads among neighbor shards. However, once the local averages are the same as the global average, where the global average is the balanced load, the values in the transfer vectors take another meaning, they represent the amount of load to transfer among neighbor shards to get a balanced workload, they are the solution to the load balancing problem.

At the conclusion of Algorithm 1, if $\Delta_i[j]$ is positive, shard i must send to shard j some load. If negative, shard i must received some load from shard j. The behavior of Algorithm 1 is illustrated in Table 1 using an extreme case of initial load imbalance. It consists of an instance of 5 shards 0, 1, 2, 3, 4, with initial loads 0, 26, 0, 0, 0.

In Table 1, given an "Iters" value i, the "Shards" columns list the workload of each shard while the "Transfer vectors" columns provide the values of the two relevant entries of each transfer vector, for example $\Delta_0[4]$ $\Delta_0[1]$ is the transfer vector of shard 0. At iteration 0 the algorithm computes the transfer vectors for the loads obtained from the previous epoch. For example, shard 0 receives no load from shard 4 but receives 8.58 load from shard 1 as shown in $\Delta_0[4]$ $\Delta_0[1]$. We can see also that at iteration 0 shard 1 send 8.58 load to shards 0 and 3. As a consequence of the transfers computed at iteration 0, the loads of shards 0 and 2 is 8.58 at iteration 1. We can also see that each workload at iteration 1 is

Table 1. Illustration of Algorithm 1

Iters	Shards					Transfer vectors									
	0	1	2	3	4	$\Delta_0[4]$	$\Delta_0[1]$	$\Delta_1[0]$	$\Delta_1[2]$	$\Delta_2[1]$	$\Delta_2[3]$	$\Delta_3[2]$	$\Delta_3[4]$	$\Delta_4[3]$	$\Delta_4[0]$
0	0	26	0	0	0	0.00	−8.58	8.58	8.58	−8.58	0	0	0	0	0
1	8.58	8.84	8.58	0	0	2.83	−8.66	8.66	8.66	−8.66	2.83	−2.83	0	0	−2.83
2	5.83	8.66	5.83	2.83	2.83	3.82	−9.60	9.60	9.60	−9.60	3.82	−3.82	0	0	−3.82
3	5.77	6.79	5.77	3.82	3.82	4.46	−9.93	9.93	9.93	−9.93	4.46	−4.46	0	0	−4.46
4	5.46	6.12	5.46	4.46	4.46	4.79	−10.26	10.26	10.26	−10.26	4.98	−4.98	0	0	−4.98
5	5.35	5.69	5.35	4.69	5.35										
⋮															
8						5.16	−10.37	10.37	10.37	−10.37	5.16	−5.16	0	0	−5.16
9	5.21	5.24	5.21	5.16	5.16										
⋮															
14						5.19	−10.39	10.39	10.39	−10.39	5.19	−5.19	0	0	−5.19
15	5.20	5.20	5.20	5.19	5.19										

the average of the workloads at iteration 0 (modulo some rounding errors) and that the average loads at iteration 1 differ among shards, thus the load among the shards is not balanced.

The diffusion process starts with the transfers computed at iteration 1 where shard 0 and 2 sends respectively 2.83 load to shard 4 and shard 3. This is diffusion of the load originally in shard 1 to shards which are not neighbors of shard 1. A similar diffusion process takes place with the transfer vectors. Eventually the average consensus algorithm converges to balanced loads, such as in iteration 15. At this point, as in iteration 14, transfer vectors store the value of the load that must be transferred from each shard to its neighbors such that each shard starts the next epoch with a balanced workload.

Algorithm 1 computes the load that must be transferred, but it does not actually migrate the accounts that actualize these workload transfers. The next step consists to migrate accounts for which the sum of the processing times is equivalent to the transfer values computed by Algorithm 1. During this step, each shard i must find $|\mathcal{N}_i|$ subsets of accounts such that the sum of the processing times of the accounts in a subset is equal to a positive $\Delta_i[j]$, $j \in \mathcal{N}_i$. Computing these subsets is equivalent to solve the subset sum problem, a problem that is known to be NP-Hard. We use a heuristic to compute these subsets. The accounts in a shard i are sorted in decreasing order of their processing time. Then the list is traversed from the largest processing time to the smallest one, when the processing time of an account is smaller than a current positive $\Delta_i[j]$ value, the corresponding account is selected to be migrated to shard j.

It may not be possible for all shards to complete the accounts migration in one round. The load of some shards could be smaller than the positive values it needs to transfer to its neighbors. In an extreme case, the workload might be concentrated in one shard i (such as shard 1 in Table 1), thus the migration of accounts in the first round is only possible from shard i to its neighbors. In such extreme load imbalance, the migration of accounts may only be completed after several rounds, where in each round accounts migration only partially fulfill the vector transfers. However, according to [28], it can be shown that the theoretical number of rounds is bounded by the diameter of the network, which is consistent with a lower-bound result in [7], in practice we have observed that the number of rounds is much smaller.

3.2 Centralized Algorithms

In this section we introduce two algorithms for the independent task scheduling problem on multiprocessors. The canonical formulations of this problem and load-balancing are very similar. Algorithms for load-balancing are used for solving the task scheduling and vice-versa. Assuming accounts are independent, the load-balancing problem for sharded blockchain easily conforms to the independent task scheduling problem.

The algorithms described in this section are well-known approximation algorithms for the static independent task scheduling problem: the Longest Processing Time (LPT) and the MULTIFIT algorithm. The two algorithms are proposed as centralized solutions. They have been adapted to solve the dynamic load-balancing problem for sharded blockchains.

Longest Processing Time. Longest Processing Time solves load-balancing for shards by first sorting the accounts in decreasing order of their respective processing times. In that order, the algorithm sequentially assigns accounts to the shard with the smallest load so far.

The easiest way to implement the sorting phase of this algorithm in a distributed computing environment is to send the processing time of each account to a single network node. This node sorts the processing times and migrates accounts to shard if their assignment differs from the one they had in the previous epoch. According to [9], LPT has an approximation factor of $(\frac{4}{3} - \frac{1}{3m})$ (about 1.58) of the optimal solution. In [5], the authors report a new approximation factor of $(\frac{4}{3} - \frac{1}{3(m-1)})$. The asymptotic running time of LPT is $O(n \log n + n \log m)$ (n = number of shards, m = number of accounts).

MULTIFIT. MULTIFIT, first introduced in [11], is a second approximation algorithm for solving the independent task scheduling problem. MULTIFIT repetitively solves a bin packing problem. A set of n bins (shards) are assigned with a same capacity $C = \frac{1}{2}(A = \frac{\sum_{i=1}^{m} c_i}{n} + B = 2\frac{\sum_{i=1}^{m} c_i}{n})$ (m = number of tasks, c_i = processing time of task i). The tasks are sorted in decreasing order of their respective processing times. In that order, the tasks are assigned into a first bin until a c_i overload the capacity C of the current bin. Then task t_i is assigned to the next bin, this continue until all the tasks are assigned in the n bins, or until all the n bins are full. If all the tasks have been assigned then a new bin packing round starts with a lower capacity $C' = \frac{1}{2}(A + (B = C))$ for each bin. If not all the tasks have been assigned in the previous round, then the new round starts with an increased capacity $C' = \frac{1}{2}((A = C) + B)$. This bin packing problem is solved for k rounds. The value of B in the last round is the length of the longest schedule. The asymptotic running time of MULTIFIT is $O(n \log n + kn \log m)$, and the approximation factor is $1.22 + (\frac{1}{2})^k$. Compared to LPT, MULTIFIT has a slightly worst asymptotic running time but it has a tighter approximation factor.

The repeat rounds of solving the bin packing problem can only be executed by a single node. Thus the implementation of MULTIFIT for the shard load-balancing problem follows a similar pattern as for the LPT algorithm. After the last round, bins are mapped to shards, the content (accounts) of each bin is migrated to the respective mapped shard, if needed.

4   Experimentation

This section provides preliminary results from tests which simulate the sharding of three account-based blockchains: Ethereum, Zilliqa and Binance Smart Chain. Data for the simulations are obtained from crawling the following transactions tracker websites: Ethereum (https://etherscan.io/), Zilliqa (https://viewblock.io/zilliqa) and Binance Smart Chain (https://bscscan.com). Ethereum 2.0, a sharded version of Ethereum, is only known at the moment of writing this paper as protocol, thus here we use Ethereum transactions. The transactions tracker website for Zilliqa does not list with which shard transactions and accounts are associated. Binance Smart chain is a non-sharded account-based blockchain.

Table 2. Number of accounts and transactions per blockchains

	Ethereum	Binance	Zilliqa
Number of transactions	44511	49989	49980
Number of accounts	21548	5507	12487
Number of shards	10	10	10
Accounts per shard (approximately)	2155	551	1249

The data collected per transaction are the following ones: transaction hash id, block hash id where the transaction is indexed, the account source, the account destination, the time the transaction processing started, the amount of asset transferred in the transaction and the transaction fees. We have tracked 44511 Ethereum transactions, 49989 Binance transactions and 49980 Zilliqa transactions. Table 2 lists the number of source accounts for the transactions that have been crawled.

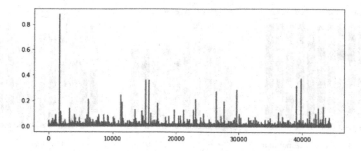

Fig. 1. Transaction fees for the 44522 Ethereum transactions

4.1 Sharding Simulations

The simulations only implement the most basic components of account-based shard designs such as the assignment of accounts and transactions to shards. The processing time of Ethereum transactions is reflected into the transaction fees expressed in "gwei", one gwei = 0.000000001 ETH [21], the native Ethereum currency. The fees that users must pay for processing a transaction are largely based on the number of CPU cycles spent to verify the transaction and some other marginal considerations such as the size of the transaction. The transaction fees expressed in ETH, as reported on the tracking website, is used here as a measure of transaction processing times. Processing times for Zilliqa and Binance transactions are also based on transaction fees in a similar way.

Prior to run our load-balancing algorithms, we have excluded some outlier transactions from the crawled data. An outlier transaction is one for which the transaction fees are disproportionally high compared to the vast majority of the transactions. Figure 1 illustrates this filtering process for Ethereum transactions where those transactions that cost more than 0.2 ETH are excluded, 11 transactions have been filtered out. Outlier transactions obscure at the present stage the comparative analysis of the performance trends for the three load-balancing algorithms. The number of transactions listed in Table 2 is after removing outliers.

We simulate 10 shards for each of the three blockchain platforms. In account-based blockchains, shards are constituted by assigning each account to a shard. The assignment of accounts to shards can be done by users. Alternatively, accounts can be assigned by the sharded platform which statically assigns accounts to shards. Our simulation follows the last approach. The accounts from the crawled transactions are assigned randomly and as evenly as possible to shards. This initial distribution of accounts per shards is reported in the first column of charts in Fig. 2, in which the first, second and third rows stand respectively for results from Ethereum, Binance and Zilliqa. This figure shows that the number of accounts per shard is distributed uniformly across all shards and blockchain platforms. The second column of charts in Fig. 2 reports the number of transactions per shard given the initial random assignment of accounts to

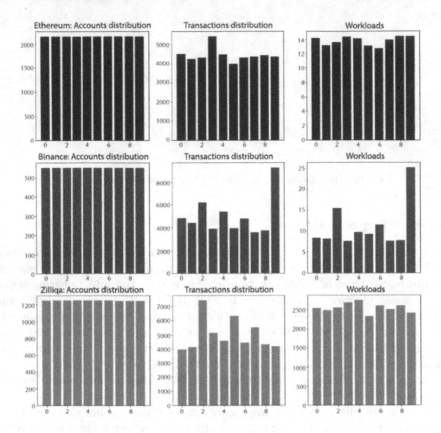

Fig. 2. Shards workload and transaction distributions

shards. The number of transactions per shard varies substantially for Binance and Zilliqa.

The simulation runs for 2 epochs. For the first epoch, the transactions fees of transactions originating from a same account are summed up together, this sum represents the processing time of the corresponding account. The sum of the processing time of the accounts assigned to a same shard is computed, which is the workload of the shard. The third column in Fig. 2 reports the initial workload of each shard. For Binance, the imbalance in the number of transactions for the last shard translates in an imbalance in the workload for this same shard.

4.2   Numerical Results

The three algorithms described in Sect. 3 are executed at the end of epoch 1 using as input the processing time of each account as calculated for epoch 1. According to the outputs of these algorithms, accounts are re-assigned to shards. Then the workload of each shard is computed in the same way as for epoch 1 using the same set of transactions (after re-assigning accounts). Figures 3, 4 and 5 report

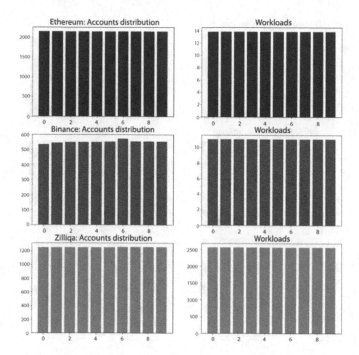

Fig. 3. Shards workload-balance using LPT algorithm

the computed workloads of epoch 2 respectively for the LPT, MULTIFIT and the diffusion algorithm. The first column of charts in each of these three figures report the number of accounts per shard while the second column of charts report the workload per shard.

The LPT algorithm in Fig. 3 shows an excellent workload-balance per shard as well as quite even distributions of the accounts per shard for the three sharded blockchain simulations. For the MULTIFIT algorithm, Fig. 4, we can see that the workload of shard 10 for Ethereum and Zilliqa is off a little bit. However, the striking aspect of the MULTIFIT solutions is the irregular distribution of accounts per shard. MULTIFIT sorts accounts in decreasing order of their transaction fees. Thus the first accounts fill the bins (shards) with relatively few accounts while at the low end of the sorted accounts, many accounts are needed to fill a single bin. As a result, the number of accounts in the last shards is substantially larger than those in beginning shards.

Last, Fig. 5, reports the performance of the diffusion algorithm. The load-balance for the diffusion algorithm appears as good as LPT. This indicates that workloads are as good as the approximation factor of the LPT algorithm. Although not shown in the charts, the second phase of the diffusion algorithm was completed in one round for all the tests.

The number of tests reported in this section is too small to draw general conclusions about these algorithms. Further tests are required over larger sets

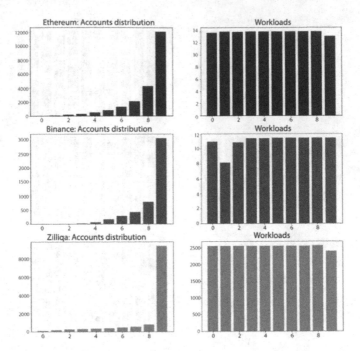

Fig. 4. Shards workload-balance using MULTIFIT algorithm

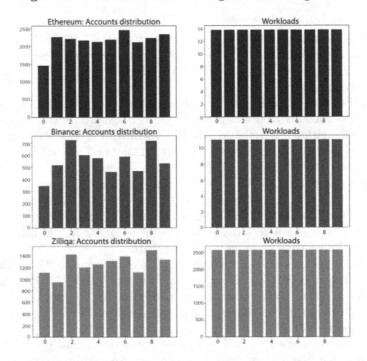

Fig. 5. Shards workload-balance using the diffusion algorithm

of crawled transactions, over several epochs, each epoch using a different set of crawled transactions. Computing times of each load-balancing algorithm will have to be recorded and compared. Intuitively, we expect the diffusion algorithm to have better computing times as it is a parallel algorithm. Outlier transactions would have to be included in the test sets to evaluate the robustness of the algorithms.

However, based on Ethereum online statistical data available at https://bitinfocharts.com/ethereum, it is not expected that significant divergences will exist between our preliminary results and results of more extensive simulations. Statistics about Ethereum data extracted from Table 2 and Fig. 2 conform with the daily reported statics on the above website. The number of initiated transactions per account is about 2 in Table 2, similar to the daily average number of initiated transactions per account. Transaction fees in Fig. 2, which are on average about 0.003 ETH per transaction, are quite similar to the daily average transaction fees reported for Ethereum. Simulations conducted with a large sample of Ethereum transactions are likely to have similar workload distributions as the distributions in our current tests. Unfortunately, similar online statistics are not available for Binance and Zilliqa.

4.3 Discussion

Load-balancing adds vulnerabilities and computing/communication overheads to the operation of sharded blockchains. The diffusion algorithm mitigates some of these issues, it is fast as it runs in parallel, it is not susceptible to single point of failure issues such as DoS attacks or other attacks aiming at controlling the load balancing protocol executed by a single node and migration of accounts is also performed in parallel and is local which likely cost less in terms of communication. Centralized algorithms such LPT and MULTIFIT rank low under these criteria, account processing times must be forwarded to a single server, account scheduling to shards is computed sequentially, and the migration of accounts is again under the control of a single blockchain node. However literature reports possible slow convergence speed for consensus algorithms and possible poor load-balancing solutions for the diffusion algorithm.

The convergence speed of diffusion and consensus algorithms has been thoroughly analyzed in several publications. Convergence depends on several parameters, two of them, the diameter of the network and the initial load imbalance can be briefly discussed here. The diameter of a network is the maximum distance between a pair of nodes, for example the diameter of a ring network is $\frac{n}{2}$, the diameter of a line network is $n-1$. In Table 1 for a ring network of 5 nodes, the system converges to a state where the difference between the largest and smallest load is smaller/equal 1 after only 5 iterations. The same network topology with 10 nodes requires 17 iterations to achieve the same degree of load balance, 51 iterations with 20 nodes. Among regular graphs, ring networks have a relatively large diameter, it is well documented that consensus algorithms embedded in such network have slow convergence. In regular graphs, the diameter depends on the degree of the nodes, thus sharded blockchains with large number of shards will have to be embedded in networks where shards are more highly interconnected.

The distribution of the initial loads has a lesser impact on the convergence behavior of the diffusion algorithm. Among the three shard blockchains that have been tested, Zilliqa has the largest initial load imbalance in absolute values (because of the differences in scale, 1 ZIL ≈ ETH0.00003182 according to https://www.coingecko.com/en/coins/zilliqa/eth). Our tests in Figs. 3, 4 and 5 have been run using the absolute values in Fig. 2 where the difference between the largest and the smallest load in Zilliqa is closed to 300 units, compared to 9 for Binance and 3 for Ethereum. Zelliqa required 36 iterations to reach a load balancing solution with a difference smaller than 1 among all the shards, while Binance required 12 iterations and Ethereum only 4 iterations. If we force the load imbalance to be smaller/equal to 0.03, which is the average cost of an Ethereum transaction, Ethereum needs 22 iterations and Binance 36. Considering much shorter epochs and 20 shards, literature [29] reports initial load imbalances up to 65% for sharded simulations of Ethereum. Using a similar initial Ethereum load imbalance for 10 shards, an unbalance <1 is achieved after 10 iterations and 52 iterations for an unbalance smaller/equal to 0.03.

There are two main reasons that may cause our implementation of the diffusion algorithm to return poor load-balancing solutions for sharded blockchains. One issue is Algorithm 1 can divide the load into infinitesimal values while account processing times are coarser, indivisible values. In some cases there may not exist a combination of the account processing times corresponding to the transfer vector values computed by Algorithm 1. Thus the heuristic that computes subsets of accounts to migrate across shards may produce effective loads that differ with the values computed by the diffusion load-balancing algorithm. The second issue is related with the locality of the migration process in the diffusion algorithm. Each shard can only migrate its accounts to its neighbors. Centralized algorithms such as LPT and MULTIFIT can schedule any account in the system to migrate to any shard. Thus centralized algorithms have a theoretical advantage. This is why in the paper we compare the load-balancing solutions based on the diffusion algorithm with solutions obtained using approximation algorithms such LPT and MULTIFIT. These comparisons ensure that the effective loads resulting from the diffusion algorithm are within the approximation factors of the best centralized algorithms.

5 Conclusion

This paper proposes a consensus-based dynamic load-balancing algorithm for account-based sharded blockchains. The proposed algorithm is fully distributed, load-balance and accounts migration is computed and executed locally and periodically by each shard. Simulation tests of sharded blockchains have been run for three well-known blockchain platforms. Load-balancing solutions of the distributed algorithm were as good or better than two other centralized approximation algorithms that have approximation factors for solutions no further apart than 22% above the optimal solution.

Load-balancing in isolation may not by itself translate into overall improvement sharded blockchain performance. User account migration in some account-based permissionless sharded blockchain platforms incur fees which users may refuse to pay. Accounts migration may also increase the number of cross-shard transactions, increasing latency from transactions initiation to confirmation. In fact, load-balancing interacts or is bound by several design parameters of sharded blockchain. One possible future work direction will be to include new constraints in the current optimization model to represent some of these design parameters interacting with load-balancing in sharded blockchains. Such constraints will model for example increase latencies for cross-shard transactions or network communication costs for migrating accounts across shards.

Acknowledgement. This work was funded by Gia Lam Urban Development and Investment Company Limited, Vingroup and supported by Vingroup Innovation Foundation (VINIF) under project code VINIF.2019.DA07.

References

1. Al-Bassam, M., Sonnino, A., Bano, S., Hrycyszyn, D., Danezis, G.: ChainSpace: a sharded smart contracts platform. CoRR abs/1708.03778 (2017). http://arxiv.org/abs/1708.03778
2. Amiri, M.J., Agrawal, D., Abbadi, A.E.: Sharper: sharding permissioned blockchains over network clusters. CoRR abs/1910.00765 (2019). http://arxiv.org/abs/1910.00765
3. Buterin, V.: On sharding blockchains FAQs (2020). https://eth.wiki/sharding/Sharding-FAQs
4. Chen, H., Wang, Y.: SSChain: a full sharding protocol for public blockchain without data migration overhead. Pervasive Mob. Comput. **59**, 101055 (2019)
5. Croce, D.F., Scatamacchia, R.: The longest processing time rule for identical parallel machines revisited. J. Sched. **23**(2), 163–176 (2020)
6. Cybenko, G.: Dynamic load balancing for distributed memory multiprocessors. J. Parallel Distrib. Comput. **7**(2), 279–301 (1989)
7. Dai, H.K., Toulouse, M.: Lower-bound study for function computation in distributed networks via vertex-eccentricity. SN Comput. Sci. **1**(1), 10:1–10:14 (2020). https://doi.org/10.1007/s42979-019-0002-3
8. Doyle, J.: Load balancing and rate limiting based algorithms for improving cloud computing performance. Trinity College Dublin (2012). https://books.google.com.vn/books?id=_ifroAEACAAJ
9. Graham, R.L.: Bounds for certain multiprocessing anomalies. Bell Syst. Tech. J. **45**(9), 1563–1581 (1966). https://doi.org/10.1002/j.1538-7305.1966.tb01709.x
10. Jadbabaie, A., Lin, J., Morse, A.: Coordination of groups of mobile autonomous agents using nearest neighbor rules. IEEE Trans. Autom. Control **48**(6), 988–1001 (2003). https://doi.org/10.1109/TAC.2003.812781
11. Coffman, E.G., Jr., Garey, M.R., Johnson, D.S.: An application of bin-packing to multiprocessor scheduling. SIAM J. Comput. **7**(1), 1–17 (1978)
12. Kokoris-Kogias, E., Jovanovic, P., Gasser, L., Gailly, N., Syta, E., Ford, B.: OmniLedger: a secure, scale-out, decentralized ledger via sharding. In: 2018 IEEE Symposium on Security and Privacy (SP), pp. 583–598 (2018). https://doi.org/10.1109/SP.2018.000-5

13. Król, M., Ascigil, O., Rene, S., Sonnino, A., Al-Bassam, M., Rivière, E.: Shard scheduler: object placement and migration in sharded account-based blockchains. CoRR abs/2107.07297 (2021). https://arxiv.org/abs/2107.07297
14. Liu, Y., et al.: Building blocks of sharding blockchain systems: concepts, approaches, and open problems. CoRR abs/2102.13364 (2021). https://arxiv.org/abs/2102.13364
15. Luu, L., Narayanan, V., Zheng, C., Baweja, K., Gilbert, S., Saxena, P.: A secure sharding protocol for open blockchains. In: Proceedings of the 2016 ACM SIGSAC Conference on Computer and Communications Security, CCS 2016, pp. 17–30. Association for Computing Machinery, New York (2016). https://doi.org/10.1145/2976749.2978389
16. Manuskin, A., Mirkin, M., Eyal, I.: Ostraka: secure blockchain scaling by node sharding. CoRR abs/1907.03331 (2019). http://arxiv.org/abs/1907.03331
17. Nguyen, L.N., Nguyen, T.D.T., Dinh, T.N., Thai, M.T.: OptChain: optimal transactions placement for scalable blockchain sharding. In: 2019 IEEE 39th International Conference on Distributed Computing Systems (ICDCS), July 2019. https://doi.org/10.1109/icdcs.2019.00059. http://dx.doi.org/10.1109/ICDCS.2019.00059
18. Okanami, N., Nakamura, R., Nishide, T., et al.: Load balancing for sharded blockchains. In: Bernhard, M. (ed.) FC 2020. LNCS, vol. 12063, pp. 512–524. Springer, Cham (2020). https://doi.org/10.1007/978-3-030-54455-3_36
19. Qing, C., Guo, B., Shen, Y., Li, T., Zhang, Y., Zhang, Z.: A secure and effective construction scheme for blockchain networks. Secur. Commun. Netw. **2020**, 8881881:1–8881881:20 (2020). https://doi.org/10.1155/2020/8881881
20. Stegos, A.G.: STEGOS: a platform for privacy applications (2019). https://stegos.com/docs/stegos-whitepaper.pdf
21. Ethereum Team: Gas and fees (2021), https://ethereum.org/en/developers/docs/gas/
22. Harmony Team: Harmony: Technical whitepaper, version 2.0 (2020). https://harmony.one/whitepaper.pdf
23. The Zilliqa Team: The zilliqa technical whitepaper (2017). https://docs.zilliqa.com/whitepaper.pdf
24. Tsitsiklis, J., Bertsekas, D., Athans, M.: Distributed asynchronous deterministic and stochastic gradient optimization algorithms. IEEE Trans. Autom. Control **31**(9), 803–812 (1986). https://doi.org/10.1109/TAC.1986.1104412
25. Vicsek, T., Czirók, A., Ben-Jacob, E., Cohen, I., Shochet, O.: Novel type of phase transition in a system of self-driven particles. Phys. Rev. Lett. **75**, 1226–1229 (1995)
26. Wang, G., Shi, Z.J., Nixon, M., Han, S.: SoK: sharding on blockchain. In: Proceedings of the 1st ACM Conference on Advances in Financial Technologies, AFT 2019, pp. 41–61. Association for Computing Machinery, New York (2019). https://doi.org/10.1145/3318041.3355457
27. Wang, J., Wang, H.: Monoxide: Scale out blockchains with asynchronous consensus zones. In: 16th USENIX Symposium on Networked Systems Design and Implementation (NSDI 2019), pp. 95–112. USENIX Association, Boston, February 2019. https://www.usenix.org/conference/nsdi19/presentation/wang-jiaping
28. Watts, J., Taylor, S.: A practical approach to dynamic load balancing. IEEE Trans. Parallel Distrib. Syst. **9**(3), 235–248 (1998). https://doi.org/10.1109/71.674316
29. Woo, S., Song, J., Kim, S., Kim, Y., Park, S.: GARET: improving throughput using gas consumption-aware relocation in ethereum sharding environments. Clust. Comput. **23**(3), 2235–2247 (2020)
30. Xiao, L., Boyd, S.P., Kim, S.J.: Distributed average consensus with least-mean-square deviation. J. Parallel Distrib. Comput. **67**(1), 33–46 (2007)

31. Yu, G., Wang, X., Yu, K., Ni, W., Zhang, J.A., Liu, R.P.: Survey: sharding in blockchains. IEEE Access **8**, 14155–14181 (2020). https://doi.org/10.1109/ACCESS.2020.2965147
32. Zamani, M., Movahedi, M., Raykova, M.: RapidChain: scaling blockchain via full sharding. In: Proceedings of the 2018 ACM SIGSAC Conference on Computer and Communications Security, CCS 2018, pp. 931–948. Association for Computing Machinery, New York (2018). https://doi.org/10.1145/3243734.3243853
33. Zochowski, M.: The logos network (2018). https://medium.com/logos-network

Neighboring Information Exploitation for Anomaly Detection in Intelligent IoT

Thien-Binh Dang[1], Duc-Tai Le[3], Moonseong Kim[2]([✉]),
and Hyunseung Choo[3]([✉])

[1] Department of Electrical and Computer Engineering, Sungkyunkwan University,
Seoul, South Korea
dtbinh@skku.edu
[2] Department of IT Convergence Software, Seoul Theological University,
Bucheon, South Korea
moonseong@stu.ac.kr
[3] College of Computing and Informatics, Sungkyunkwan University,
Seoul, South Korea
{ldtai,choo}@skku.edu

Abstract. The identification of anomalies has become increasingly important for the security of sensory data gathering in the intelligent Internet of Things (iIoT). The current approaches might not be applied to the general cases of anomalies, i.e., both long-term and short-term anomalies, as well as not be suitable with real-time applications such as natural disaster monitoring and early warning systems. To address this challenge, this paper proposes a comprehensive approach, named DWT-PCA Anomaly Detection (DAD) to detect both long- and short-term anomalies. DAD bases on the combination of Discrete Wavelet Transform (DWT) and Principal Component Analysis (PCA) to improve the system performance. In particular, we first utilize the DWT to extract approximation coefficients and detail coefficients from the input data which are capable to highlight long-term and short-term anomalies, respectively. We then exploit the spatial-temporal correlation of neighboring sensors by applying PCA on these coefficients to obtain a high detection accuracy. Numerical experiments based on the real dataset of Intel Berkeley Research reveal that the proposed scheme obtains higher accuracy and a lower false-positive rate on detecting three typical anomalies: drift, noise, and outlier, comparing to existing schemes.

Keywords: Intelligent IoT · Wireless sensor network · Security · Sensory data · Anomaly detection

This research was supported by the MSIT, Korea, under the Grand Information Technology Research Center support program (IITP-2021-2015-0-00742), and the ICT Creative Consilience program (IITP-2021-2020-0-01821) supervised by the IITP and was supported by the NRF grant funded by the Korea government (MSIT) (No. NRF-2020R1A2C2008447).

T. K. Dang et al. (Eds.): FDSE 2021, LNCS 13076, pp. 260–271, 2021.
https://doi.org/10.1007/978-3-030-91387-8_17

1   Introduction

In the intelligent Internet of Things (iIoT), data collected from sensor networks are highly vulnerable to random faults and cyber-attacks, resulting the anomalies. In fact, the limitations on the size and cost make the capability of sensors scaled down, such as weak computational speed, small memory capacity, limited energy and restricted communication bandwidth [1,2]. In practice, most of abnormal events tamper the data of a victim sensor in three forms: drift, outlier, and noise as shown in Fig. 1, which can be described as:

- Drift anomaly: a series of measurements which drift away from the normal measurements.
- Outlier: an isolated measurement which is significantly deviates from the other normal measurements.
- Noise: a series of measurements which experience unexpectedly high variation.

These three form of anomalies categories can be classified into long-term anomaly (drift) and short-term anomaly (noise, outlier) based on the continuum of abnormal measurements. Short-term anomaly appears discretely in short periods of time; for instance, sensors with battery failure and other hardware or connection failures often produce sudden changes in sensed measurements (outlier anomaly) greater than expected, or sensors having low battery voltage often produce noisy measurements (noise anomaly).

Point-based anomaly detection schemes [3–7] which handle each measurement individually can deal well with short-term anomalies by applying statistical, classification or clustering methods. The authors in [3] proposed an isolation based distributed outlier detection framework by exploiting the information of neighboring sensors. Outlier score, which is used to make decision, is calculated based on the distances between every two measurements. In this work, the authors only focus on outlier anomaly. A similar k-nearest-neighbor-based (kNN-based) scheme for WSNs was presented in [4]. In this work, the authors overcome the lazy-learning issue of conventional kNN-based scheme which occasionally generates unaffordable communication overhead by using intuition hypergrid.

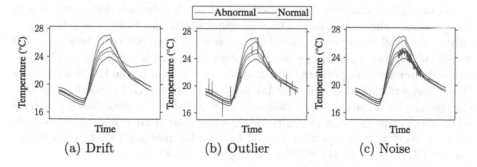

(a) Drift   (b) Outlier   (c) Noise

Fig. 1. Three typical anomalies in iIoT.

Authors in [5] presented two schemes: centered hyperspherical and hyperellip-soidal SVM to identify anomaly in two distinct methods, namely centralization and dispersion. While the former achieves excellent detection accuracy, it has significant data redundancy and a restricted potential for distributed deployment in WSNs. The latter is more suitable for distributed implementation and provides comparable detection accuracy while requiring less communication overhead than the former. A clustering-base scheme which is proposed by Sutharshan Rajasegarar et al. in [6]. The proposed technique reduces communication cost by first do the clustering on an individual sensor data and then combining these clusters before transmitting the shorten description to other nodes. Sutharshan Rajasegarar et al. also proposed another approach in [7] which can adapt to the changes of non-stationary environments while receiving streaming data. Despite the above approaches can deal well with the short-term anomaly, they encounter the difficulty in detecting the long-term anomaly which slowly appears in the data for a long period of time.

On the contrary to short-term anomaly, long-term anomaly exposes a relative change in sensory data. In long-term anomaly, an individual measurement can be normal but when a collection of measurements occur together, this collection is abnormal and usually displays a long-term abnormal pattern; for example, due to a calibration mistake in a sensor, a displacement value is added to the actual observed data, or a jammed sensor may send a constant stream of bytes into the network for a certain duration of time. [8,9]. Long-term anomaly is identical to collective anomaly which is defined in [10]. In order to detect long-term anomaly, segment-based detection schemes [11–13] (or window-based detection schemes in some works of literature) analyze the correlation among consecutive measurements. Segment-based detection approaches can overcome the disadvantages of point-based approaches by analyzing the set of successive data instead of a single measurement. In [11], Miao Xie et al. proposed a segment-based anomaly detection scheme with approximated sample covariance matrix (SAD). By using the data segments of nearest sensors with a strong assumption that these sensors disclose identical data patterns at the same period, SAD estimate the prediction variances utilized for anomaly detection. Furthermore, it lowers communication cost by providing summary data descriptions of sensors to the cluster head rather than raw data. In a similar manner, [12] presented an outlier detection scheme based on sliding windows. The experimental results showed that even though the proposed scheme obtained a high specificity and sensitivity, the scheme can only detects well point anomalies. Another work of Miao Xie et al. was proposed in [13], which focused on a specific long-term anomaly that arises in a collection of nearby sensors and persists over time. A combination of the Kullback-Leibler (KL) divergence method and kernel density estimation (KDE) was developed in this system for anomaly identification. The distributed segment-based recursive KDE checks the global probability density function (PDF), and the difference between every two time periods is used for decision making. Due to an approximation DL divergence, the communication cost is reduced significantly.

While some of the previous researches perform well with short-term anomaly, some work well with long-term anomaly. A potential research direction is the development of a new detection mechanism that can identify both of these anomaly forms with a high accuracy and a low false positive rate. As illustrated in Fig. 1, there are five solid lines representing sensory data of five neighboring sensors in which the red line is the data of the abnormal sensor and the blue line is the data of normal ones. Due to the continuum of short-term and long-term anomalies, discrete wavelet transform (DWT) is an applicable approach to handle both of them based on its capacity to recognize the overall trend as well as the transient in data. Furthermore, because of the linear relationships that exist across geographically proximate sensors, nearby data should show a fairly similar pattern throughout the same time period. Principal component analysis (PCA) is used to project the data into a normal subspace and an abnormal subspace in order to leverage the spatial correlations. While the dominating variance in the original data is contained in the normal subspace, the anomalous data is plainly visible in the abnormal space, which is beneficial for anomaly identification. To adopt the advantages of both DWT and PCA, we are motivated to combine DWT and PCA to develop a DWT-PCA Anomaly Detection (DAD) scheme which is able to detect both short-term and long-term anomalies efficiently. Finally, we evaluate the proposed scheme by comparing to SAD by a wide range of numerical experiments on the well-known IBRL dataset [16].

2 Propose Scheme

2.1 The Overall Process

DAD has two phase: offline training phase and online testing phase. In the offline training phase (Fig. 2), DWT first is applied on the training data of M sensors separately to obtain the approximation and detail coefficients. These coefficients are collected into $L + 1$ matrices $\mathbf{D}_1, \mathbf{D}_2, \cdots, \mathbf{D}_L$, and $\mathbf{A}_L$. While $\mathbf{A}_L$ can capture the long-term pattern of the data, abrupt changes in the data will show clearly in $\mathbf{D}_1, \mathbf{D}_2, \cdots, \mathbf{D}_L$. Thus, $\mathbf{D}_1, \mathbf{D}_2, \cdots, \mathbf{D}_L$, is suitable to detect short-term anomaly and $\mathbf{A}_L$ is appropriate with long-term anomaly detection. Next, the $L + 1$ detection thresholds for short- and long-term anomalies are estimated by applying PCA on these $L = 1$ coefficients matrices. In the online testing phase (Fig. 3), the same process is applied on testing data to obtain $\mathbf{D}_1, \mathbf{D}_2, \cdots, \mathbf{D}_L$, and $\mathbf{A}_L$. Then the square prediction errors (SPE) of these coefficients are calculated based on parameters obtained from the training phase. If the SPE of the data is greater than the corresponding detection thresholds, the data is marked as anomaly, otherwise it is considered as normal data. The detail calculation steps are shown in the Data Processing Model subsection.

2.2 Data Processing Model

A matrix $\mathbf{X} \in \mathbb{R}^{N \times M}$ is meant to hold the normalized training data, with each column $\mathbf{x}_i$ of $\mathbf{X}$ containing N measurements taken from one of M sensors (i :

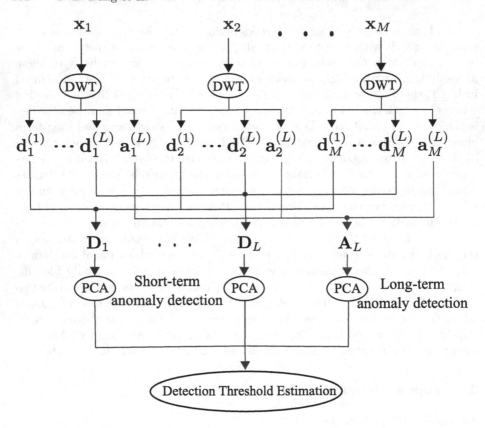

Fig. 2. Offline training phase.

$1..M$). The DWT is apply on each column $\mathbf{x}_i$ to obtain $\mathbf{d}_i^{(1)}, \mathbf{d}_i^{(2)}, \cdots, \mathbf{d}_i^{(L)}$, and $\mathbf{a}_i^{(L)}$. The DWT decomposition tree is presented in Fig. 4. The input data $\mathbf{x}_i$ of the decomposition process is a vector of N measurements representing the collected data of sensor i. $\mathbf{f}_{hp}$ and $\mathbf{f}_{lp}$ are high-pass and low-pass filters. DWT has a maximum predefined decomposition level L. The output coefficients at level ℓ ($\ell = 1..L$) are obtained as:

$$\mathbf{a}_i^{(\ell)} = (\mathbf{a}_i^{(\ell-1)} * \mathbf{f}_{lp}) \downarrow 2, \tag{1}$$

$$\mathbf{d}_i^{(\ell)} = (\mathbf{a}_i^{(\ell-1)} * \mathbf{f}_{hp}) \downarrow 2, \tag{2}$$

where $\mathbf{a}_i^{(\ell)}$ and $\mathbf{d}_i^{(\ell)}$ are the approximation coefficients and the detail coefficients at level ℓ of sensor i, respectively[1]. Vectors $\mathbf{a}_i^{(\ell)}, \mathbf{d}_i^{(\ell)} \in \mathbb{R}^{1 \times N/2^{\ell}}$. In DWT family, we choose Haar filter [17] where $\mathbf{f}_{lp} \triangleq \frac{1}{\sqrt{2}} [1 \; 1]$ and $\mathbf{f}_{hp} \triangleq \frac{1}{\sqrt{2}} [1 \; -1]$ because it is the fastest one and it is adequate to analyze sensory data.

---

[1] $\mathbf{a}_i^{(\ell)} = \mathbf{x}_i$ where $l = 0$.

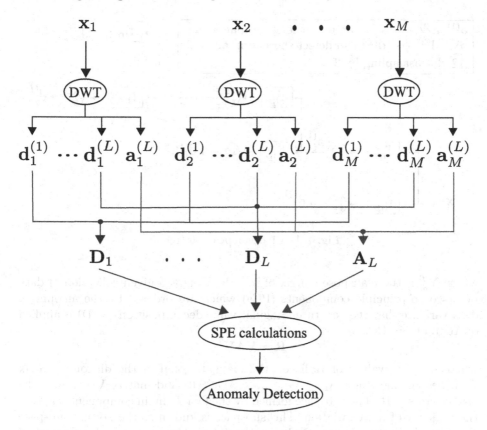

Fig. 3. Online testing phase.

The approximation and detail coefficients can be used to detect the anomalies in an individual sensor; i.e., while the approximation coefficients represent the trend of the data which is useful for detecting long-term anomalies, low-pass filter pulls out the transient in detail coefficients of data which clearly shows the abrupt changes of short-term anomalies. However, the accuracy can be improved by leveraging the spatially proximal sensors' linear correlations. Thus, we apply PCA on these coefficients for a better detection accuracy.

Let $\mathbf{Y} \in \{\mathbf{A}_L, \mathbf{D}_1, \cdots, \mathbf{D}_L\}$, The data matrix $\mathbf{Y}$ is normalized to zero-mean and scaled to unit variance. Let $\mathbf{Y}_s$ be the normalized data, $\mathbf{Y}_s$ can be expressed as:

$$\mathbf{Y}_s = \left(\mathbf{Y} - \bar{\mathbf{Y}}\right) \mathbf{D}^{-\frac{1}{2}}, \tag{3}$$

where $\bar{\mathbf{Y}} = \mathbf{1}(\frac{1}{N}\mathbf{1}^T\mathbf{Y})$ and $\mathbf{D} = \frac{1}{N-1}\left[\left(\mathbf{Y} - \bar{\mathbf{Y}}\right)^T \left(\mathbf{Y} - \bar{\mathbf{Y}}\right)\right] \circ \mathbf{I}_M$, with $\circ$ denoting the Hadamard multiplication and $\mathbf{I}_M$ is the identity matrix. The co-variance matrix of matrix $\mathbf{Y}_s$ is calculated as:

$$\mathbf{R} = \frac{1}{M}\mathbf{Y}_s^T\mathbf{Y}_s, \tag{4}$$

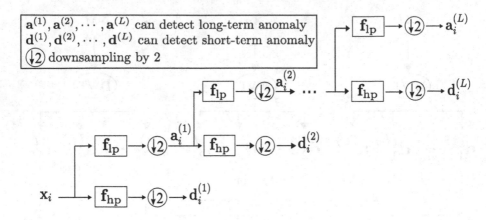

Fig. 4. DWT decomposition tree.

where $\mathbf{Y}_s^T$ is the transpose matrix of $\mathbf{Y}_s$. PCA maps a given collection of data points onto principle components (PCs) which are ordered by the amount of data variance that they capture. Singular value decomposition (SVD) is applied on $\mathbf{R}$ to obtain PCs as:

$$\mathbf{R} = \mathbf{V}\boldsymbol{\Lambda}\mathbf{V}^T, \tag{5}$$

where the eigenvalues of $\mathbf{R}$ lie on the main diagonal of the diagonal matrix $\boldsymbol{\Lambda}$ in descending order ($\lambda_1 \geq \lambda_2 \geq ... \geq \lambda_M \geq 0$) and matrix $\mathbf{V}$ contains the eigenvectors of $\mathbf{R}$. The subspace formed by the first K main components capture the majority of data variation. This subspace is known as the normal subspace $\mathcal{S}_{no}$. The abnormal subspace $\mathcal{S}_{ab}$ is constructed by the remaining $M - K$ principal components. The detection of anomalies is based on $\mathcal{S}_{ab}$ which can be obtained by the decomposition on $\mathbf{Y}_s$ as:

$$\mathbf{Y}_s = \hat{\mathbf{Y}} + \tilde{\mathbf{Y}}, \tag{6}$$

where $\hat{\mathbf{Y}} = \mathbf{Y}_s\hat{\mathbf{C}}$ and $\tilde{\mathbf{Y}} = \mathbf{Y}_s\tilde{\mathbf{C}}$. $\hat{\mathbf{C}}$ and $\tilde{\mathbf{C}}$ are the projection matrices of $\mathcal{S}_{no}$ and $\mathcal{S}_{ab}$. $\hat{\mathbf{C}}$ and $\tilde{\mathbf{C}}$ are respectively defined as:

$$\hat{\mathbf{C}} \triangleq \hat{\mathbf{P}}\hat{\mathbf{P}}^T, \tag{7a}$$

$$\tilde{\mathbf{C}} \triangleq \tilde{\mathbf{P}}\tilde{\mathbf{P}}^T. \tag{7b}$$

While $\hat{\mathbf{P}}$ is constructed by the first K eigenvectors corresponding to the first K largest eigenvalues ($\lambda_1, \lambda_2, \cdots, \lambda_K$), $\tilde{\mathbf{P}}$ is constituted by the rest $M - K$ eigenvectors of $\mathbf{R}$. $\mathbf{P} = [\hat{\mathbf{P}}, \tilde{\mathbf{P}}]$ is the loading matrix. Because anomalies often cause a large shift in $\tilde{\mathbf{Y}}$, Square Prediction Error (SPE) is a useful measure for spotting anomalous data. SPE statistics of $\tilde{\mathbf{Y}}$ and the detection threshold δ^2 [18] are calculated as:

$$\mathbf{SPE} = diag(\mathbf{Y}_s\tilde{\mathbf{C}}\tilde{\mathbf{C}}^T\mathbf{Y}_s^T), \tag{8}$$

$$\delta^2 = \theta_1\left[\frac{C_\alpha\sqrt{2\theta_2h_0^2}}{\theta_1} + 1 + \frac{\theta_2h_0(h_0 - 1)}{\theta_1^2}\right]^{\frac{1}{h_0}}, \tag{9}$$

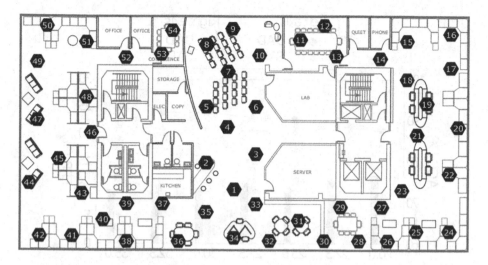

Fig. 5. Sensors allocation in IBRL [16].

$$h_0 = \frac{2\theta_1\theta_3}{\theta_2^2},\tag{10}$$

$$\theta_i = \sum_{j=K+1}^{L} \lambda_j^i, (i = 1...3),\tag{11}$$

where λ_j is the eigenvalue associated with the j-th eigenvector and C_α is the standard normal deviation corresponding to the confident level α of standard normal distribution. An measurement is considered as an anomalous data if its $SPE > \delta^2$.

3   Performance Evaluation

3.1   Experimental Dataset and Evaluation Method

To evaluate the performance of DAD, we use the well-known IBRL dataset [16]. The dataset is a collection of light, voltage, humidity, and temperature measurements which are collected from 54 Mica2Dot sensors in the duration from February 28th to April 25th, 2004. Without loss of generality, we select temperature measurements from eleven sensors whose IDs from 22 to 32 for evaluation. These sensors are close to each other and have a strong spatial correlation as shown in Fig. 5. For offline training phase, we use the data of Feb 29th. The data of March 1st is used for online testing phase.

Because the IBRL dataset contains only normal data, we use the anomaly model in [19] to artificially generate and inject drift, noise, and outlier anomalies into it for testing. The performance of our scheme (DAD) and the referred scheme (SAD) is then evaluated using accuracy (ACC) and false positive rate

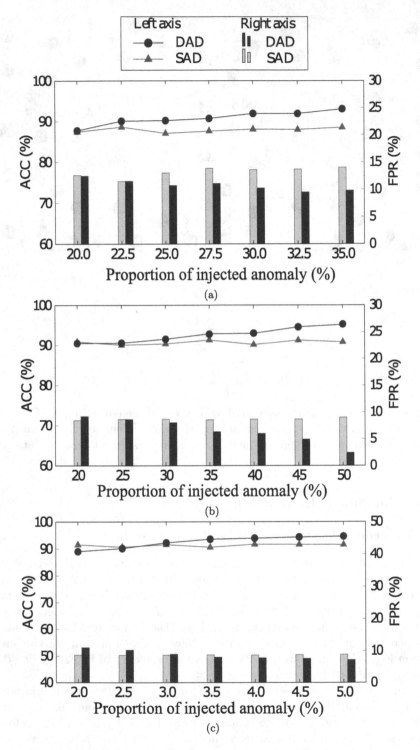

Fig. 6. (a), (b), and (c) are detection results of noise, drift, and outlier anomalies.

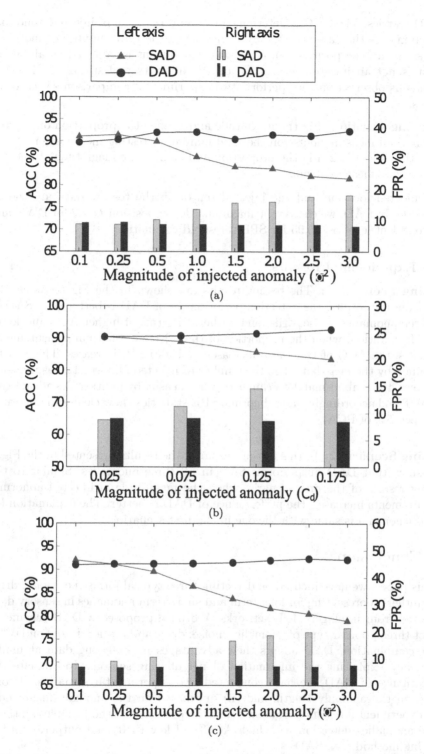

Fig. 7. (a), (b), and (c) are detection results of noise, drift, and outlier anomalies.

(FPR) metrics. The ACC is determined by the proportion of injected abnormal measurements that are successfully identified to the total number of measurements and a false positive caused if an measurement is detected as abnormal but it is not an injected one. Based on the quantity and size of the injected anomalous observations, we perform two experimental comparison scenarios as follows:

- Testing scenario 1: Fix the magnitude and change the proportion of injected abnormal measurements (on the total number of testing measurements).
- Testing scenario 2: Fix the proportion and change the magnitude of injected abnormal measurements.

The injected location and the injected sensor in the testing data are chosen randomly. In DAD, we choose the maximum decomposition $L = 2$ for DWT and the confident level $\alpha = 0.95$ for SPE threshold calculation.

3.2   Experimental Results

Testing Scenario 1: The testing results are shown in the Fig. 6. As can be seen from the results, the detection performance of DAD is better than SAD in all three anomalies: noise, drift, and outlier, in terms of higher ACC and lower FPR. In addition, when the proportion of the injected abnormal measurements increases, the ACC of DAD also increases and its FPR decreases. This can be explained by the fact that when the number of injected abnormal measurements increases, these abnormal data can be captured easily by multi-scales advantages of DWT and accordingly their abnormal SPE statistics show clearer in abnormal subspace $\mathcal{S}_{ab}$ of PCA.

Testing Scenario 2: In this testing scenario, the results presented in the Fig. 7 also show that DAD outperforms SAD in all three anomalies. Similar to the testing results of the first scenario, when the magnitude of injected abnormal measurements increases, the performance of DAD is better. The explanation for this phenomena is same with the one in the first scenario.

4   Conclusion

In this paper, we have focused on detecting three typical forms (i.e., noise, drift, and outlier) representing for long-term and short-term anomalies in sensory data collected from intelligent IoT networks. We have proposed a DAD scheme to detect three popular type of anomalies: noise, drift, and outlier. Based on DWT-PCA combination, DAD adopts their advantages in analyzing data at multi-scales and exploiting the information of neighboring sensors. To evaluate the performance of DAD, we have conducted two experimental scenarios: (i) one based on the anomaly length, and (ii) one based on the anomaly magnitude. The experimental results have shown that the proposed method enables all those three anomalies detection with high ACC and low FPR, and outperforms the existing method, i.e., SAD.

References

1. Yu, J.Y., Lee, E., Oh, S.R., Seo, Y.D., Kim, Y.G.: A survey on security requirements for WSNs: focusing on the characteristics related to security. IEEE Access **8**, 45304–45324 (2020)
2. Sen, J.: A survey on wireless sensor network security. arXiv preprint arXiv:1011.1529 (2010)
3. Wang, Z.M., Song, G.H., Gao, C.: An isolation-based distributed outlier detection framework using nearest neighbor ensembles for wireless sensor networks. IEEE Access **7**, 96319–96333 (2019)
4. Xie, M., Hu, J., Han, S., Chen, H.H.: Scalable hypergrid k-NN-based online anomaly detection in wireless sensor networks. IEEE Trans. Parallel Distrib. Syst. **24**(8), 1661–1670 (2012)
5. Rajasegarar, S., Leckie, C., Bezdek, J.C., Palaniswami, M.: Centered hyperspherical and hyperellipsoidal one-class support vector machines for anomaly detection in sensor networks. IEEE Trans. Inf. Forensics Secur. **5**(3), 518–533 (2010)
6. Rajasegarar, S., Leckie, C., Palaniswami, M., Bezdek, J.C.: Distributed anomaly detection in wireless sensor networks. In: 2006 10th IEEE Singapore international conference on communication systems, pp. 1–5. IEEE (2006)
7. Moshtaghi, M., Leckie, C., Karunasekera, S., Bezdek, J.C., Rajasegarar, S., Palaniswami, M.: Incremental elliptical boundary estimation for anomaly detection in wireless sensor networks. In: 2011 IEEE 11th International Conference on Data Mining, pp. 467–476. IEEE (2011)
8. Noshad, Z., et al.: Fault detection in wireless sensor networks through the random forest classifier. Sensors **19**(7), 1568 (2019)
9. Wen, S., Xiang, Y., Zhou, W.: A lightweight intrusion alert fusion system. In: 2010 IEEE 12th International Conference on High Performance Computing and Communications (HPCC), pp. 695–700. IEEE (2010)
10. Rajasegarar, S., Leckie, C., Palaniswami, M.: Anomaly detection in wireless sensor networks. IEEE Wirel. Commun. **15**(4), 34–40 (2008)
11. Xie, M., Hu, J., Guo, S.: Segment-based anomaly detection with approximated sample covariance matrix in wireless sensor networks. IEEE Trans. Parallel Distrib. Syst. **26**(2), 574–583 (2014)
12. Yu, Y., Zhu, Y., Li, S., Wan, D.: Time series outlier detection based on sliding window prediction. Math. Probl. Eng. **2014** (2014)
13. Xie, M., Hu, J., Guo, S., Zomaya, A.Y.: Distributed segment-based anomaly detection with Kullback-Leibler divergence in wireless sensor networks. IEEE Trans. Inf. Forensics Secur. **12**(1), 101–110 (2016)
14. Vuran, M.C., Akan, Ö.B., Akyildiz, I.F.: Spatio-temporal correlation: theory and applications for wireless sensor networks. Comput. Netw. **45**(3), 245–259 (2004)
15. Jindal, A., Psounis, K.: Modeling spatially correlated data in sensor networks. ACM Trans. Sens. Netw. (TOSN) **2**(4), 466–499 (2006)
16. Intel Berkeley research lab. http://db.csail.mit.edu/labdata/labdata.html
17. Haar, A.: Zur theorie der orthogonalen funktionensysteme. Math. Ann. **69**(3), 331–371 (1910). https://doi.org/10.1007/BF01456326
18. Jackson, J.E., Mudholkar, G.S.: Control procedures for residuals associated with principal component analysis. Technometrics **21**(3), 341–349 (1979)
19. Dang, T.B., Le, D.T., Nguyen, T.D., Kim, M., Choo, H.: Monotone split and conquer for anomaly detection in IoT sensory data. IEEE Internet Things J. (2021)

Feature Representation of AutoEncoders for Unsupervised IoT Malware Detection

Huu Noi Nguyen, Van Cuong Nguyen, Nguyen Ngoc Tran, and Van Loi Cao$^{(\boxtimes)}$

Le Quy Don Technical University, Hanoi, Vietnam
{noi.nguyen,cuongpd,ngoctn,loi.cao}@lqdtu.edu.vn

Abstract. The feature representation of AutoEncoders (AEs) has been widely used for unsupervised learning, particularly in cybersecurity domain, and demonstrated promising performance. However, deeply investigations of the feature learner for the task of IoT attack detection in unsupervised learning have not been carried out yet. In this paper, we study the feature representation of AEs in combination with a subsequent clustering-based technique like Self-Organizing Maps (SOM) for unsupervised learning IoT attack detection. This aims to get insight into the characteristics of the AE learners in the tasks of unsupervised IoT detection such as identifying unknown/new IoT attacks and transfer learning. To highlight the behavior of AE-based learners, a feature reduction like Principle Component Analysis (PCA) is used to construct a feature space for facilitating SOM. The proposed models are investigated and assessed extensively by a number of experiments and analyses on the NBaIoT dataset. The experimental results highly suggest that AEs should be used for transferring models as training data is highly un-balanced and includes IoT attacks being similar to Benign. If the training data seems to be balanced, and contains IoT attacks being significantly deviated from Benign, the feature reduction like PCA is more preferable.

Keywords: AutoEncoders · Self-organizing maps · IoT malware · IoT anomaly detection · Transfer learning

1 Introduction

IoT technology has played a vital role in enhancing smart applications in Human's life, such as smart home, healthcare, transportation, education [7,24]. However, the rapid growth, diversity, large-scale nature of IoT systems have posed new security challenges [1]. IoT attack methods have been developed increasingly and diversely, such as on-board attacks, security gateway attacks, control server attacks, eavesdropping attacks. One of the prominent attacks can be known as BotNet Mirai. It is a special type of botnet that has recently caused large-scale DDoS attacks by exploiting vulnerabilities on IoT devices [17]. Therefore, the traditional IoT security solutions might be insufficient to keep up with the rapid and diverse evolution of IoT attacks to protect such systems [17].

© Springer Nature Switzerland AG 2021
T. K. Dang et al. (Eds.): FDSE 2021, LNCS 13076, pp. 272–290, 2021.
https://doi.org/10.1007/978-3-030-91387-8_18

Machine learning (ML), particularly deep learning (DL), has currently widely employed in many different security problems, such as classification and detection of malicious codes, detection of network attacks, prediction the anomalous behavior of network data [14,27]. Anomaly detection approaches has been commonly used in cyber-security domain in which cyberattacks and malicious actions can be modelled as anomalies while normal behaviors and activities are treated as normal [20,28,29]. Machine learning-based anomaly detection methods can rely on training data to construct detection models. These models will be then used to distinguish network traffic into "normal" or "anomaly". Thanks to the development in ML recently, advanced techniques can deal with some challenges in cybersecurity, particular IoT security, such as federated learning, transfer learning and meta-learning.

Unsupervised and semi-supervised learning techniques are considered as typical approaches for developing cyberattack detection models [22]. This is because these approaches can deal with the shortage of cyberattack data and their labels, and can leverage the ability in identifying unknown/new malicious activities. Recently, deep neural networks (DNNs), such as Deep Belief Networks (DBNs), deep AutoEncoders (AEs), have been employed to represent the original data into useful feature spaces for eliminating subsequent detection/classification methods [5,8,21,29]. Amongst DNNs, AE-based methods are usually employed for developing feature learning models in both semi-supervised and unsupervised learning manner. An AutoEncoder is a feed-forward neural network consisting of Encoder and Decoder. The objective of AEs is to learn to reproduce the input data at its the output layer. The middle hidden layer, typical a bottleneck layer, can project the input data into lower and more meaningful feature space (called latent feature space) [4,12,28]. When applying for identifying anomalies in cybersecurity, AEs have been widely used to construct detection models from one class of data, particular the normal data, and have shown the state-of-the-art performance. This suggests that AEs have a powerful in representation data, prefer to learn data within a single distribution. In addition, the latent representation of AEs also employed to represent multiple classes of data for facilitating subsequent clustering methods in unsupervised learning, such as in [21]. In cybersecurity domain, to the best Of our knowledge, no work has investigated why AEs can learn to represent multiple classes (normal and anomaly classes), and which kinds of training data are AEs preferable.

This study aims to get insight into the latent feature space of AEs in representing multiple classes of data (normal data and IoT Malware) for subsequent unsupervised learning methods. In other words, this attempts to answer to the questions that why the feature space of AEs can benefit unsupervised learners despite of the fact that AEs are considered to prefer learning single class?. When learning multiple classes (normal data and IoT Malware), what kinds of IoT Malware groups can benefit AE learners? To do these, the feature space of AEs will be learnt from training datasets with different rations of IoT Malware and normal data. We also observe the characteristics of AEs when using different IoT Malware groups for training. Self-Organizing Maps are chosen as the subsequent clustering method. Beside these, Principle Component Analysis (PCA) is investigated together with AEs in order to highlight the behaviors of the AE learner. PCA attempts to project data into a new coordinate system so that the data are as much separated as possible [13]. It means that PCA keeps features which make data the most separable. Thus, PCA will perform well on data in which the classes of the data are very separated from each others. Our investigations will be carried out on the NBaIoT dataset [18], and focused on evaluating unkown/new Malware detection and the transferring ability between different devices and IoT Malware groups. The contributions of this paper can be listed as follows:

- We introduce approaches to deeply investigate the latent feature space of AEs on IoT attack detection in the unsupervised learning manner.
- We carry out extensive experiments to evaluate the hybrid AEs and SOMs, and comparisons to the models using PCA instead of AEs or without AEs.
- Unknown/new IoT attack detection and transfer learning are also examined.

The rest of paper is organized as follows. Section 2 and 3 briefly present some related works and background, respectively. Our proposed models are described in details in Sect. 4. Experiments, results and discussions are presented and illustrated in Sects. 5, 6. Section 7 draws some conclusion and future directions about results and discussion.

2   Related Works

In this section, we discuss recent works which are used for detecting IoT anomalies. Most of approaches used are based on unsupervised/semi-supervised learning methods. We discuss the use of AEs in particular. Next, we also carry out some studies that tend to solve the classification problem using the unsupervised technique.

AE-based methods have been widely used as a feature learner for latent representation [3–5,8,10,18,25,28]. This latent is later used for anomaly detection models. The latent feature representation can be learned in different manners like supervised learning [18,28], semi-supervised learning [5], and unsupervised learning [8,10]. The common way is as follow, all methods are trained on the training data (usually one-class learning), and when the training process is done, only the encoder is used for further stages. This encoder tries to learn the latent features from the input, and this latent is next used in the detection models constructed by traditional machine learning methods. Since the dimension of latent data is decreased and much less than that of the input data, then the classification models are mostly faster than previous.

Recently Cao et al. [5] introduced two regularized AEs, namely SAE and DVAE for capture the normal behaviors of network data. These regularizers AEs are attempted to put normal data towards a small region at the origin of the latent feature space, which can result in reserving the rest of the space for anomalies occurring in the future. These regularized AEs were designed to overcome the problem of identifying anomalies in high-dimensional network data. The latent representation of SAE and DVAE was then used for enhancing simple one-class classifiers. In a supervised manner, Vu et al. [28] proposed Multi-distribution VAE (MVAE) to represent normal data and anomalous data into two different regions in the latent feature space of VAE. Originally, Variational EutoEncoders (VAEs) learn to map input data into a standard Gaussian distribution $\mathcal{N}(0,1)$ in its middle hidden layer. The proposed model was evaluated on two publicly network security datasets, and it produces promising performance.

In unsupervised manner, Gustavo et al. [6] explore the power of SOM for multi-label classification. Since the SOM has ability to map input instances to a map of neurons. After performing SOM, similar instances are grouped in the same class. Also using SOM, Andreas Rauber et al. [23] presented the LabelSOM approach, which can automatically label and train the SOM with the features of the most relevant input data in a particular cluster. In [26] J. Tian et al. proposed a method that can improve SOM for anomaly detection. For a given test data, all the neighbors are identified by using the k-nearest neighbor algorithm. The Euclidean distance was used to measure the distance between the test data observation and the centroid of the neighbors.

This work attempt to investigate the latent representation of AEs for IoT malware detection tasks in an unsupervised manner. This means that we use AEs to learn a latent representation for both normal and IoT malware without labels. The latent representation then facilitate SOMs for discovering clusters.

3   Background

3.1   AutoEncoder

An AutoEncoder is a feed-forward neural network that attempts to reconstruct the original input data at the output layer [2,11]. The traditional AE is used for dimensional reduction and feature learning. The AE architecture is shown in Fig. 1 consists of two parts, Encoder, Decoder, connecting by its bottleneck layer. The hidden layer h that described a code used to represent the input [9]. The encoder function f is used to learn the input and represented as *code*, the decoder function g is used to reconstruct the data from the encoded representation.

Mathematically, given data x with no-labels and the function f for encoder and function g for decoder. Then we have the following equations:

$$z = f(x) = a_e(wx + b) \tag{1a}$$
$$\hat{x} = g(z) = g(f(x)) = a_d(w'.f(x) + b') \tag{1b}$$

where a_e and a_d are the activation functions of the encoder and decoder, $\hat{x}$ is x's reconstruction. The reconstruction loss function (e.g. squared loss error) is to minimize the difference between the input x and the output $\hat{x}$.

$$L(x, \hat{x}) = \|x - \hat{x}\|^2 \tag{2}$$

The AEs have many applications, such as image compression, image denoising, feature extraction, image generation, sequence to sequence prediction and recommendation systems.

3.2   Principle Component Analysis

Dimensionality reduction is one of the important techniques in Machine Learning. Dimensionality reduction is finding a function that takes the input of a data point $x \in R^D$ (D is the number of dimensions) and create a new data point $z \in R^K$ with $K < D$.

One of the simplest dimensionality reduction algorithms is based on a linear model. This method is called Principal Component Analysis (PCA) [13,30]. This method is based on the observation that the data are not normally distributed randomly in space, but are often distributed near-certain special lines/faces. PCA considers a special case when those special faces are linear in sub-spaces. Some modern PCA algorithms are Kernel PCA, Sparse PCA, Nonlinear PCA, Robust PCA.

3.3   Self-Organizing Maps

SOM Introduction: The Self-Organizing Maps (SOM) [15,16] are self-organizing neural networks that are able to map similar instances to a group in a map, then each neuron is placed next to each other. This map provides a mapping from a high-dimensional input space to lower-dimensional output space (usually two dimension). SOMs apply competitive learning as opposed to error-correction learning (such as back-propagation with gradient descent), and they use a neighborhood function to preserve the topological properties of the input space. In other words, competitive learning is an unsupervised learning method, and it is most suitable to illustrate the appropriateness of learning from a single-layer neural network.

SOMs have been successfully applied in a number of fields, such as identification, data clustering and text prediction. The data types include sound, image and text. More details of SOMs will be described in the followed subsection.

Mapping Process: On the training SOM, the winning neuron is archived using Euclidean distance, which is represented in Eq. 3.

$$d_j(x) = \sqrt{\sum_{i=1}^{A}(x_i - w_{ji})^2} \tag{3}$$

where x is the attribute vector of the instance and w_j is the weight vector of the jth neuron, A is the number of attributes of an instance.

When the winning neuron is obtained, its weights will be adjusted to approximate it to the instance. Since then, a map of its neighborhood is defined. The process is continued as follow: the weights of each neighborhood also is updated, and approximate to the winning neuron; a good choice for finding for the neighborhood is using Gaussian function 4, where $h_{j,i}$ is the neighborhood of the winning neuron i, while j is the older winning neuron, the distance $d_{j,i}$ is a distance between neurons, the σ defines the spreading of neighborhoods [6].

$$h_{j,i} = exp(-\frac{d_{j,i}^2}{2\sigma^2}) \tag{4}$$

The weight update process is given by Eq. 5, where w_j is the weight vector, x is input instance, η is a learning rate.

$$\Delta w_j = \eta h_{j,i}(x - w_j) \tag{5}$$

Finally, the weight vector at iteration $(t+1)$ is updated by Eq. 6.

$$\Delta w_j(t+1) = w_j(t) + \eta h_{j,i}(x - w_j(t)) \tag{6}$$

The training process of the algorithm is shown in Algorithm 1. Firstly, the weight matrix is initialized by generating randomly or using PCA method. Secondly, the winning neuron is selected from the neuron grid Ω by using the distance metric (e.g. Euclidean). Finally, the weights of all related neurons will be updated using Eq. 6 [6]. The process is terminated when the network is converged, and the weights in the map might have the same distribution as the input vectors. It means, after the finite number of iterations, the inputs are placed in appropriate positions in the Kohonen network (another name of Self-organizing maps). In Algorithm 1, X is a dataset, q is a number of instances, a is a number of attributes, l is a number of labels and n is a number of neurons.

Algorithm 1. The SOM algorithm

Input: $X = [q, (a + l)]$, e: number of epochs
Output: $W = [n, a]$
Main loop
  for $i \leftarrow 1$ to e **do**
    Initialize weight matrix W
    for $j \leftarrow 1$ to q **do**
      $o(x_j) = argmin_k \|x_j - w_k\|$, $k \in \Omega$
      $w_k(i+1) = w_k(i) + \eta(i)h_{k,o(x_j)}(i)(x_j(i) - w_k(i))$
    end for
  end for
  return W

---

4   Hybrid AEs and SOMs for IoT Malware Detection

This section presents our proposed hybrid for IoT Malware detection. The proposed approach consists of two phases as shown in Fig. 1:

Phase 1: The feature space of an AE is learnt from unlabeled data (noraml data and IoT Malware). The feature space is in a lower dimension, and can reveal more meaningful features.

Phase 2: A SOM plays as a subsequent clustering-based technique that works directly on the feature space of the AE. The trained feature space will help the SOM discover and label clusters associated with normal data and IoT malware. These cluster labels are used for classifying. Another feature reduction methods like PCA are also involved in our experiments presented in Sect. 5. In other words, a PCA is used instead of AEs in the hybrid for highlighting the performance of AEs.

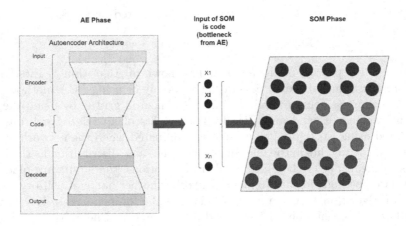

Fig. 1. The system architecture

4.1   Latent Representation of AEs

We aim to represent the original data into a lower dimension and more meaningful feature space. Many AE-based structures have been developed from the original AE. In this work, we employ an ordinary AE with a 3-layer in its encoder and a 3-layer in its decoder. The AE architecture is shown in Fig. 1.

The AE is trained in the unsupervised learning manner. This means that the training dataset is a unlabelled dataset consisting of both normal data and IoT Malware. To examine the effectiveness of the AE feature space in unsupervised learning task, we create four versions of the training datasets with four different level of unbalances (ratio r) between IoT Malware and normal data on each IoT Malware group. The values of r are manually chosen as 0.01, 0.1, 1.0, and 5.0. To investigate the behavior of AEs on different IoT Malware groups, the normal data is combined with each of IoT Malware groups to generate training datasets for each IoT device. For training process, a training dataset is passed into an AE model with the objective of minimizing the difference between the original data and its reconstruction. The process is terminated if the training error is satisfied an early-stopping criteria. Once the training terminated, the decoder part is discarded while the encoder part is kept as a feature learning component facilitating the subsequent clustering-based technique.

4.2 SOM-Based Clustering Algorithm

Given instance x_i, and the mission is to classify this instance into an appropriate class. Firstly, all classes are represented by binary vector v_i. The jth position in this vector is corresponding to the jth class (c_j). If the instance x_i is a member of class c_j, then the $v_{i,j}$ has value 1, and 0 vice versa [6].

The test instance is mapped to labels-map by a trained SOM. Once its closest neurons found, a vector (called prototype vector) is obtained by averaging the class vectors of the training instances mapping to this neuron, notated as $\bar{v}$. It is the classification probability of the test instance falling into each class. The formulation of the prototype vector is shown in Eq. 7. The S_n is the training set which mapped to neuron n, while the $S_{n,j}$ is the training instances which mapped to the neuron n and classified to class c_j.

$$\bar{v}_{n,j} = \frac{|S_{n,j}|}{|S_n|} \tag{7}$$

To map instance n to class c_j, a *predetermined threshold* is set, typically 0.5, on the $\bar{v}_{n,j}$. The position whose the $\bar{v}_{n,j}$ value is not smaller than the *threshold* receives the value 1, and 0 otherwise.

The SOM-based clustering model is shown in Algorithm 2 [6]. The sample is classified using the labels-map which received after training SOM (Algorithm 1). A label c_j will be assigned to the neuron if most of samples mapping to that neuron has the label c_j. It means that the $\bar{v}_{n,j}$ is greater than the *threshold*. In case if no label is assigned to this neuron then the algorithm function will assign the most common label in the dataset. The steps are as follow:

Step 1: The winning neuron is selected from the neuron grid Ω;
Step 2: Get training instances mapped to winning neuron;
Step 3: Calculate prototype vector;
Step 4: Compare the prototype vector to *threshold*;
Step 5: Get the appropriate class.

In the Algorithm 2 q is the number of training instances, a is the number of attributes dataset, l is the number of labels, m is the number of instances in the testing set, W is the weight matrix and P is the prediction matrix.

5 Experiments

In this section, we present a set of experiments to evaluate the feature space of AEs in unsupervised learning IoT Malware detection. To do this, we employ a SOM method as the subsequent clustering-based method. In order to highlight the characteristics of AEs, the SOM working with other feature reduction

Algorithm 2. SOM-based clustering algorithm

Input: $X^{train} = [q, (a + l)]$, $W = [n, a]$
Output: P
Main loop
   for $j \leftarrow 1$ to m **do**
      $o(x_j^{test}) = argmin_k \|x_j^{test} - w_k\|$, $k \in \Omega$
      $T \leftarrow$ instances mapped to $o(x_j^{test})$;
      $\bar{v}_j \leftarrow$ average of the label vectors from T;
      $x_j^{test} \leftarrow x_j^{test} + \bar{v}_j$
      $p_j \leftarrow \bar{v}_j$
      p_j compares to *threshold*
   end for
   return P

method like PCA and the stand-alone SOM are assessed in comparison to the hybrid AEs and SOMs. We use the terms AESOM and PCASOM to refer to the hybrid an AE and a SOM, the hybrid PCA and a SOM respectively. The description of the IoT dataset, hyper-parameter settings, and evaluation metrics are presented in the rest of this section.

We design three main experiments to extensively investigate the characteristics of the AE feature space for unsupervised learning IoT Malware detection. The experiments consists of *(1) IoT data analysis; (2) the ability in identifying unknown/new IoT attacks; and (3) the ability in transferring learning.* The details are as follows,

- IoT data analysis attempts to get insight into the characteristics of three classes of data: benign, Gafgyt, and Mirai,
- Unknown/new IoT attacks is to evaluate our proposed model on unknown/new IoT attacks from the same IoT device,
- Transfer learning is to evaluate our proposed model on the data from different devices.

All experiments are implemented in Python using Keras[1], Scikit-learn[2] and Minisom[3] frameworks.

5.1 Datasets

The NBaIoT dataset[4] was introduced by Y. Meidan et al. [18]. It contains data samples collected from nine different IoT devices, as described in Table 1. For each device, the two most popular kinds of IoT Malware are launched such as Mirai and BASHLITE (or Gafgyt) for generating Malware data together with benign data. These devices can be categorized into four main groups: doorbell,

---

[1] https://keras.io/.
[2] https://scikit-learn.org/.
[3] https://github.com/JustGlowing/minisom.
[4] https://archive.ics.uci.edu/ml/datasets/detection_of_IoT_botnet_attacks_N_BaIoT.

thermostat, monitor and camera/webcam. Each record in the dataset consists of 115 features extracted by using Kitsune [19]. In addition, the NBaIoT dataset contains two main different groups of IoT attack types, namely Mirai and Gafgyt. Each group of IoT attacks consists of many sub-classes of attack types. However, our experiments focus on classifying benign and IoT attacks, thus Mirai and Gafgyt are examined instead of sub-classes of IoT attacks.

Table 1. The nine IoT datasets in NBaoIoT

Device ID	Device name	Type	Benign	Gafgyt	Mirai
D1	Danmini_Doorbell	Doorbell	49548	652100	316650
D2	Ecobee_Thermostat	Thermostat	13113	512133	310630
D3	Ennio_Doorbell	Doorbell	39100	316400	
D4	Philips_B120N10_Baby_Monitor	Monitor	175240	312273	610714
D5	Provision_PT_737E_Security_Camera	Camera	62154	330096	436010
D6	Provision_PT_838_Security_Camera	Camera	98514	309040	429337
D7	Samsung_SNH_1011_N_Webcam	Webcam	52150	323072	
D8	SimpleHome_XCS7_1002_WHT_ Security_Camera	Camera	46585	303223	513248
D9	SimpleHome_XCS7_1003_WHT_ Security_Camera	Camera	19528	316438	514860

In our experiments, we employ two groups of IoT devices: doorbell (D1 and D3) and camera (D5, D6 and D8) as shown in Table 1. For the training phase, we use one type of IoT Malware together with legitimate data on each IoT device. The resulting models are then utilized to evaluate the other IoT Malware type on the same/different devices. For more details, we carry out two groups of scenarios for doorbell and camera as shown in Table 2. Thermostat, webcam and monitor types do not include in our experiments because they have only device such as D2, D4 and D7, respectively.

We divide the dataset into two parts: train (70%) and test (30%). The early-Stopping technique is employed to avoid over-fitting. For each IoT device, all benign data is used for training because it is much less than the amount of IoT Malware. The IoT Malware data for training is randomly selected from the original IoT Malware data with ratios of $0.01, 0.1, 1.0, 5.0$ in comparison to the amount of the benign data, respectively. The relative ratio between IoT attacks and benign data is signed as r.

5.2 Parameters Settings

Firstly, we set up the size of the encoded layer (bottleneck) for the AE and PCA. We choose the ratio between the size of the encoded layer and the original feature space as 0.25. Thus, the encoded layer size is 29 ($0.25 * 115$). The sizes of other hidden layers are set as 0.75, 0.5, 0.33, 0.25 (the encoded layer) of the input layer's size, respectively. In training phase, we split the training data into training set (80%) and validation set (20%). The optimization method is Adam with the loss function of Mean Squared Error (MSE). The maximum number of

Table 2. The training and testing scenarios

Group	Scenarios	Training	Testing Same device but different attacks	Testing Different devices
1	1.1	D1: Gafgyt	D1: Mirai	D3: Gafgyt
	1.2	D1: Mirai	D1: Gafgyt	D3: Gafgyt
	1.3	D3: Gafgyt		D1: Gafgyt, Mirai
2	2.1	D5: Gafgyt	D5: Mirai	D6: Gafgyt, Mirai D8: Gafgyt, Mirai
	2.2	D5: Mirai	D5: Gafgyt	D6: Gafgyt, Mirai D8: Gafgyt, Mirai
	2.3	D6: Gafgyt	D6: Mirai	D5: Gafgyt, Mirai D8: Gafgyt, Mirai
	2.4	D6: Mirai	D6: Gafgyt	D5: Gafgyt, Mirai D8: Gafgyt, Mirai
	2.5	D8: Gafgyt	D8: Mirai	D5: Gafgyt, Mirai D6: Gafgyt, Mirai
	2.6	D8: Mirai	D8: Gafgyt	D5: Gafgyt, Mirai D6: Gafgyt, Mirai

epochs is 50 (with the early-stopping method) and the batch size is 200. The *tanh* activation function is used in all layers.

The size of the bottleneck (coded size) in PCA is set as the AE. All other PCA hyper-parameters are set to default values. For training SOM, we use tuning techniques to search for the best hyper-parameters. The hyper-parameters are σ, *learning rate* η. Once these parameters found, SOM is trained to build the label-maps and determine the outlier percentage.

5.3 Evaluation Metrics

We utilize Area Under the Curve (AUC) for evaluating the performance of our proposed methods. To estimate AUC, the true positive rate (TPR) and false positive rate (FPR) need to be calculated first by the following formulas.

$$TPR = \frac{TP}{TP + FN} \qquad FPR = \frac{FP}{FP + TN} \tag{8}$$

By plotting TPR against FPR, we received the Receiver Operating Characteristic curve (ROC curve) at a number of thresholds. AUC can be calculated as the entire area underneath the ROC curve. AUC provides an aggregate measure of performance across all possible classification thresholds.

6   Results and Discussion

In this section, we describe the experimental results and provide discussions. All experiments are performed as described in Table 2. The details about results are shown in the followed tables and figures in this section. In the next subsections, we present the reports and associated analysis for IoT data analysis, unknown IoT attack detection and transfer learning, respectively.

6.1   IoT Data Analysis

Mirai[5] and Gafgyt (also known as BASHLITE)[6] are all botnets that developed to inject IoT devices and generate the DDoS traffic from that bots. Gafgyt is a branch of Mirai which is developed upon the existing source code. The difference is that Gafgyt focuses on generating DDoS traffic, while Mirai focuses on making the victim become a bot.

This analysis aims to show some statistics and visualization of Gafgyt and Mirai against Benign to show how they are deviate from Benign. Thus, we first use the scale model built from Benign for normalizing both Mirai and Gafgyt. The metrics of *mean, median* and *standard deviation (std)* are calculated over the normalized versions of Benign, Gafgyt and Mirai as presented in Table 3. Statistics values on the metrics are the smallest for Benign, the second smallest for Gafgyt and the largest for Mirai on all devices shown in Table 3. This suggest an assumption that Gafgyt may share some common aspects with Benign while the attributes of Mirai tend to be significantly different from Benign.

Table 3. Statistics on Benign, Gafgyt and Mirai data

Device	Dataset	Mean	Median	Std	Device	Dataset	Mean	Median	Std
D1	Benign	0.14	0.02	0.17	D6	Benign	0.10	0.03	0.18
	Gafgyt	16.40	0.05	123.90		Gafgyt	1.49	0.00	10.36
	Mirai	**45.74**	**0.14**	**223.91**		Mirai	**2.71**	**0.03**	**13.45**
D3	Benign	0.08	0.03	0.13	D8	Benign	0.12	0.00	0.21
	Gafgyt	4.32	0.01	32.65		Gafgyt	0.47	0.00	3.15
	Mirai	–	–	–		Mirai	**1.72**	0.00	**7.53**
D5	Benign	0.09	0.02	0.18					
	Gafgyt	2.22	0.01	17.11					
	Mirai	**4.05**	**0.04**	**21.70**					

To confirm this assumption, we visualize the normalized versions of Gafgyt and Mirain against Benign in a coordinate system. In Fig. 2, we use two pairs of features of these data on the devices D1 and D5. The visualization of Mirai,

---

[5] https://en.wikipedia.org/wiki/Mirai_(malware).
[6] https://en.wikipedia.org/wiki/BASHLITE.

Gafgyt and Benign are colored by green, orange and blue respectively. It is clear that Mirai deviates significantly from Benign while some parts of Gafgyt appear in and around Benign on the two pairs of the observed features.

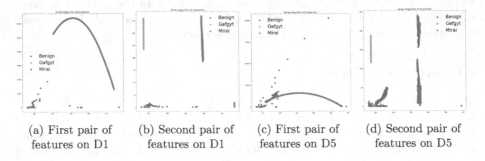

| (a) First pair of features on D1 | (b) Second pair of features on D1 | (c) First pair of features on D5 | (d) Second pair of features on D5 |

Fig. 2. Data visualization of D1 and D5 (Color figure online)

Based on the above considerations, we would draw a conclusion that Mirai deviates significantly from both Benign and Gafgyt whereas Gafgyt and Benign may share some common features. Therefore, traditional machine learning-based methods can achieve good performance on Mirai, while it may suffer from identifying Gafgyt from Benign. In addition, this analysis results also suggest that a training data consisting of Benign and Gafgyt may benefit AE learners while PCA may prefer the combination of Benign and Mirai. To verify this suggestion, we will do a number of analyses on the results of the unknown/new IoT attack ability and the transfer learning ability in the next subsections.

6.2 Unknown/New IoT Attack Detection

This subsection will provide some analyses on the ability to detect unknown/new IoT attacks from our proposed models. For all the experiments in this subsections, the models are trained on training data including Mirai, and evaluated on training including Gafgyt, and vice versa. The experimental results are shown in Table 4 and also in Fig. 3. The device D3 does not include in the experiment because it misses Mirai. For each device, we report the AUC values produced by the three models on each setting of the training data, and also the mean and median over the four different settings of training data. The highest values amongst three methods on each setting of training data are highlighted in the gray color.

It can be seen from Table 4 that AESOM seem to perform better than two others on some highly unbalanced settings, such as $r = 0.01$ and 0.1 while PCA-SOM prefers balanced training data. When $r = 0.01$ and 0.1, Benign dominates in training data. Thus, training data can be considered as a single class which will benefit the AE learners. On the other hand, PCA can yield good AUC values on $r = 1.0$. In the balanced case, training data will have two equal classes, PCA

Table 4. Unknown/new attack detection results on D1, D5, D6 and D8

Device	Models	Data Ratio r				Metrics		Data Ratio r				Metrics	
		0.01	0.10	1.00	5.00	mean	median	0.01	0.10	1.00	5.00	mean	median
		Train on Gafgyt - test on Mirai						Train on Mirai - test on Gafgyt					
D1	AESOM	0.67	0.62	0.81	0.99	0.77	0.74	0.53	0.95	0.84	0.97	0.82	0.90
	PCASOM	0.79	0.71	0.90	1.00	0.85	0.85	0.87	0.98	0.98	0.86	0.92	0.92
	SOM	0.91	0.67	0.67	0.80	0.77	0.74	0.64	0.98	0.97	0.99	0.89	0.97
D5	AESOM	0.68	0.58	0.80	0.92	0.75	0.74	0.95	0.87	0.96	0.64	0.85	0.91
	PCASOM	0.64	0.67	0.82	0.97	0.78	0.74	0.64	0.98	0.65	0.98	0.81	0.82
	SOM	0.51	0.75	0.74	0.87	0.72	0.74	0.63	0.97	0.98	0.97	0.89	0.97
D6	AESOM	0.57	0.66	0.66	0.81	0.67	0.66	0.63	0.86	0.84	0.94	0.82	0.85
	PCASOM	0.68	0.67	0.68	0.92	0.74	0.68	0.97	0.56	0.65	0.98	0.79	0.81
	SOM	0.65	0.68	0.68	0.68	0.67	0.68	0.95	0.95	0.82	0.96	0.92	0.95
D8	AESOM	0.61	0.73	0.69	0.74	0.69	0.71	0.58	0.96	0.95	0.95	0.86	0.95
	PCASOM	0.60	0.66	0.59	0.75	0.65	0.63	0.91	0.95	0.97	0.65	0.87	0.93
	SOM	0.54	0.66	0.82	0.97	0.75	0.74	0.50	0.64	0.67	0.95	0.69	0.66

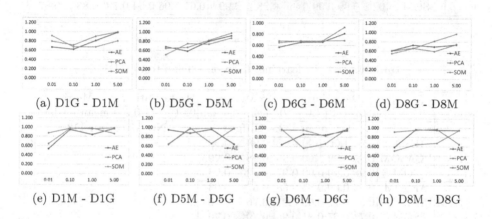

(a) D1G - D1M   (b) D5G - D5M   (c) D6G - D6M   (d) D8G - D8M

(e) D1M - D1G   (f) D5M - D5G   (g) D6M - D6G   (h) D8M - D8G

Fig. 3. Unknown/new attack detection results on the same device

can find a coordinate system in which the training data can be highest separated. Moreover, the Fig. 3 illustrates some interesting characteristics of training data. All models trained on the training data of Benign and Gafgyt yield AUC values very similar to each others on the four ratios. While the AUC values of the models trained on the combination of Benign and Mirai tend to be deviated from each others.

6.3 Transfer Learning

This subsection aims to evaluate the ability to transfer models to different IoT devices. The models trained on a device with a IoT attack group can be used to detect IoT attacks on other devices. We examine two groups of transferring model knowledge such as doorbell (D1 and D3) and camera (D5, D6, D8). The detailed results are presented in the following subsections.

Transfer Learning on the Doorbell Devices. There are two devices D1 and D3 in the doorbell. Firstly, the models are trained on Gafgyt and Mirai of the device D1, and tested on Gafgyt of the device D3 respectively. We then train the models on Gafgyt of the device D3, and test on Mirai and Gafgyt of the device D1, each. The experimental results are reported in Tables 5 and 6. The highest AUC values over three models are indicated in the gray color. Note that Mirai does not include in the data of the device D3.

Table 5. Transfer learning: train on D1 and test on D3

Test	Models	Data Ratio r				Metrics		Data Ratio r				Metrics	
		0.01	0.1	1.0	5.0	mean	median	0.01	0.1	1.0	5.0	mean	median
		Train on D1 - Gafgyt						Train on D1 - Mirai					
D3G	AESOM	0.98	0.99	0.99	0.95	0.98	0.98	0.61	0.93	0.90	0.71	0.79	0.81
	PCASOM	0.94	0.99	0.99	0.94	0.97	0.97	0.62	0.97	0.95	0.81	0.84	0.88
	SOM	0.98	0.99	0.99	0.95	0.98	0.99	0.61	0.96	0.94	0.79	0.83	0.87

Table 6. Transfer learning: train on D3 and test on D1

Test	Models	Data Ratio r				Metrics		Data Ratio r				Metrics	
		0.01	0.1	1.0	5.0	mean	median	0.01	0.1	1.0	5.0	mean	median
		Train on D3 - Gafgyt						No Mirai on D3					
D1G	AESOM	0.94	1.00	1.00	0.98	0.98	0.99						
	PCASOM	0.99	0.99	1.00	0.95	0.98	0.99						
	SOM	0.98	0.99	0.99	0.97	0.98	0.98						
D1M	AESOM	0.56	0.69	0.85	0.58	0.67	0.63						
	PCASOM	0.68	0.72	0.68	0.95	0.76	0.70						
	SOM	0.71	0.67	0.81	0.95	0.79	0.76						

When training on Gafgyt, AESOM often outperform the two others on identifying both Gafgyt and Mirai from other devices. This can be seen from the first row on each devices in Tables 5 and 6. The reason can be that the AE learners may prefer to learn from data points that are not too separated from each others. Regard to the aspect, based on the analysis presented in Sect. 6.1, the combination of Benign and Gafgyt can be considered better than the training data consisting of Benign an Mirai.

On the other hand, PCASOM often yields good AUC values the balanced case ($r = 1.0$), particularly outperforms other models when training data includes Mirai. This suggests that PCA can create a better feature space for the case when training data consists of two highly separated classes. SOM working on the original data also produces the highest values in few cases.

Transfer Learning on Camera Devices. The camera group consists of the devices D5, D6, and D8. D5 and D6 are the two device versions of the same brand, while D8 is from anther brand. In this experiment, we use the data from one device for training and those of the two others for testing. The experimental results are shown in Tables 7, 8 and 9.

Table 7. Transfer learning: train on D5 and test on D6 and D8

Test	Models	Data Ratio r				Metrics		Data Ratio r				Metrics	
		0.01	0.10	1.00	5.00	mean	median	0.01	0.10	1.00	5.00	mean	median
		Train on D5 - Gafgyt						**Train on D5 - Mirai**					
D6G	AESOM	0.99	1.00	1.00	0.96	0.98	0.99	0.64	0.62	0.84	0.94	0.76	0.74
	PCASOM	0.98	1.00	1.00	0.96	0.98	0.99	0.55	0.97	0.91	0.93	0.84	0.92
	SOM	0.96	1.00	0.99	0.95	0.98	0.98	0.85	0.97	0.65	0.92	0.84	0.88
D6M	AESOM	0.67	0.66	0.88	0.94	0.79	0.78	0.97	0.99	0.99	0.97	0.98	0.98
	PCASOM	0.67	0.69	0.68	0.93	0.74	0.69	0.96	1.00	1.00	0.97	0.98	0.98
	SOM	0.52	0.68	0.70	0.94	0.71	0.69	1.00	1.00	0.99	0.95	0.98	0.99
D8G	AESOM	0.99	0.99	0.77	0.70	0.86	0.88	0.61	0.61	0.68	0.86	0.69	0.65
	PCASOM	0.97	0.99	0.95	0.72	0.91	0.96	0.55	0.97	0.91	0.69	0.78	0.80
	SOM	0.96	0.99	0.97	0.95	0.97	0.97	0.85	0.94	0.58	0.91	0.82	0.88
D8M	AESOM	0.81	0.65	0.68	0.70	0.71	0.69	0.74	0.74	0.52	0.82	0.71	0.74
	PCASOM	0.66	0.67	0.63	0.67	0.66	0.66	0.70	0.94	0.73	0.71	0.77	0.72
	SOM	0.60	0.66	0.65	0.95	0.71	0.66	0.72	0.73	0.77	0.93	0.79	0.75

Table 8. Transfer learning: train on D6 and test on D5 and D8

Test	Models	Data Ratio r				Metrics		Data Ratio r				Metrics	
		0.01	0.10	1.00	5.00	mean	median	0.01	0.10	1.00	5.00	mean	median
		Train on D6 - Gafgyt						**Train on D6 - Miria**					
D5G	AESOM	0.98	1.00	1.00	0.99	0.99	0.99	0.93	0.85	0.85	0.89	0.88	0.87
	PCASOM	0.99	1.00	1.00	0.98	0.99	1.00	0.96	0.96	0.95	0.97	0.96	0.96
	SOM	1.00	1.00	1.00	0.99	1.00	1.00	0.97	0.97	0.64	0.97	0.89	0.97
D5M	AESOM	0.65	0.71	0.91	0.94	0.80	0.81	0.95	0.99	0.98	0.98	0.98	0.98
	PCASOM	0.66	0.68	0.85	0.89	0.77	0.76	0.99	1.00	1.00	0.99	0.99	0.99
	SOM	0.66	0.68	0.68	0.73	0.69	0.68	0.98	1.00	1.00	1.00	0.99	1.00
D8G	AESOM	0.98	0.99	0.97	0.94	0.97	0.98	0.93	0.84	0.83	0.56	0.79	0.84
	PCASOM	0.99	0.99	0.94	0.76	0.92	0.97	0.78	0.96	0.84	0.95	0.88	0.90
	SOM	0.99	0.99	0.99	0.88	0.96	0.99	0.95	0.97	0.64	0.84	0.85	0.90
D8M	AESOM	0.62	0.80	0.81	0.81	0.76	0.80	0.96	0.78	0.80	0.66	0.80	0.79
	PCASOM	0.66	0.72	0.69	0.68	0.69	0.69	0.56	0.98	0.89	0.76	0.80	0.82
	SOM	0.63	0.63	0.66	0.72	0.66	0.65	0.73	0.75	0.79	0.71	0.74	0.74

Again, AESOM seems to be perform better than PCASOM and SOM on almost devices when training data including Gafgyt. This can be seen from the

Table 9. Transfer learning: train on D8 and test on D5 and D6

Test	Models	Data Ratio r				Metrics		Data Ratio r				Metrics	
		0.01	0.10	1.00	5.00	mean	median	0.01	0.10	1.00	5.00	mean	median
		Train on D8 - Gafgyt						**Train on D8 - Miria**					
D5G	AESOM	0.92	0.99	0.99	0.92	0.95	0.95	0.53	0.96	0.93	0.83	0.81	0.88
	PCASOM	0.84	0.97	0.93	0.93	0.92	0.93	0.51	0.97	0.96	0.72	0.79	0.84
	SOM	0.95	0.98	0.99	0.69	0.90	0.97	0.51	0.96	0.98	0.77	0.80	0.86
D5M	AESOM	0.63	0.66	0.68	0.79	0.69	0.67	0.66	0.94	0.89	0.84	0.83	0.87
	PCASOM	0.64	0.66	0.65	0.78	0.69	0.66	0.83	0.95	0.96	0.75	0.87	0.89
	SOM	0.64	0.68	0.68	0.59	0.65	0.66	0.79	0.83	0.96	0.78	0.84	0.81
D6G	AESOM	0.92	0.99	0.99	0.87	0.94	0.95	0.52	0.96	0.91	0.78	0.79	0.85
	PCASOM	0.84	0.98	0.91	0.89	0.90	0.90	0.51	0.96	0.96	0.75	0.80	0.86
	SOM	0.95	0.98	0.98	0.74	0.92	0.97	0.51	0.96	0.98	0.71	0.79	0.84
D6M	AESOM	0.63	0.67	0.68	0.73	0.68	0.67	0.66	0.94	0.88	0.80	0.82	0.84
	PCASOM	0.64	0.66	0.63	0.75	0.67	0.65	0.83	0.95	0.96	0.79	0.88	0.89
	SOM	0.64	0.68	0.68	0.64	0.66	0.66	0.79	0.83	0.96	0.73	0.83	0.81

first row on each device, particularly on mean values. However, the Table 9 does not demonstrate this point clearly. This is because the device D8 was produced from a different brand to the devices D5 and D6. But in that case, the dataset is more "balanced" compare to other values of 0.01 and 0.1. Similarly to the analysis on transferring model between D1 and D3, PCASOM and SOM stand alone tend to produce high AUC values when training data is not too highly unbalanced.

In IoT networks, the number of IoT devices is massive, the IoT protocols are also diverse, and the ratio between Benign and IoT Malware is varied. Applying the trained model on a installed device to new devices is valuable. Based on the discussion in this section, the feature representation of AEs are highly recommended to be used for transfer learning when training data is highly unbalanced, and includes IoT attacks being similar to Benign. On the other hand, if the training data seems to be balanced, and contains IoT attacks being significantly deviated from Benign, the feature reduction like PCA is more preferable.

7   Conclusions and Future Work

This work aims to examine intensively and extensively the feature representation of AEs in unsupervised IoT Malware detection tasks such as identifying unknown/new attacks and transferring model knowledge to other IoT devices. Thus, the hybrid AEs and SOMs is proposed to investigations. To highlight the behavior of AE-based learners, we observe the use of a feature reduction like PCA for facilitating SOM, and the case of using SOM stand alone. The proposed models are evaluated extensively by a number of experiments and analyses on the NBaIoT dataset. The experimental results highly recommend that AEs can be used for transferring knowledge and identifying unknown/new IoT attacks for highly unbalanced data cases and IoT attacks being similar to Benign. For balanced data cases and IoT attacks being significantly deviated from Benign, the feature reduction like PCA is more preferable. Moreover, data consisting of IoT attacks being similar to Benign tends to help different methods yielding similar performance. The work of investigating other feature representation models and developing better feature learning models is postponed to the near future. In future work, we will also evaluate the latent feature space of AEs on other metrics, such as the computational cost.

References

1. Abomhara, M., Køien, G.M.: Cyber security and the internet of things: vulnerabilities, threats, intruders and attacks. J. Cyber Secur. Mob. 65–88 (2015)
2. Bourlard, H., Kamp, Y.: Auto-association by multilayer perceptrons and singular value decomposition. Biol. Cybern. 291–294 (1988). https://doi.org/10.1007/BF00332918
3. Bui, T.C., Cao, V.L., Hoang, M., Nguyen, Q.U.: A clustering-based shrink autoencoder for detecting anomalies in intrusion detection systems. In: 2019 11th International Conference on Knowledge and Systems Engineering (KSE), pp. 1–5. IEEE (2019)
4. Cao, V.L., Nicolau, M., McDermott, J.: A hybrid autoencoder and density estimation model for anomaly detection. In: Handl, J., Hart, E., Lewis, P.R., López-Ibáñez, M., Ochoa, G., Paechter, B. (eds.) PPSN 2016. LNCS, vol. 9921, pp. 717–726. Springer, Cham (2016). https://doi.org/10.1007/978-3-319-45823-6_67
5. Cao, V.L., Nicolau, M., McDermott, J.: Learning neural representations for network anomaly detection. IEEE Trans. Cybern. **49**(8), 3074–3087 (2018)
6. Colombini, G.G., de Abreu, I.B.M., Cerri, R.: A self-organizing map-based method for multi-label classification. In: 2017 International Joint Conference on Neural Networks (IJCNN), pp. 4291–4298. IEEE (2017)
7. Dastjerdi, A.V., Buyya, R.: Fog computing: helping the internet of things realize its potential. Computer **49**(8), 112–116 (2016)
8. Erfani, S.M., Rajasegarar, S., Karunasekera, S., Leckie, C.: High-dimensional and large-scale anomaly detection using a linear one-class SVM with deep learning. Pattern Recogn. **58**, 121–134 (2016)
9. Goodfellow, I., Bengio, Y., Courville, A.: Deep Learning. MIT Press, Cambridge (2016)
10. Hawkins, S., He, H., Williams, G., Baxter, R.: Outlier detection using replicator neural networks. In: Kambayashi, Y., Winiwarter, W., Arikawa, M. (eds.) DaWaK 2002. LNCS, vol. 2454, pp. 170–180. Springer, Heidelberg (2002). https://doi.org/10.1007/3-540-46145-0_17
11. Hinton, G.E., Zemel, R.S.: Autoencoders, minimum description length, and Helmholtz free energy. Adv. Neural. Inf. Process. Syst. **6**, 3–10 (1994)
12. Japkowicz, N., Myers, C., Gluck, M., et al.: A novelty detection approach to classification. In: IJCAI, vol. 1, pp. 518–523. Citeseer (1995)
13. Jolliffe, I.: Principal component analysis. Encycl. Stat. Behav. Sci. **30**(3), 487 (2002)
14. Jordan, M.I., Mitchell, T.M.: Machine learning: trends, perspectives, and prospects. Science **349**(6245), 255–260 (2015)

15. Kohonen, T.: The self-organizing map. Proc. IEEE **78**(9), 1464–1480 (1990)
16. Kohonen, T.: Essentials of the self-organizing map. Neural Netw. **37**, 52–65 (2013)
17. Kolias, C., Kambourakis, G., Stavrou, A., Voas, J.: DDoS in the IoT: Mirai and other botnets. Computer **50**(7), 80–84 (2017)
18. Meidan, Y., et al.: N-baiot–network-based detection of IoT botnet attacks using deep autoencoders. IEEE Pervasive Comput. **17**(3), 12–22 (2018)
19. Mirsky, Y., Doitshman, T., Elovici, Y., Shabtai, A.: Kitsune: an ensemble of autoencoders for online network intrusion detection. arXiv preprint arXiv:1802.09089 (2018)
20. Nguyen, T.D., Marchal, S., Miettinen, M., Fereidooni, H., Asokan, N., Sadeghi, A.R.: Dïot: a federated self-learning anomaly detection system for IoT. In: 2019 IEEE 39th International Conference on Distributed Computing Systems (ICDCS), pp. 756–767. IEEE (2019)
21. Nguyen, V.Q., Nguyen, V.H., Le-Khac, N.-A., Cao, V.L.: Clustering-based deep autoencoders for network anomaly detection. In: Dang, T.K., Küng, J., Takizawa, M., Chung, T.M. (eds.) FDSE 2020. LNCS, vol. 12466, pp. 290–303. Springer, Cham (2020). https://doi.org/10.1007/978-3-030-63924-2_17
22. Pang, G., Shen, C., Cao, L., Hengel, A.V.D.: Deep learning for anomaly detection: a review. ACM Comput. Surv. (CSUR) **54**(2), 1–38 (2021)
23. Rauber, A.: LabelSOM: on the labeling of self-organizing maps. In: IJCNN 1999. International Joint Conference on Neural Networks. Proceedings (Cat. No. 99CH36339), vol. 5, pp. 3527–3532. IEEE (1999)
24. Ray, S., Jin, Y., Raychowdhury, A.: The changing computing paradigm with internet of things: a tutorial introduction. IEEE Design Test **33**(2), 76–96 (2016)
25. Song, C., Liu, F., Huang, Y., Wang, L., Tan, T.: Auto-encoder based data clustering. In: Ruiz-Shulcloper, J., Sanniti di Baja, G. (eds.) CIARP 2013. LNCS, vol. 8258, pp. 117–124. Springer, Heidelberg (2013). https://doi.org/10.1007/978-3-642-41822-8_15
26. Tian, J., Azarian, M.H., Pecht, M.: Anomaly detection using self-organizing maps-based k-nearest neighbor algorithm. In: PHM Society European Conference, vol. 2 (2014)
27. Tsai, C.W., Lai, C.F., Chiang, M.C., Yang, L.T.: Data mining for internet of things: a survey. IEEE Commun. Surv. Tutor. **16**(1), 77–97 (2013)
28. Vu, L., Cao, V.L., Nguyen, Q.U., Nguyen, D.N., Hoang, D.T., Dutkiewicz, E.: Learning latent distribution for distinguishing network traffic in intrusion detection system. In: ICC 2019–2019 IEEE International Conference on Communications (ICC), pp. 1–6. IEEE (2019)
29. Vu, L., Nguyen, Q.U., Nguyen, D.N., Hoang, D.T., Dutkiewicz, E.: Deep transfer learning for IoT attack detection. IEEE Access **8**, 107335–107344 (2020)
30. Wold, S., Esbensen, K., Geladi, P.: Principal component analysis. Chemom. Intell. Lab. Syst. **2**(1–3), 37–52 (1987)

Machine Learning and Artificial Intelligence for Security and Privacy

Potential Threat of Face Swapping to eKYC with Face Registration and Augmented Solution with Deepfake Detection

Trong-Le Do[1,3], Mai-Khiem Tran[1,2,3], Huy H. Nguyen[4,5],
and Minh-Triet Tran[1,2,3]([✉])

[1] University of Science, VNU-HCM, Ho Chi Minh City, Vietnam
{dtle,tmkhiem,tmtriet}@hcmus.edu.vn
[2] John von Neumann Institute, VNU-HCM, Ho Chi Minh City, Vietnam
[3] Vietnam National University, Ho Chi Minh City, Vietnam
[4] Graduate University for Advanced Studies, SOKENDAI, Hayama, Japan
[5] National Institute of Informatics, Tokyo, Japan
nhhuy@nii.ac.jp

Abstract. It is necessary to develop an efficient and secure mechanism to verify customers digitally for various online transactions. Integrating biometric solutions into the online user registration and verification processes is a promising trend for electronic Know Your Customer (eKYC) systems. However, Deepfake or face manipulation techniques may become a threat for eKYC with face authentication. In this paper, we introduce this potential attack of Deepfake on eKYC by swapping and manipulating faces between source and target faces. We then propose to augment the security for current eKYC systems with Deepfake detection. We conduct the experiments on the 10K video clips in the private test of Deepfake Detection Challenge 2020, and our method, following the Capsule-forensics approach, achieves the Logloss score of 0.5189, among the top 6% best results among the 2114 teams worldwide. This result demonstrates that our deepfake detection algorithm can be a promising method to provide extra protection for eKYC solutions with face registration and authentication.

Keywords: eKYC · Faceswap · Deepfake detection

1 Introduction

KYC, or Know Your Customer, are strategies for a financial organization to provide a better understanding and insight into customer engagement [23]. This cohesion makes it possible for financial organizations to identify, verify, and analyze the probabilities of risks when working with different clients. With technology advancement, eKYC is introduced as an electronic customer identification aiming to simplify procedures and documents, creating convenience for

© Springer Nature Switzerland AG 2021
T. K. Dang et al. (Eds.): FDSE 2021, LNCS 13076, pp. 293–307, 2021.
https://doi.org/10.1007/978-3-030-91387-8_19

customers. This is a high-accuracy electronic identification solution approach to digitize the customer experience, increase conversion rates, and reduce risks for businesses.

In eKYC, it is essential to link a digital identity, *e.g.* an online account of an e-wallet, with the true identity of an individual. In the registration process for digital identity, besides the legal documents, *e.g.* citizen ID card, passport, or driving license, the user is usually required to prove his or her true identity via face registration. This task is to prevent the wrong association between digital and real identities. As face registration or authentication can be tricked by photos or recorded video clips of the true user's face, the liveness of a face should be considered and integrated into such security processes [21].

Deepfake [22] refers to synthetic media, *e.g.* photos, audio, or video clips, using advanced algorithms in machine learning and artificial intelligence to manipulate or generate visual and audio content. Face swapping [25], or manipulation [28] are among the common and well-known demonstrations of Deepfake techniques [22]. We can easily notice the widespread use of face swap or manipulation in with social enthusiasm, from new commercialized applications for iPhones and Android smartphones, *e.g.* Reflect, Reface, to various Youtube channels with video clips synthesized with DeepFaceLab utility, *etc.*.

Deepfake techniques can be used to swap a target person's face into a photo or video of a source person. Hence, it can become a potential threat to trick security tasks [4,22], including eKYC registration and authentication processes. Without the existence of a real true person to register a digital identity, it would be possible to synthesize a fake face video of that person to bind that identity to an online account illegally [21]. Even if the eKYC solution employs the challenge that requests a user to perform some sequence of facial actions, *e.g.* smile, eye blink, turn left, look up, *etc.*, fake face video of a target person can also be synthesized with insignificant delay under the controlled of a source person. Thus, in this paper, we introduce a potential threat of using Deepfakes to trick eKYC systems based on face registration and authentication.

Various methods have been proposed to differentiate between real and fake video clips of human faces to meet the urgent need in society [4,7,15,20]. From 2019 to 2020, Facebook partners with industry leaders and academic experts to organize Deepfake Detection Challenge (DFDC [12]) to encourage individuals and organizations to develop solutions for identifying potentially fake face videos. In this challenge, our method achieves the result in the top 6% worldwide, with the Logloss score of 0.5189 on 10K video clips in the private test set of DFDC [12]. We inherit the idea of Capsule-forensics approach [24] employ integrate Capsule layers [30] for fake image/video detection. In this paper presents our implementation for Deepfake detection and proposes our solution to prevent the potential attack on eKYC with face swapping by integrating Deepfake detection into eKYC platforms.

The content of this paper is organized as follows. In Sect. 2, we briefly review the overview of eKYC and biometric factors, then we present the related work on Deepfake and Deepfake detection. In Sect. 3, we introduce a potential threat to eKYC systems by using face swapping and manipulation to generate fake face video in realtime to trick the registration and authentication processes of eKYC. We then propose in Sect. 4 our solution to enhance eKYC solutions with Deepfake detection techniques. The experiments of our Deepfake detection methods are discussed in Sect. 5. Finally, we conclude and present open problems for future work in Sect. 6.

2   Related Work

2.1   Know Your Customer

Depending on the progress of the digitization process and the level of potential risk, some or all of the intermediary steps involved in the KYC process may be automated. eKYC, or electronic KYC is a concept of compliance with electronic KYC guidelines [23]. This form of KYC has become a de facto standard today. eKYC provides for financial organizations not only customers' social information but initial information about the customer's biometrics [35].

For a regular eKYC system, citizen portrait identification or any similar static 2D recognition method has the same weakness as written signature, in the sense that it is static and easily forgeable. One metric often used in judgment is the perceived "liveness" of the content [21]. Ideally, this metric should be zero for static images, and the opposite for a face captured live [13].

The two prominent methods for identification and authentication being currently adopted are Face ID and Video ID. These methods employ new and emerging technologies such as Machine Learning to complete the task.

Face identification data acquisition techniques capture the human face at multiple angles, giving both photometric (colors) and geometric (surface) information [9]. Video identification is a step up from face identification, requiring the user to perform an action and the result is a video [21]. This action commonly involves the user saying something then pick up and hold the identity card in front of the camera. This way, even more information can be analyzed from the content, such as gait, body identification, and proposed identity card. Video identification has the benefit that it is very natural to the user while giving more information at the same time.

However, recently popular techniques, such as Deepfake [2], which is currently state-of-the-art in the subject of Face manipulation and face swapping, can generate realistic-looking faces that confuse even humans [6]. Deepfake requires visual images of a person to generate ("target person"), and the movements of the controlling person. It then proceeds to generate the target person's face on the controlling person. This ability poses a threat to all existing face-based visual identification and authentication systems, as well as real-time face identification challenges.

2.2   Deepfake Generation and Detection

Fake videos can be generated/synthesized by different methods with rapid advancement to reach photo-realistic level [22], *e.g.* FaceSwap [1], Face2Face [33], DeepFakes [2] and NeuralTextures [34]. DeepFaceLab [26] provides an easy to use open source framework that anyone (amateur) can use to make deepfake videos.

The common pipeline for swapping faces from a source image to a target image usually consists of the following steps. First, the faces of two persons are extracted from both source and target images. Some methods require face normalization, facial landmark detection, and pose estimation. Finally, the face segment in the source image is transferred and blended into the corresponding segment in the target image. Deepfake generation can not only transferring facial features from a source to a target face, but also head pose or facial movement [16] or expression [33].

In practice, some methods require training (finetune) the network to well-adapt to a pair of source and target persons. This approach is appropriate to create high-quality Deepfake media. Using a pre-trained network for generic faceswap can be done, but the results may not be as good as the finetuned one. However, some recent methods no longer require training with new faces [25] as in early techniques [3]. Refinement with few-shot technique can also be applied for quality enhancement [38].

Several datasets have been released to evaluate different Deepfake detection methods. FaceForensics++ [28] and DeepFake Detection Challenge (DFDC) Dataset [12] are among the common datasets for facial forgery detection.

To detect fake images, researchers usually focus on artifacts resulting from face tampering, such as from face-swapping [39], face morphing [27,38], splicing attacks [5], or Deepfake methods [18].

Most of the current methods for face manipulation detection in videos are CNN-based per-frame classification. Besides, inconsistency in facial movement, including eye blinking [19], head pose movement [37], or facial expressions [4], can also be exploited for the detection of fake videos. Several methods use CNN, 3D CNN [10], or RNN [29] to detect such inconsistency and exploit temporal dependencies in Deepfake video.

3 Deepfake Attack on eKYC Systems

3.1 Regular Registration Process for eKYC

We consider a typical registration process for an eKYC system, as illustrated in Fig. 1. There are two main inputs for this process, legal documents and bio-

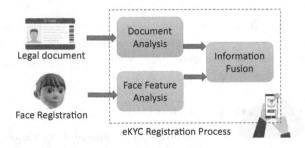

Fig. 1. Overview of a regular registration process in eKYC system.

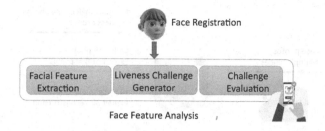

Fig. 2. Regular face feature analysis component.

metric information, which is usually facial features. Document Analysis component extracts information about the person registering an online identity, *e.g.* full name, date of birth, address, citizenship *etc.*; while Face Feature Analysis extracts facial features for account registration. Information Fusion component provides extra necessary steps to link information between the two analysis components. For example, face verification may be employed in the registration process of some eKYC systems to confirm that the live face of the person who is registering the account actually matches the face in the legal document.

In Fig. 2, we present the main modules for a regular face feature analysis component. Apart from the facial feature extraction to register a user's face, it is crucial to verify the liveness of that face during the registration process. The Live Challenge Generator produces a sequence of requests R for specific facial gestures that a user should perform to confirm the liveness. Each request in the challenge R is selected from a known challenge face gesture set S, including smile, left/right eye blink, right eye blink, open mouth, turn head left/right/up/down, *etc.*. The number and order of requests in a challenge R can be pre-defined or generated adaptively depending on the confidence evaluation and environment of the user.

3.2   Attack Registration Process with Deepfake

In this section, we present two scenarios for an adversary to trick the face registration process of an eKYC system. In scenario 1, as shown in Fig. 3, the registration process requires any customer to capture and submit a number of photos of his/her face (see the left part), which allows for 3D reconstruction of the customer's face.

The essence of this challenge is to create multiple differentiators between real and live data versus generated and/or collected data to present as few difficulties to legit customers as possible while being very hard to complete for an adversary. These differentiators could be, for example, the perceived liveness index of the photos, the difference between the requested face angle versus the received face angle, the consistency among the photos, or the amount of time to produce an answer. A legit customer will have no difficulties accomplishing this challenge by taking pictures of his/her face in a reasonably short amount of time. This was not the case for adversarial methods. Methods that are based on a large collection

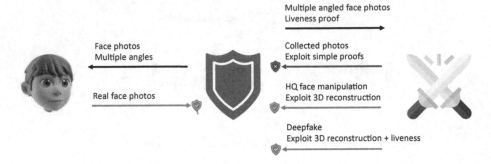

Fig. 3. Scenario 1: attack on face registration with a photo

of a victim's data rely on the chance that there exists a sequence of face photos, with minimal difference in requested angles presented in the challenge.

Nevertheless, if the incentive is high enough, a malicious party could exploit the 3D reconstruction process by manually manipulating the collected photos to adapt to the requested angle. This method satisfies the differentiator established in the difference between the requested angle and the received face angle. However, it is likely time-consuming and thus is differentiated by the time differentiator and failing the challenge.

Figure 4 presents another scenario where the registration process requires new customers to film a video proof. In the video, a user is requested to perform some actions, such as holding an ID card and tilt his/her head. This method is commonly used by not only banks but also exchanges for customers that are willing to trade in high volume.

Video-based proofs further enhance the consistency and the amount of data available to analyze, while making minimal changes to the customer experience. Owing to the advance of ubiquitous mobile technology developments, the process of taking video proof differs in little with taking multiple photos. Similar to taking multiple photos, this approach generates multiple photos from the video and 3D reconstruction of the customer's face is possible and can be performed.

By using video, various movements contained within can be easier analyzed for discrepancies. Likewise, any legit customer can complete this challenge with

Fig. 4. Scenario 2: attack on face registration with a live video.

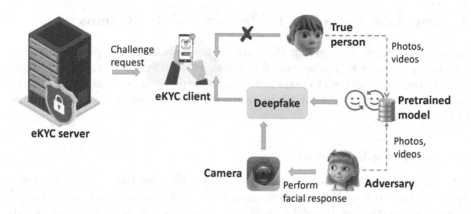

Fig. 5. Scenario 3: attack with Deepfake.

little difficulty, while it is unlikely that there exists a pre-recorded video that an adversary possesses that matches the requested head poses.

However, common to the two aforementioned methods, now exists a threat from Deepfake. For an adversary that uses a controlling person and possession of a target face, Deepfake has shown to be very flexible in generating faces on the controlling person using the target face. With a resourceful adversary party, the time required to perform face swapping can be reduced to real time, and thus render the time differentiator ineffective. The quality of targets that Deepfake can generate is good enough that it even allows 3D reconstruction to take place.

In Fig. 5, we propose a potential attack on the face registration process in eKYC system using Deepfake techniques. From face photos and/or video clips collected from a true person B, we can train a model to learn how to synthesize/swap face from B to a face motion video of an adversary A. This training phase can be time-consuming but can be done offline.

During the face registration process in an eKYC application, an eKYC server generates a sequence of facial gesture challenge requests to check the liveness of a use. Instead of having the true person B to actually perform face gestures according to requests from eKYC server, an adversary A can impersonate B by capturing his/her face as a sequence of frames, which are then fed into a Deepfake system to synthesize a corresponding video with the target face of B. This sequence of frames can be generated with low latency in response to the request from the eKYC server.

Our experiment can generate a Deepfake video with a length of 10 s in less than 5 s with a system with two 2080Ti GPUs. Currently, we aim to create high-quality fake video clips, so we use a desktop implementation. If we want to further boost the Deepfake video generation, we may employ knowledge distillation technique to train a smaller student network for Deepfake face image synthesis from a larger teacher network. We also plan to utilize mobile apps for a Deepfake video generation in future work.

4   Deepfake Detection to Protect eKYC Systems

In this section, we first present the common framework for Deepfake detection, and our strategies to realize this framework with different components. Then we propose to integrate Deepfake detection module as an add-on component to enhance the security of an eKYC system with face registration and authentication.

4.1   Deepfake Detection

Deepfake detection has become an attractive problem for many researchers and practitioners worldwide. Therefore, more than 2000 teams already participated in the competition of Deepfake Detection Challenge (2019-2020). Despite different techniques and strategies of implementation, we can find the common approach for current Deepfake detection methods, as illustrated in Fig. 6. There are two main modules in the framework. The first module extracts facial features while the second one is for real/fake image classification. In our implementation, we employ different backbone for feature extraction, *e.g.* VGG19 [31], Xception [8], *etc..* For classification, we consider the classical approach with FC layer(s) and also experiment the CapsuleNetwork [24,30].

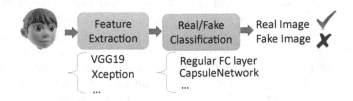

Fig. 6. Common framework for Deepfake detection.

In our first implementation, we use a VGG19-like network for feature extraction. As shown in Fig. 7, we consider up to three levels of details in a face, thus we use the output after the 3rd MaxPooling layer of the VGG-like network as the face feature for classification. For a deeper analysis of face, we may use features after the 4th or 5th MaxPooling layer. We do not employ the feature flattening steps as we want to preserve much face information. In our future work, we also consider combining features output from different levels of details, such as after each MaxPooling layer, to utilize more information about the face.

To better learn how to classify between a real and fake face image, we employ CapsuleNetwork [30] for the classification task, as illustrated in Fig. 8. This approach can be considered as the implementation following the idea of Capsuleforensics, proposed by Nguyen et al. in [24], for detecting fake image and video. In the traditional approach for convolutional neural networks, features extracted are usually flattened, then processed through one or several fully connected layers, and finally evaluated by a softmax one. CapsuleNetwork utilizes the idea to

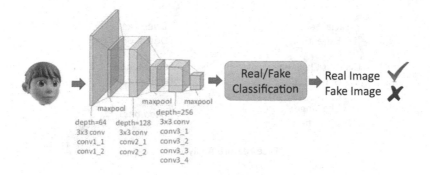

Fig. 7. Face feature extraction with VGG19 in Deepfake detection.

mimic the brain's operation, *i.e.* the brain should employ a mechanism for "routing" low level visual information to what it believes is the best capsule, a nested set of neural layers, for handling it. Dynamic routing, or routing by agreement, is considered to be superior than the current mechanism like max-pooling that routes based on the strongest feature detected in the lower layer.

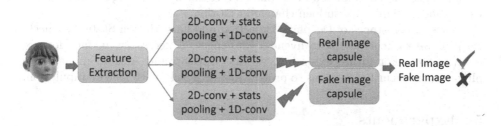

Fig. 8. Deepfake detection with CapsuleNetwork.

To train a Deepfake detection system with two main components, as shown in Fig. 6, we divide this task into two phases. In phase 1, we reuse pretrained weights for the feature extraction module and freeze them, then only train the classification module. By this way, we expect to initialize the feature extraction module with good-enough weights, and focus on training the second module. In phase 2, we unfreeze the feature extraction module, and finetune the whole network.

4.2 Enhanced eKYC with Deepfake Detection

To strengthen the security for eKYC systems, particularly with face registration and authentication, we propose to integrate several components to validate face photos and photo sequences, as represented in Fig. 9. The Deepfake Detection module is responsible for verifying if a face image is real or fake. In practice, we

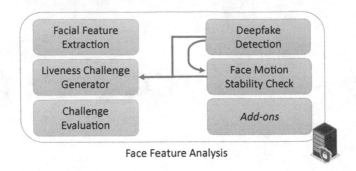

Fig. 9. Enhanced face feature analysis component with Deepfake detection.

have a sequence of face images returned from a mobile eKYC client, responding to a face gesture request from eKYC server. Therefore, we should check if that face image sequence is real and reliable or not.

In our experiment, we select a threshold $\theta = 0.8$ to be the minimum percentage of real photos in a sequence to confirm that the sequence is trustworthy. This task is solved by Face Motion Stability Check. This module is also responsible for checking the continuity and stability of visual data to trigger notification when there is significant sudden change in motion.

With the assistance of Deepfake Detection and Face Motion Stability Check, we propose to generate more liveness challenges, if necessary, in face feature analysis step. For future extension, we also develop and Add-on manager to plug-in more security functions to protect future attacks with face manipulation.

5 Experiments

We conduct our proposed solutions for Deepfake detection on DFDC dataset in the Deepfake Detection Challenge (2019-2020). The DFDC dataset includes 100K videos of 3426 paid actors [12], and the length of each clip is 10 s. The public test set has 4K video clips with 214 unique subjects, and 2K Deepfake video clips (50%). The private test set consists of 10K video clips with 5K Deepfake videos. To evaluate the robustness of proposed methods, the private test set has unseen content collected from various sources. Figure 10 shows some examples from the DFDC dataset.

For face detection, we use several models like DSFD [17], PyramidBox [32], RetinaFace [11] with pretrained weights (from WIDER FACE [36] and/or FDDB [14]). We choose PyramidBox to run based on observations about the accuracy as well as the calculation speed of the model. We use the architecture of PyramidBox with the modules Low-level feature pyramid network, Data-anchor-sampling, PyramidAnchors, and context prediction module pretrained on the WIDER FACE set with the confidence threshold of a face to be 0.8. For each bounding box of a detected face, we scale this bounding box to a size of 1.3 times

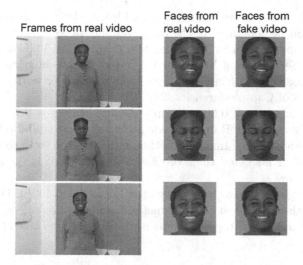

Fig. 10. Examples from DFDC dataset.

Table 1. Experimental results on DFDC datasets.

No.	Team Name	Logloss	Medal
1	Selim Seferbekov	0.42798	Gold
2	WM	0.42842	Gold
3	NtechLab	0.43452	Gold
...	...	...	Gold
14	drupal	0.46060	Gold
15	Ryan Wong	0.46099	Silver
...	...	...	Silver
113	wonderful CV	0.51289	Silver
114	[ods.ai] Catalyst	0.51400	Bronze
...	...	...	Bronze
128	**Our**	**0.51880**	**Bronze**
...	...	...	Bronze
226	Ibako	0.53787	Bronze
500	MeeCreeps	0.54567	
1000	Bowen Pei	0.61560	
1500	Ken Krige	0.69314	
2000	Marcus Lin	2.30784	
2114	20/20 vision	17.45439	

so that the real/fake face discriminator model learns the surrounding features of the face that can be generated in the face swapping process.

In our implementations for Deepfake detection, we use VGG19-like network, as presented in Sect. 4.1, and Xception. We employ two techniques for classification, namely classical fully connected layers, and CapsuleNetwork [30], folloiwing the original idea of Capsule-forensics [24].

We use learning rate of 0.001, Adam optimizer with $\beta = 0.9$, and CrossEntropyLoss. We also employ Random color jitter and horizontal flip for data augmentation.For each video frame, our method evaluates the score to measure that frame to be real or fake, and the score for the whole video is the average score of all video frames.

In the public leaderboard, we achieve the Logloss of 0.40214. For the official leaderboard with private test set, our method achieves Logloss of 0.51880 with Bronze Medal, among the top 6% teams in the challenge (Table 1).

6   Conclusion

Common financial relationships are moving towards a fully automated eKYC process with the help of newer technologies such as Machine Learning, while financial entities that offer services that could involve high value or high trust transactions may opt for a hybrid automation process that involves humans. Face registration and authentication are commonly used in eKYC systems, but Deepfake techniques can be used to trick such processes.

Therefore, in this paper, we introduce a potential threat for eKYC solutions by using face-swapping to synthesize/generate in real-time sequence of face images in response to facial gesture challenges from eKYC systems. Then we present our method for detecting Deepfake videos and propose its application to augment the security for eKYC systems with Deepfake detection.

The experimental results of our method on a challenging dataset DFDC from the Deepfake Detection Challenge (2019-2020) demonstrate that our solution achieves promising results, among the top 6% of all teams worldwide in the challenge. Currently, we are focusing our effort to improve our solution for Deepfake detection by setting attention to important facial blending regions, and the temporal consistency of photos.

Acknowledgment. Trong-Le Do and Mai-Khiem Tran were funded by Vingroup Joint Stock Company and supported by the Domestic Master/ PhD Scholarship Programme of Vingroup Innovation Foundation (VINIF), Vingroup Big Data Institute (VINBIGDATA), code VINIF.2019.ThS.22 and VINIF.2020.ThS.JVN.06, respectively.

References

1. Faceswap (2017). https://github.com/MarekKowalski/FaceSwap
2. Deepfake (2018). https://github.com/deepfakes/faceswap

3. Terrifying high-tech porn: Creepy deepfake videos are on the rise (2018). https://www.foxnews.com/tech/terrifying-high-tech-porn-creepy-deepfake-videos-are-on-the-rise
4. Agarwal, S., Farid, H., Gu, Y., He, M., Nagano, K., Li, H.: Protecting world leaders against deep fakes. In: Proceedings of the IEEE/CVF Conference on Computer Vision and Pattern Recognition (CVPR) Workshops (June 2019)
5. Bappy, J.H., Simons, C., Nataraj, L., Manjunath, B.S., Roy-Chowdhury, A.K.: Hybrid lstm and encoder-decoder architecture for detection of image forgeries. IEEE Trans. Image Process. **28**(7), 3286–3300 (2019)
6. Bhattacharjee, S., Mohammadi, A., Anjos, A., Marcel, S.: Recent advances in face presentation attack detection. In: Marcel, S., Nixon, M.S., Fierrez, J., Evans, N. (eds.) Handbook of Biometric Anti-Spoofing. ACVPR, pp. 207–228. Springer, Cham (2019). https://doi.org/10.1007/978-3-319-92627-8_10
7. Bonettini, N., Cannas, E.D., Mandelli, S., Bondi, L., Bestagini, P., Tubaro, S.: Video face manipulation detection through ensemble of CNNs. In: 2020 25th International Conference on Pattern Recognition (ICPR), pp. 5012–5019 (2021)
8. Chollet, F.: Xception: deep learning with depthwise separable convolutions. In: 2017 IEEE Conference on Computer Vision and Pattern Recognition (CVPR), pp. 1800–1807. IEEE, Honolulu, HI (July 2017)
9. Costa-Pazo, A., Vazquez-Fernandez, E., Alba-Castro, J.L., González-Jiménez, D.: Challenges of face presentation attack detection in real scenarios. In: Marcel, S., Nixon, M.S., Fierrez, J., Evans, N. (eds.) Handbook of Biometric Anti-Spoofing. ACVPR, pp. 247–266. Springer, Cham (2019). https://doi.org/10.1007/978-3-319-92627-8_12
10. Davletshin, A.: (2020). https://github.com/ntech-lab/deepfakedetection-challenge
11. Deng, J., Guo, J., Ververas, E., Kotsia, I., Zafeiriou, S.: RetinaFace: single-shot multi-level face localisation in the wild. In: 2020 IEEE/CVF Conference on Computer Vision and Pattern Recognition (CVPR), pp. 5202–5211. IEEE, Seattle, WA, USA (June 2020)
12. Dolhansky, B., et al.: The deepfake detection challenge dataset. CoRR abs/2006.07397 (2020)
13. Hernandez-Ortega, J., Fierrez, J., Morales, A., Galbally, J.: Introduction to face presentation attack detection. In: Marcel, S., Nixon, M.S., Fierrez, J., Evans, N. (eds.) Handbook of Biometric Anti-Spoofing. ACVPR, pp. 187–206. Springer, Cham (2019). https://doi.org/10.1007/978-3-319-92627-8_9
14. Jain, V., Learned-Miller, E.: FDDB: a benchmark for face detection in unconstrained settings. Technical report UM-CS-2010-009, University of Massachusetts, Amherst (2010)
15. Khalid, H., Woo, S.S.: Oc-FakeDect: classifying deepfakes using one-class variational autoencoder. In: Proceedings of the IEEE/CVF Conference on Computer Vision and Pattern Recognition (CVPR) Workshops (June 2020)
16. Kim, H., et al.: Deep video portraits. ACM Trans. Graph. (TOG) **37**(4), 163 (2018)
17. Li, J., et al.: DSFD: dual shot face detector. In: 2019 IEEE/CVF Conference on Computer Vision and Pattern Recognition (CVPR), pp. 5055–5064. IEEE, Long Beach, CA, USA (June 2019)
18. Li, L., et al.: Face x-ray for more general face forgery detection. In: 2020 IEEE/CVF Conference on Computer Vision and Pattern Recognition (CVPR), pp. 5000–5009 (2020)
19. Li, Y., Chang, M.C., Lyu, S.: In ictu oculi: exposing ai created fake videos by detecting eye blinking. In: 2018 IEEE International Workshop on Information Forensics and Security (WIFS), pp. 1–7 (2018)

20. Li, Y., Lyu, S.: Exposing deepfake videos by detecting face warping artifacts. In: IEEE Conference on Computer Vision and Pattern Recognition Workshops (CVPRW) (2019)
21. Marcel, S., Nixon, M.S., Fiérrez, J., Evans, N.W.D. (eds.): Handbook of Biometric Anti-Spoofing - Presentation Attack Detection, Second Edition. Advances in Computer Vision and Pattern Recognition, Springer, Heidelberg (2019)
22. Mirsky, Y., Lee, W.: The creation and detection of deepfakes: a survey. ACM Comput. Surv. **54**(1) (2021)
23. Mondal, P.C., Deb, R., Huda, M.N.: Transaction authorization from know your customer (kyc) information in online banking. In: 2016 9th International Conference on Electrical and Computer Engineering (ICECE), pp. 523–526 (2016)
24. Nguyen, H.H., Yamagishi, J., Echizen, I.: Capsule-forensics: using capsule networks to detect forged images and videos. In: ICASSP 2019–2019 IEEE International Conference on Acoustics. Speech and Signal Processing (ICASSP), pp. 2307–2311. IEEE, Brighton, United Kingdom (May 2019)
25. Nirkin, Y., Keller, Y., Hassner, T.: Fsgan: subject agnostic face swapping and reenactment. In: Proceedings of the IEEE/CVF International Conference on Computer Vision (ICCV) (October 2019)
26. Perov, I., et al.: Deepfacelab: a simple, flexible and extensible face swapping framework (2020)
27. Raghavendra, R., Raja, K.B., Venkatesh, S., Busch, C.: Transferable deep-cnn features for detecting digital and print-scanned morphed face images. In: 2017 IEEE Conference on Computer Vision and Pattern Recognition Workshops (CVPRW), pp. 1822–1830 (2017)
28. Rossler, A., Cozzolino, D., Verdoliva, L., Riess, C., Thies, J., Niessner, M.: FaceForensics++: learning to detect manipulated facial images. In: 2019 IEEE/CVF International Conference on Computer Vision (ICCV), pp. 1–11. IEEE, Seoul, Korea (South) (October 2019)
29. Sabir, E., Cheng, J., Jaiswal, A., AbdAlmageed, W., Masi, I., Natarajan, P.: Recurrent convolutional strategies for face manipulation detection in videos. In: IEEE Conference on Computer Vision and Pattern Recognition Workshops, CVPR Workshops 2019, Long Beach, CA, USA, 16–20 June 2019, pp. 80–87. Computer Vision Foundation/IEEE (2019)
30. Sabour, S., Frosst, N., Hinton, G.E.: Dynamic routing between capsules. In: Proceedings of the 31st International Conference on Neural Information Processing Systems, pp. 3859–3869. NIPS 2017, Curran Associates Inc., Red Hook, NY, USA (2017)
31. Simonyan, K., Zisserman, A.: Very deep convolutional networks for large-scale image recognition. In: Bengio, Y., LeCun, Y. (eds.) 3rd International Conference on Learning Representations, ICLR 2015, San Diego, CA, USA, 7–9 May 2015, Conference Track Proceedings (2015). http://arxiv.org/abs/1409.1556
32. Tang, X., Du, D.K., He, Z., Liu, J.: Pyramidbox: a context-assisted single shot face detector. In: Proceedings of the European Conference on Computer Vision (ECCV) (September 2018)
33. Thies, J., Zollhofer, M., Stamminger, M., Theobalt, C., Niessner, M.: Face2face: real-time face capture and reenactment of rgb videos. In: Proceedings of the IEEE Conference on Computer Vision and Pattern Recognition (CVPR) (June 2016)
34. Thies, J., Zollhöfer, M., Nießner, M.: Deferred neural rendering: image synthesis using neural textures. ACM Trans. Graph. **38**(4), 1–12 (2019)
35. Wang, J.S.: Exploring biometric identification in fintech applications based on the modified tam. Financ. Innov. **7**(1), 1–24 (2021)

36. Yang, S., Luo, P., Loy, C.C., Tang, X.: WIDER FACE: a face detection benchmark. In: 2016 IEEE Conference on Computer Vision and Pattern Recognition (CVPR), pp. 5525–5533. IEEE, Las Vegas, NV, USA (June 2016)
37. Yang, X., Li, Y., Lyu, S.: Exposing deep fakes using inconsistent head poses. In: ICASSP 2019–2019 IEEE International Conference on Acoustics, Speech and Signal Processing (ICASSP), pp. 8261–8265 (2019)
38. Zakharov, E., Shysheya, A., Burkov, E., Lempitsky, V.: Few-shot adversarial learning of realistic neural talking head models. In: 2019 IEEE/CVF International Conference on Computer Vision (ICCV), pp. 9458–9467. IEEE (October 2019)
39. Zhou, P., Han, X., Morariu, V.I., Davis, L.S .: Two-stream neural networks for tampered face detection. In: Proceedings of the IEEE Conference on Computer Vision and Pattern Recognition (CVPR) Workshops (July 2017)

Spliced Image Forgery Detection Based on the Combination of Image Pre-processing and Inception V3

Trung-Tri Nguyen[1] and Kha-Tu Huynh[2]([✉])

[1] Industrial University of Ho Chi Minh City, Ho Chi Minh City, Vietnam
[2] International University, Vietnam - Vietnam National University, Ho Chi Minh City, Ho Chi Minh City, Vietnam
hktu@hcmiu.edu.vn

Abstract. The paper proposes a method to detect the splicing in images which the spliced regions are created from other images. Our model is built to classify images with fake or not in a set of images. The image classification aims to save time when collecting and removing the fake images in a huge dataset and skip the step of searching the forged areas if there is no forgery in images. Previous studies mainly focused on the finding forged areas on the interfered image. The image preprocessing step and Inception V3 are combined to improve the detection. The objective of preprocessing step is to highlight the cropped area of the spliced images. The proposed model is experimented on Columbia Uncompressed Image Splicing Detection Evaluation Dataset and our built-in dataset with the average accuracy of spliced image detection up to 93.7%. The high accuracy demonstrates the effectiveness of the proposed method and is also a new contribution to the field of image forensics.

Keywords: Inception V3 · Image Forgery Detection (IFD) · Splicing · Enhancement · Image classification

1 Introduction

Image forensics is one of the areas of applied image processing. When the images that we see every day are interfered by such modern tools and technology that we cannot confirm the originality of the photos, image authentication has become a challenge. The interference is positive for beauty images, art photos, but in certain cases, counterfeiting causes effects. There are many ways to protect the integrity of images such as watermarking, digital signatures, but these methods will know the original image in advance and protect that original image using a security code. However, if the original image information is not known in advance, for any image, determining the integrity is a problem for the field of blind image detection. Copy-move and Splicing are two prominent cases of blind image detection. Many images that are faked from the same image (called Copy-Move) or spliced from another image (called Splicing) have challenged researchers in image forensics field. In the last 5 years, there are about 455 publications related to detection

© Springer Nature Switzerland AG 2021
T. K. Dang et al. (Eds.): FDSE 2021, LNCS 13076, pp. 308–322, 2021.
https://doi.org/10.1007/978-3-030-91387-8_20

of fake images in the form of copy-move and 245 publications related to detection of fake images in the form of splicing in IEEE Xplore and Science Direct. In this paper, the authors propose a method to determine whether an image is fake or not in the form of splicing using a deep learning model. This method is a new contribution and highly applicable.

The main contributions of the paper are:

1. Detect the spliced images using the combination of preprocessing and the model of Inception V3 [1].
2. Build a new dataset and simulating the proposed model using Columbia Uncompressed Image Splicing Detection Evaluation Dataset [2] and the new one to prove the effectiveness of the proposed model.

The introduction is followed by literature review, problem statement, the proposed model for image forgery detection with related theories, simulation results and conclusion.

2 Literature Review

Image forgery is done in many manipulations by many different tools and technologies in which Copy-Move is commonly used because the cut-pasting of information on the same image will be difficult to detect due to the structure and distribution of pixels on the same image is easier to achieve a higher correlation and manipulations and post processing is not too complicated, while splicing from one image to another requires more complicated processing techniques. According to the statistics mentioned in the introduction, in 5 years, the number of publications related to Copy-Move has nearly doubled the number of publications related to Splicing. Most of the techniques for detecting images due to splicing are based on image properties [3] or cameras properties [4, 5]. Researchers focus on image regions that are captured by resampling [6, 7] or double compression or heterogeneity in blur [8], sharpness [9], image smoothness, noise [10], quantization level on image [11], moment-based features [12] and illuminant [13], ... In the case of relying on cameras features, the published algorithms measure the camera's characteristic parameters to find the inconsistency of the spurious areas compared with the rest of the image. In recent years, the development of deep learning has brought many breakthroughs to the Splicing detection problem in which many methods have used neural networks [14–17] as the core of the solutions.

The paper proposes a new approach by applying the Inception model to detect if images are faked. The authors do not go into depth in identifying the forged regions, but we classify the fake images as a whole by evaluating images based on classification. Determining which images are faked in a huge set of images is complicated because there are hundreds of forgery operations, and our solution solves this challenge.

3 Problem Statement

With an arbitrary image in a dataset of the original and spliced images, the requirements of the problem are:

- Identify whether that image is forged or not. This step is useful and effective to remove the original image from a huge image dataset before finding and locating the forged regions in images. Previous research in image forgery detection often skips the image classification and only solves problem with the forgery location directly.
- Detect fake images by splicing with the reliable accuracy.

Based on the problem requirements, authors have come up with solution to solve the problem as follows:

- Choose the reliable datasets including both original and spliced images to train the model. There are many datasets for image forgery detection but most of them are copy-move, and a few of splicing. To increase the number of images to train, we also capture more natural images in the dataset. The more data used for training, the better the accuracy will be.
- Train the model using Inception V3. Inception V3 is candidate of model because of its fast trained speed and effective computation.
- Apply the preprocessing of enhancement and contrast images before passing them to the model of Inception V3 to update the dataset and improve the detection.
- Implement the proposed method for both images with and without splicing and evaluate the achieved results.

4   Proposed Method

The general block diagram of the proposed method is shown in the Fig. 1 and the details of blocks are presented in the following parts.

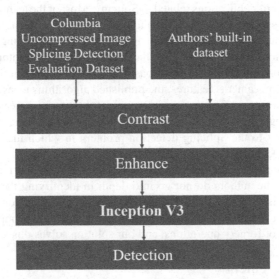

Fig. 1. The general block diagram of the proposed method.

4.1 The Architecture of Inception V3

Inception V3 is the next version of Inception V2 which is developed from Inception V1. However, Inception V2 is not popular and not commonly used so Inception V3 is known as the successor of Inception V1, with 24M parameters. Therefore, this section will give a brief introduction to the architecture of Inception V1 before considering Inception V3.

Inception V1. [18] Most neural network architectures use filters with sizes of 11×11, 5×5, 3×3 to 1×1. That of combining these filters into the same block produces a new architecture called Inception block architecture with better effectiveness in trained process. The architecture of Inception V1 including stem and Inception module are represented in Fig. 2.

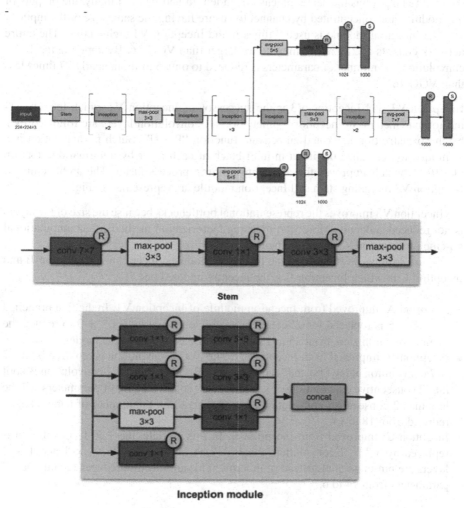

Fig. 2. The architecture of Inception V1.

The Inception module (see Fig. 2) consists of 4 parallel branches. Filters with sizes of $1 \times 1, 3 \times 3, 5 \times 5$ respectively are applied in the Inception Module to extract various features from cognitive regions with different sizes.

At the terminal of branches 1, 2, 4 from the top, a 1×1 convolution is used on each pixel as a fully connection to reduce the channel depth and number of parameters of the model. Example: In the previous layer, we have width $\times$ height $\times$ channels $= 12 \times 12 \times 256$. Applying 32 filters with the size of 1×1 will not change width, height but depth down to 32, then the size of the output shape is $12 \times 12 \times 32$. In the next layer, when doing the convolution over the full depth, the filters will be 32 instead of 256 in depth. Hence the number of parameters is significantly reduced. At the third branch from the top, the data dimension is reduced by a 3×3 max-pooling layer and then applying a 1×1 filter to change the number of channels. The branches are applied padding and stride so that the output has the same dimensions in length and width. Finally, the outputs of the modules are concentrated by channel to ensure having the same size of the input.

The Inception module is used 7 times in the Inception-V1 architecture. The entire network consists of 22 Layers, more twice larger than VGG-16. By applying the 1×1 convolution, the number of parameters is reduced to only 5 million, nearly 27 times less than VGG-16.

Inception V3. [1] Inception V3 is the successor of Inception-V1 which includes 24 million parameters. In Inception V3 network, all convolution layers are followed by a batch normalization layer and an activate function "ReLU". Batch normalization is a technique to normalize the input in mini-batch at each layer by a normal distribution of $N(0, 1)$ which improves the algorithm training process faster. The architecture of Inception V3 including stem and Inception module are represented in Fig. 3.

Inception V3 improves the representational bottlenecks because the size of the layers is not reduced suddenly. Moreover, by using factorization methods, the computational scheme of Inception V3 is more efficient.

Figure 3 shows 03 versions of the Inception modules: Inception-A, Inception-B and Inception-C in which Inception-C is the recently updated one.

- Inception-A: improved from Inception module of Inception V1. In the first branch, a layer 5×5 is replaced by 02 consecutive convolutional layers 3×3 to reduce the number of parameters from 25 to 18 and increase the depth of the model.
- Inception-B: improved from Inception-A. The 3×3 convolution is replaced by the 7×7 convolution at the first and second branch. Besides, the 7×7 convolution is split into 2 consecutive convolutions 7×1 and 1×7 so the number of parameters will be less than 2 consecutive convolutions 3×3. Therefore, the number of parameters is reduced from 18 to 14.
- Inception-C: improved from Inception-B. In this module, the 7×1 convolution is replaced by a 3×1 convolution, the 1×7 by a 1×3 and the 3×1 and 1×3 layers are put in parallel instead of in a row. This architecture reduces the number of parameters from 14 to 6.

In addition, Inception V3 uses two architectures called Reduction-A and Reduction-B to reduce the data dimensions.

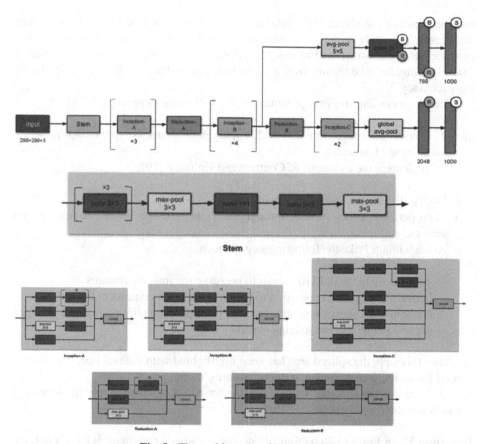

Fig. 3. The architecture of Inception V3.

4.2 Spliced Image Forgery Detection

This section demonstrates the method using preprocessing and Inception V3 architectural model to determine whether an image has been tampered by splicing. The image forgery detection before locating the forged area saves processing time and can be applied in case of authenticity validation in a huge image database.

Preprocessing. For images that have been confirmed with tampering, the next step is to locate the forged area. Our research focuses on the forged images in the form of splicing, that is, the forged region is concatenated from another image. To be able to detect the forgery accurately, the authors performed a pre-processing step with the aim of highlighting the cropped area. Based on features of splicing manipulations, recognition techniques have usually analyzed two basic features: edge extraction and image enhancement-based extraction.

For simple splicing techniques, transforming the input image upon the edge extraction is quite effective to the detection. However, for the complicated or higher splicing techniques, there is no differences between the edges of the object being spliced are and the edges of the object in the image, so edge extraction is not good solution. For that reason, our study does not use edge detection, but focuses on using image quality enhancement by Contract and Enhance for spliced image before putting it into the trained dataset to train the fake region positioning model. The method has been developed with high accuracy.

The preprocessing steps are presented by the following three steps:

Step 1: Convert all images into 224 × 224 images to reduce computation time while preserving image features.
Step 2: Implement the technique of Contract and Enhance [19].

- Firstly, declare the Python Pillow library.
- Then perform a loop running through each image in the set of real and spliced images.
- At each image, take the following steps in turn:

    – Analyze image with RGB format to preserve the image's features.
    – Perform the Contract operation to increase the color contrast of the image.
    – Perform the Enhance operation in order to improve the light difference between the real image and the splicing object.

After this step, the spliced area has been highlighted with the real one. This step is crucial for image feature detection during training model.
Step 3: Save the preprocessed images into the trained dataset and apply the Inception V3 to detect the forged images.

Inception V3 in Forgery Detection. To determine whether an image is fake or not, the authors have built a 10-step algorithm as follows:

Step 1: Define the training input data consisting of original and splicing.
Step 2: Resize the input image to 224 × 224.
Step 3: Divide the dataset into training set and testing set with the ratio 3:1. The dataset is updated from images which are preprocessed by contrast and enhance manipulations.
Step 4: Declare the using model Inception V3; Use ReLU activation function to reduce the computation cost; Use Softmax activation function to increase the accuracy in classification.
Step 5: Implement Checkpoint declaration to save training results.
Step 6: Initialize Optimizer with Learning Rate = 0.0001.
Step 7: Compile Model with Loss Function as Crossentropy.
Step 8: Training the model with Batch Size = 64, Epochs number = 100 and Validation number = 0.2.
Step 9: Evaluate the accuracy of the trained model.
Step 10: Testing with arbitrary images.

5 Simulation Results

This section introduces the datasets used to train and test, the environment of experiments, and presents the simulation results achieved for identifying the original or forged images. The proposed model is built to train and test these images on the CPU Ryzen 7 4800H, RAM 16 GB, Card: 1650Ti.

Datasets. Authors use dataset which is composed of 02 datasets: Columbia Uncompressed Image Splicing Detection Evaluation Dataset and images created by the authors. Total images are used to detect the forgery is 2629.

Experiment. In this section, the authors run the simulation to detect the image forgery in the case of using Inception V3 with preprocessing. The implementation aims to confirm that the accuracy is greatly improved only if there is preprocessing before detecting. This is also a suggestion for researchers to develop methods for identifying image forgery, the role of image preprocessing also plays an important and useful role. We use Inception V3 model after the preprocessing as shown in Table 1.

The training figure and an example of prediction are shown in Fig. 4 and Fig. 5, respectively.

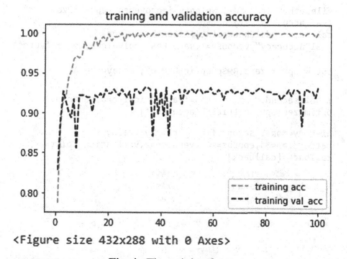

<Figure size 432x288 with 0 Axes>

Fig. 4. The training figure

From Fig. 4, the red dashed line is collection of value oscillating around 93%. This shows that the accuracy of the model is more than 93%. Actually, with experimental results, the average accuracy is 93.7%

In Fig. 5, the prediction percentage of the image belonging to the NoForgery class is 4.63^{-05}% and the Splicing class is 99%. This prediction result confirms that the input image 1 belongs to Splicing class and it also means that the prediction is exact.

Table 1. Steps for detecting the forged image in programming.

Step	Programming
1	`listObj = ['Original', 'Splicing']`
2	`IMG_SIZE = 224`
3	`x_train,x_test,y_train,y_test =` `train_test_split(X,Y,test_size=0.25,random_state=42)`

4	```# create the base pre-trained model```
	`base_model = InceptionV3(weights='imagenet', include_top=False)`
	
	`# add a global spatial average pooling layer`
	`x = base_model.output`
	`x = GlobalAveragePooling2D()(x)`
	`# let's add a fully-connected layer`
	`x = Dense(1024, activation='relu')(x)`
	`# and a logistic layer -- let's say we have 2 classes`
	`predictions = Dense(2, activation='softmax')(x)`
	
	`# this is the model we will train`
	`model_second = Model(inputs=base_model.input, outputs=predictions)`

5	`filepath="FirstModel-weights-improvement-epoch:02d}-`
	`{val_accuracy:.2f}.hdf5"`
	`callback=ModelCheckpoint(filepath,monitor=`
	`'val_accuracy",verbose=1,save_best_only=True,mode= 'auto')`

6	`opt = optimizers.RMSprop(lr=0.0001, decay=1e-6)`
7	`model_second.compile(loss='categorical_crossentropy',` `optimizer=opy, metrics=['accuracy']`

8	`history=model_second.fit(x_train,y_train,`
	`batch_size=64,epochs=100,verbose=2,validation_split=0.2,`
	`callback=[callback]`

9	`y_hat = model_second.predict(x_test)`
	`y_train_pre = np.argmax(y_hat, axis=1)`
	`y_val_label = np.argmax(y_test, axis=1)`
	`# Tính accuracy: (tp + tn) / (p + n)`
	`accuracy = accuracy_score(y_val_label, y_train_pre)`
	`print('Accuracy: %f' % accuracy)`
	`# Tính precision tp / (tp + fp)`
	`precision = precision_score(y_val_label, y_train_pre, average='macro')`
	`print('Precision: %f' % precision)`
	`# Tính recall: tp / (tp + fn)`
	`recall = recall_score(y_val_label, y_train_pre, average='macro')`
	`print('Recall: %f' % recall)`
	`# Tính f1: 2 tp / (2 tp + fp + fn)`
	`f1 = f1_score(y_val_label, y_train_pre, average='macro')`
	`print('F1 score: %f' % f1)`
	
	`# Tính confusion matrix`
	`matrix = confusion_matrix(y_val_label, y_train_pre)`
	`print(matrix)`

10	`Run the program`

Input image: Splicing

Prediction:
[[4.6380803e-05 9.9995363e-01]]
Result: Splicing

Fig. 5. An example of the prediction.

Input image: Original

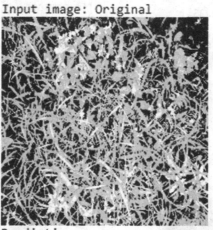

Prediction:
[[9.9999940e-01 5.6485516e-07]]
Result: NoForgery

Fig. 6. An example of the prediction.

Input image: Original

Prediction:
[[0.9649285 0.03507147]]
Result: NoForgery

Input image: Original

Prediction:
[[0.9108775 0.0891225]]
Result: NoForgery

Input image: Original

Prediction:
[[0.96302384 0.03697611]]
Result: NoForgery

Input image: Original

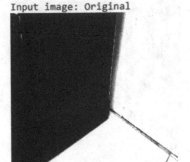

Prediction:
[[0.99833137 0.0016687]]
Result: NoForgery

Input image: Original

Prediction:
[[9.9977690e-01 2.2315579e-04]]
Result: NoForgery

Input image: Original

Prediction:
[[0.7867402 0.21325985]]
Result: NoForgery

Fig. 7. Simulation results.

Input image: Splicing

Prediction:
[[7.1552486e-06 9.9999285e-01]]
Result: Splicing

Input image: Splicing

Prediction:
[[4.945476e-04 9.995054e-01]]
Result: Splicing

Input image: Splicing

Prediction:
[[0.00951854 0.99048144]]
Result: Splicing

Input image: Splicing

Prediction:
[[0.00791391 0.9920861]]
Result: Splicing

Input image: Splicing

Prediction:
[[0.6697838 0.3302162]]
Result: NoForgery

Input image: Splicing

Prediction:
[[0.0069146 0.9930854]]
Result: Splicing

Fig. 7. continued

In Fig. 6, the prediction percentage of the image belonging to the NoForgery class is 99% and the Splicing class is $1.35^{-04}\%$. This prediction result confirms that the input image 1 belongs to NoForgery class and it also means that the prediction is exact.

Some other simulation results are also shown in Fig. 7. From the Fig. 7, we can see that the accuracy of the proposed method is relatively high. When applying in the dataset of 2629 images in which 2103 images are used for training and 526 images for testing, the average accuracy is 93.7%. This result proves that the model is a good candidate for spliced image detection.

Most of previous research mainly focused on the finding forged areas on the interfered images. This means that the input images are confirmed to be spliced already while the proposed method focuses on the image classification in huge datasets so the input images maybe original or splicing. Although the approaches are different, the objective is applied in image forensics, especially for spliced image detection. Our results can be applied to classify the forged images effectively with reliable accuracy.

To emphasize the role of pre-processing in the model, authors also implement the model without pre-processing in which only Inception V3 is applied. The average accuracy of model without pre-processing is 88% and with pre-processing is 93.7%. Their confusion matrices and training figures are shown in Table 2 and Fig. 8.

Table 2. Comparison the performance of Inception V3 with and without pre-processing.

Results	Inception V3		Pre-processing and Inception V3	
	True	False	True	False
True	75	16	140	32
False	11	128	9	447
Average	87.2%	88.8%	94%	93.3%
	88%		93.7%	

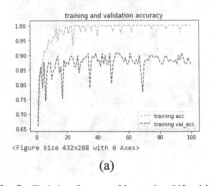
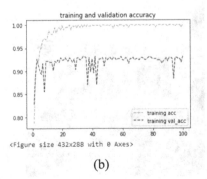

(a)   (b)

Fig. 8. Training figures of Inception V3 without pre-processing (a) and with pre-processing (b).

The validation accuracy in Fig. 8(a) is around the value of 88% and that in Fig. 8(b) is around the value of 93.7%. This proves that the model of Inception V3 with pre-processing is more effective than the model of Inception V3.

6 Conclusion

The paper aims to bring another aspect of image forgery detection based on deep learning algorithms. Factually, with the average accuracy of 93.7% in spliced image detection, our method proves that input image pre-processing combined with Inception V3 is efficient and contributes a new approach for the image forensics field. In addition, previous studies usually only evaluated on published datasets of spliced images such as Columbia Uncompressed Image Splicing Detection Evaluation Dataset or CASIA. The paper combines the standard Columbia dataset and the authors' built-in dataset which images are created and reconstructed the image forgery operations in the form of splicing. These kinds of forgery are the common image faking methods on the Internet, websites and social networks. Applying Mantranet in training process is an upcoming development orientation of the paper.

References

1. Szegedy, C., Vanhoucke, V., Ioffe, S., Shlens, J., Wojna, Z.: Rethinking the inception architecture for computer vision. In: Proceedings of the IEEE Conference on Computer Vision and Pattern Recognition, pp. 2818–2826 (2016)
2. Hsu, Y.F., Chang, S.F.: Detecting image splicing using geometry invariants and camera characteristics consistency. In: 2006 IEEE International Conference on Multimedia and Expo, pp. 549–552. IEEE, July 2006
3. Ng, T.T., Chang, S.F., Sun, Q.: Blind detection of photomontage using higher order statistics. In: 2004 IEEE international symposium on circuits and systems (IEEE Cat. No. 04CH37512), vol. 5, p. V. IEEE, May 2004
4. Fang, Z., Wang, S., Zhang, X.: Image splicing detection using camera characteristic inconsistency. In: 2009 International Conference on Multimedia Information Networking and Security, vol. 1, pp. 20–24. IEEE, November 2009
5. Zhang, Z., Wang, G., Bian, Y., Yu, Z.: A novel model for splicing detection. In: 2010 IEEE Fifth International Conference on Bio-Inspired Computing: Theories and Applications (BIC-TA), pp. 962–965. IEEE, September 2010
6. Popescu, A.C., Farid, H.: Exposing digital forgeries by detecting duplicated image regions (2004)
7. Mahdian, B., Saic, S.: Blind authentication using periodic properties of interpolation. IEEE Trans. Inf. Forensics Secur. 3(3), 529–538 (2008)
8. Kakar, P., Sudha, N., Ser, W.: Exposing digital image forgeries by detecting discrepancies in motion blur. IEEE Trans. Multimed. 13(3), 443–452 (2011)
9. Wang, W., Dong, J., Tan, T.: Effective image splicing detection based on image chroma. In: 2009 16th IEEE International Conference on Image Processing (ICIP), pp. 1257–1260. IEEE, November 2009
10. Mahdian, B., Saic, S.: Detection of resampling supplemented with noise inconsistencies analysis for image forensics. In: 2008 International Conference on Computational Sciences and Its Applications, pp. 546–556. IEEE, June 2008

11. Hsu, Y.F., Chang, S.F.: Statistical fusion of multiple cues for image tampering detection. In: 2008 42nd Asilomar Conference on Signals, Systems and Computers, pp. 1386–1390. IEEE, October 2008

12. Zhang, Z., Kang, J., Ren, Y.: An effective algorithm of image splicing detection. In: 2008 International Conference on Computer Science and Software Engineering, vol. 1, pp. 1035–1039. IEEE, December 2008

13. Youseph, S.N., Cherian, R.R.: Pixel and edge-based illuminant color estimation for image forgery detection. Proc. Comput. Sci. **46**, 1635–1642 (2015)

14. Vinoth, S., Gopi, E.S.: Neural network modeling of color array filter for digital forgery detection using kernel LDA. Proc. Technol. **10**, 498–504 (2013)

15. Liu, Y., Zhao, X.: Constrained image splicing detection and localization with attention-aware encoder-decoder and atrous convolution. IEEE Access **8**, 6729–6741 (2020)

16. Liu, Y., Zhu, X., Zhao, X., Cao, Y.: Adversarial learning for constrained image splicing detection and localization based on atrous convolution. IEEE Trans. Inf. Forensics Secur. **14**(10), 2551–2566 (2019)

17. Almawas, L., Alotaibi, A., Kurdi, H.: Comparative performance study of classification models for image-splicing detection. Proc. Comput. Sci. **175**, 278–285 (2020)

18. Szegedy, C., et al.: Going deeper with convolutions. In: Proceedings of the IEEE Conference on Computer Vision and Pattern Recognition, pp. 1–9 (2015)

19. OpenCV-Python-Tutorials–OpenCV. https://docs.opencv.org/3.4/da/df6/tutorial_py_table_of_contents_setup.html

Comprehensive Analysis of Privacy in Black-Box and White-Box Inference Attacks Against Generative Adversarial Network

Trung Ha[1], Tran Khanh Dang[2(✉)], and Nhan Nguyen-Tan[3]

[1] University of Information Technology, VNU-HCM, Thu Duc City, Vietnam
trunghlh@uit.edu.vn

[2] Ho Chi Minh City University of Technology (HCMUT), VNU-HCM, Ho Chi Minh City, Vietnam
khanh@hcmut.edu.vn

[3] H2A Technology Solutions Joint-Stock Company, Bien Hoa, Dong Nai, Vietnam

Abstract. Nowadays, deep learning models have many applications in social life. Specifically, the generative adversarial network (GAN) has many applications such as multimodal image-to-image translation, text to image, image filter, image editing, stylized images, data augmentation. However, deep neural networks are vulnerable to inference attacks as they memorize information about their training data. In this study, we set up black-box and white-box attacks to comprehensively evaluate the privacy of generalization models on the LFW dataset and CIFAR dataset. In addition, we measured the leakage of private information through the parameters of the fully trained model as well as the parameter updates of the model during training. In a white box attack setup, we evaluated inference attacks against GAN by monitoring their training data samples. In the black box attack setup, we divided it into two types of black box attacks with supporting information and without supporting information. We assumed that the attacker had about 10% to 20% of the target model training dataset in the black box attack with supporting information. Finally, we concluded the relationship between the number of training epochs and the GAN properties with information leakage.

Keywords: Differential privacy · Black-box · White-box · Inference attacks · GAN

1 Introduction

In recent years, Google, Amazon, and Microsoft have provided customers with access to APIs that allow them to easily integrate machine learning models into their applications. Organizations use machine learning as a service (MLaaS) leased from Google, Amazon, and Microsoft to build complex learning models such as: classifier training, prediction, clustering. They also allow users to query models trained on their data and can pay for it. However, if malicious users are able to recover the data used to train these models, the information leak will create serious problems [1, 23]. In particular, organizations do not have much control over the type of models and training parameters used by the

© Springer Nature Switzerland AG 2021
T. K. Dang et al. (Eds.): FDSE 2021, LNCS 13076, pp. 323–337, 2021.
https://doi.org/10.1007/978-3-030-91387-8_21

machine learning platform as a service, and this can lead to overfitting (i.e., model does not generalize well outside of the data that has been trained on), which provides attackers with a useful tool for recovering training data [18].

In the last decade, research in deep learning has made great progress, especially in the field of generative models. These models are used to generate new samples from the same underlying distribution of a given training dataset. In particular, one of the popular generative models is the generative adversarial network, which provides a way to produce artificially plausible images and videos, and are used in many applications, e.g. inpainting, compression, denoising [5], image manipulation [11, 21], image super-resolution [17, 19], image translation [20, 22].

In this paper, we study the feasibility of membership inference attacks against generative adversarial networks. We consider contexts such as given access to generative adversarial networks and an individual data record, an attacker can know if a particular record is used to train the model. Member inferences on generative models can be more difficult than discriminator ones [17]. Besides that, we consider two properties of the generative adversarial networks: diversity and accuracy, which affect the process of privacy leak attack and defense.

In details, we consider both membership inference attacks in the black box and white box: firstly, the adversary can only make queries to the attacked model, i.e. the model target, and it has no access to the internal parameters. This is called "black box" in client server architecture: the attacker sends queries to the server containing the machine learning models. Secondly, he also has accessed to the parameters, this case is considered in the collaboration model. To mount the attacks, we train the GAN model on samples generated from the target model [17]; specifically, we use generative models as a method to learn information about the target generative model and thus create a local copy of the target model from which we can initiate attack. We predict if a generative model is overfit, then the GAN, which combines the generative and discriminative model, will be able to detect this overfitting, even if it is not observable, since the discriminator is trained to learn different distributions. We rely on GANs to classify real and fake records to detect differences in the records generated from the target model. For white box attacks, the attacker-trained discriminator itself can be used to measure the information leakage of the target model. In addition, we consider information leakage in the trained GAN model and how GAN attributes affect the privacy level of the model.

In general, membership inference attacks are often the starting point for further attacks. It means that the adversary first infers whether the victim's data is part of the accessed information and then other attacks (e.g. attribute inference [8, 24]) can leak additional information about the victim [27].

The next section presents the theory about GAN. Section 3 briefly reviews the work involved with GAN including condition generative adversarial networks, control generative adversarial networks, and Wasserstein generative adversarial networks. Section 4 describes the experimentation, and Sect. 5 concludes the paper.

2 Background

Generative adversarial networks are neural networks trained in an adversarial manner to generate data based on the distribution [6]. The main idea of generative adversarial networks is that there were two competing neural network models. One model is a generator, whose task is to add noise to the input data and create samples. The other model is a discriminator, the task of the model is to receive data from the generator and the data set, and to classify between the generated data and the training data set.

The generator and the discriminator work together to oppose each other. The goal for generative networks is to learn how to generate samples that are as closed as possible to the actual data from the input data and noise. The goal of the discriminative network is to learn to get better and better at the data classification that is generated with real data as illustrated in Fig. 1.

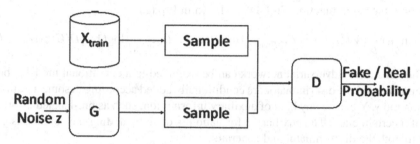

Fig. 1. Generative adversarial networks

Specifically, the output of the generator on data x is determined based on the input noise variable $p_z(z)$, then map to the space $G(z; \theta_g)$, where G is a deep learning model with parameter θ_g. The discriminator $D(x; \theta_d)$ produces the output $D(x) \in [0, 1]$, which represents the probability that x is derived from the training set rather than from the process of data generated from the generator G. The discriminator D is trained to maximize the probability of correctly labeling both the real training data and the generated data from G. The process of training the generating network G gives the way how to achieve minimization $\log(1 - D(G(z)))$. The optimization problem of generative adversarial networks is solved by the discriminator D and the generator G as shown in (1).

$$\min_{G} \max_{D} E_{x \sim P_{data}(x)} \left[\log D(x) \right] + E_{x \sim p_z(z)} \left[log(1 - D(G(z))) \right] \qquad (1)$$

First, the gradient of D is calculated to distinguish the generated data and the training data, then G is updated to make the data samples as close as possible to the real data. After several training steps, if G and D reach equilibrium, both models will reach a point where no further improvement is possible [6].

3   Related Work

3.1   Condition Generative Adversarial Networks

The generative adversarial network consisted of two 'compete' models: a generator model G that generated a data distribution and a discriminator model D that was used to distinguish a data sample from the training data and the generator G. Both G and D could be non-linear functions [7, 12].

To explore the generation of data on distribution models p_g over data generators x, the generator uses a function that maps from the predefined noise distribution $p_z(z)$ to the data space $G(z, \theta_g)$. And the discriminator $D(x, \theta_d)$ produces a scalar value that determines whether x is the data coming from the training set or not.

G and D are both trained concurrently: both of these training sets are adjusted to the parameters for G so that $\log(1 - D(G(z))$ is as small as possible (corresponding to generate data as close to the real dataset as possible). And loss function is used to adjust the parameters for D so that $\log D(X)$ is maximized (corresponding to detect larger deviations between the training data in the model and data are generated from generator G), as if they were following the min-max theoretical model of two players playing the game with the value function V(G, D) as shown in Eq. (2).

$$\min_G \max_D V(G, D) = E_{x \sim p_{data}(x)}(\log D(x)) + E_{x \sim p_z(z)}(\log(1 - D(G(z)))) \qquad (2)$$

The generative adversarial network can be extended to a conditional model if both the generator and the discriminator are conditionally added according to some additional information y. Y can be any kind of auxiliary information, such as pre-labeled information of a certain class. The auxiliary information is done by adding the conditions y as input to both the discriminator and generator.

In the generator, the input noise data predefined through the function $p_z(z)$ and y are combined to provide training for creating the generative model. In the discriminator, x and y are presented as inputs to construct the discriminant function. The objective function of minimax theory is shown in Eq. (3).

$$\min_G \max_D V(G, D) = E_{x \sim p_{data}(x)}(\log D(x|y)) + E_{x \sim p_z(z)}(\log(1 - D(G(z|y)))) \qquad (3)$$

3.2 Control Generative Adversarial Networks

The condition generative adversarial network operated on the given labels as input to the generator and the discriminator to limit them to operate under certain conditions.

The condition generative adversarial network mainly focused on generating data that were like the given actual data, rather than generating random data. As a result, it was difficult to generate samples with detailed features while condition GAN produced data with predefined important features.

The control generative adversarial network (control GAN) was composed of three modules: a generator/decoder, a discriminator and a classifier/encoder [26]. The generator in the model competed with the discriminator and classifier; at the same time, the generator was creating realistic data to fool the discriminator and classifier, while the classifier was correctly distinguishing the generated data from the real data.

The control generative adversarial network had two main advantages over existing models. First, the control generative adversarial network could be trained to focus more on specified input labels to generate samples with detailed features that were difficult for conditional GANs to generate. Second, the control generative adversarial network used an independent network to map the functions to the corresponding inputs while the discriminator performed the same function as the condition generative adversarial network and variable forms of GAN. Thus, the discriminator could focus on its own goal, which is to distinguish between fake and real samples, so that the quality of the generated samples could be improved as shown in Eq. (4).

$$
\theta_D = \arg \min\{\alpha \cdot L_D(t_D, D(x; \theta_D)) + (1 - \alpha) \cdot L_D((1 - t_D), D(G(z, l; \theta_G); \theta_D))\}
$$
$$
\theta_G = \arg \min \{\gamma_t \cdot L_C(l, G(z, l; \theta_G)) + L_D(t_D, D(G(z, l; \theta_G); \theta_D))\}
$$
$$
\theta_C = \arg \min\{L_C(l, x; \theta_C)\}
$$

(4)

where l is the binary representation of the labels marking the pattern x or as input to the generator, t_D is the label assigned to the discriminator, and α represents a parameter to the discriminator. The parameter γ_t determined the level of the generator on the labeled inputs of the generation process.

3.3 Wasserstein Generative Adversarial Networks

In the process of implementing the Wasserstein generative adversarial network model, an important component in the model was how to measure the distance between the data distribution densities, which was called the Wasserstein measurement.

The 1-Wasserstein measure is the distance between probability distributions. The measurement is defined using the transport theory concept of an optimal transport plan [15, 16]. Specifically, it supposes that we have two random variables A and B on a space X with probability distributions P_A and P_B respectively, then the 1-Wasserstein distance $d_W(P_A, P_B)$ is determined as shown in Eq. (5).

$$
d_W(P_A, P_B) = \inf_{(X,Y) \in \Pi(A,B)} E[\|X - Y\|]
$$

(5)

where $\Pi(A, B)$ is the set of all random variables with concurrent and marginal distributions of A and B respectively. Intuitively, this can be thought of as a transport problem in that the fractional probability of P_A needs to be transported optimally to the probability distribution of P_B, and the transport cost is the mass to be transported times the distance it has to travel. The 1-Wasserstein distance between two random variables is the cost of the optimal transport.

The 1-Wasserstein distance can be calculated another way by using the Kantorovich-Rubinstein duality [10], which measures the 1-Wasserstein distance between two probability distributions over space X as shown in Eq. (6).

$$
d_W(P_A, P_B) = \sup_{f:X \to R} E_{x \sim P_A}[f(x)] - E_{x \sim P_B}[f(x)]
$$

(6)

where the function f does 1-Lipschitz continuous mapping.

However, the generative adversarial network is a method that learns a distribution $G_\omega(N)$ similar to an unknown distribution H from existing samples (the samples in the training set). A common approach is to create a discriminator D_ψ that considers for each sample and the probability of that that sample produced by $G_\omega(N)$ or H.

The Wasserstein generative adversarial network was a variant of the generative adversarial network introduced by Goodfellow [6]. In the Wasserstein generative adversarial

network, the goal of the discriminator was no longer limited to find the probability that
a single sample was produced by one distribution or another, but rather to estimate the
"distance" between two distributions. Especially, this is the 1-Wasserstein distance, also
known as the earth mover distance.

$$discriminator\ loss = -E\big[D_\psi(H) - D_\psi(G_w(N))\big]$$
$$generator\ loss = E\big[D_\psi(H) - D_\psi(G_w(N))\big] \tag{7}$$

When the discriminator performs optimally, then the deviation is equal to the coefficient of the 1-Wasserstein distance between the data set H and $G_\omega(N)$ (the data set is
produced by the generator), the coefficient depends on the constant Lipschitz D_ψ. The
discriminator is trained to get better at maximizing the 1-Wasserstein distance between
the real samples (the data samples in the training set) and the fake samples (the samples
are created by the generator), and the generator trained to minimize the 1-Wasserstein
distance between the real sample and the fake sample as shown in Eq. (7). The implementation of continuous D_ψ on Lipschitz varies and includes techniques such as constraining
the neural network's weights or adding a loss penalty.

4 The Experimentation

4.1 Threat Models

When an adversary in a machine learning model aimed to infer whether a known record
was presented in the training set of a generative model. Based on the knowledge that
the adversary had attack methods divided into two methods: black box and white box
attacks. Firstly, the adversary could only make queries to the attacked target model and
had no access to the model's internal parameters. Secondly, the adversary had accessed
to the parameters of a target model. In both implementations, if the adversary knew the
size of the training set, it was not allowed to know its original data points. Variations of
the attack allowed the adversary to access some extra information. Attack accuracy was
measured by the degree of deviation of the inferred data from the set data.

In evaluating the accuracy of an attack on machine learning models, the problem
that the adversary was trying to solve, and was to distinguish the data used to train the
target model. Thus, supposing in the case that an adversary had accessed to a dataset
that they suspected contains the training data. However, the construction of the attack
did not depend on access to any data set. It was assumed that the adversary knew the
size of the training set; as part of the information including metadata in the target model
or leaked data after being attacked; but they did not know how the detailed data in the
dataset were divided in the training set and the test set.

In a black box attack setup, it was assumed that the adversary had no prior information
about the training dataset or the target model. In particular, the attack took place without
knowing the followings:

- In the target model parameters and hyper-parameters, there was no access to the
  network weights from the trained target model, nor to hyper-parameters such as
  regularization parameters or the number of epochs used for training in the model.

- In the target model architecture, the attacker had no knowledge of the target model's architecture.
- Data set used to train the target model: there was no knowledge of the data used to train the target model during the training process, because this was inferred from sampling the target model at the time of inference. In contrast, the training dataset membership inference attack on Shokri's discriminative models required some information about the dataset [18]. For example, information about the format of the dataset was used in training to generate aggregated data samples used in the attack.
- Predictive value: Shokri showed that the predictive index leaked the information used to perform membership inference attacks [18]. However, due to the nature of the generative model, an adversary could not generate a prediction score directly from the target model.

4.2 White-Box Attack

The white box attack was shown as illustrated in Fig. 2. To evaluate the attack, it was assumed that the adversary A_{wb} had access to the trained target model, namely the GAN - i.e. the generator model G_{target} and the discriminator model D_{target}. In addition, the white-box attack simulation further assumed that the attacker had access to the dataset, $X = \{x_1, ..., x_{m+n}\}$, which they suspected contains data that were known used to train the target model, where n was the size of the training set and m was the number of data that did not belong to the training set.

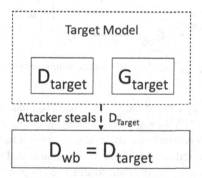

Fig. 2. The white-box attack

The target model was trained to generate patterns that were similar to the training set samples. A_{wb} created a local copy of D_{target}, which was called D_{wb}. Then, A_{wb} imported all samples $X = \{x_1, ..., x_{m+n}\}$ into D_{wb}, outputting the resulting probability vector p $= [D_{wb}(x_1), ..., D_{wb}(x_{m+n})]$. If the trained target model was overfitting on the training dataset, D_{wb} would achieve a higher confidence value for the samples, which were also part of the training set. A_{wb} sorted the prediction results of p in descending order and took the associated samples with the largest probability n as predictors of whether a data was a member of the training set.

The adversary did not need to train an attacker model; instead, they could rely on internal access to the target model, from which the attack could be launched.

4.3 Black-Box Attack Without Auxiliary Knowledge

In a black box attack setup, there was an assumption that the adversary A_{bb} did not have access to the target model parameters. Therefore, A_{bb} could not directly steal the discriminative model from the target model like in a white box attack. Furthermore, in a white box attack, the attack setup model was the generative adversarial network model. In the black box attack, it was not limited to the target model but did not include the discriminator model in the generative adversarial network. To evaluate the attack, there was assumption that the attacker had accessed to the dataset, $X = \{x_1, ..., x_{m+n}\}$, in the suspected dataset containing the data points used to train the target model, where n is the size of the training set. However, the adversary did not know how to construct the training set from X; therefore, they did not access to the labeled training data of the samples from the dataset nor could they train the model using the discriminator model approach. Instead, A_{bb} trained to build a local GAN to recreate the target model and generate a discriminator D_{bb}, the D_{bb} model could detect overfit in the target generative model G_{target}, the attack setting was modeled as illustrated in Fig. 3.

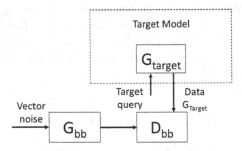

Fig. 3. The black-box attack without auxiliary

Specifically, A_{bb} locally trained the GAN (G_{bb}, D_{bb}) using queries from the target model, i.e. A_{bb} trained the GAN locally on the samples created by G_{target}. Note that this was a black box attack, so depending only on the samples returned by the target model, G_{target} could be any generated model. There was an additional assumption that A_{bb} had no knowledge and controlled over the data source used to generate the random samples generated by G_{bb}.

4.4 Black-Box Attack with Auxiliary Knowledge

In the black box attack presented in Sect. 4.3, there was an assumption that A_{bb} had no additional knowledge about the subsets of the training dataset. However, the case where an adversary could take advantage of limited additional information about the training set was not considered in the attack setup in Sect. 4.3. This was a real setup in practice. For example, knowledge of social networks was used to anonymize information on social networks [4]. Overall, adjunctive/incomplete knowledge of sensitive datasets was a common assumption in attack setups in [9, 13].

Accessing information about the training set meant that an adversary could "enhance" the probability of success when performing a black box attack. Two attack implementations were considered: generator attack and discriminator attack. In both implementations, one scenario was considered where the adversary had incomplete knowledge about the members of the test dataset, the training dataset, or both.

In the discriminator setup: the adversary trained a simple discriminator model to infer membership of the target model's training set as illustrated in Fig. 4. This was possible because the adversary now had accessed to the labels marked for the dataset as data points that belonged to the training set of the target model or not. Therefore, they did not need to train the generative model to detect overfit the target model. In this setting, two scenarios were considered:

(1) The adversary had a partial knowledge of the patterns that was used to train the target model.
(2) The adversary had limited complementary knowledge about both the training and test set samples.

In both cases, the attack method was the same: the adversary trained the model locally to detect to overfitting in the target model. In (a), the discriminator D was fed samples from this auxiliary set, including those labeled as fake and those generated by the target

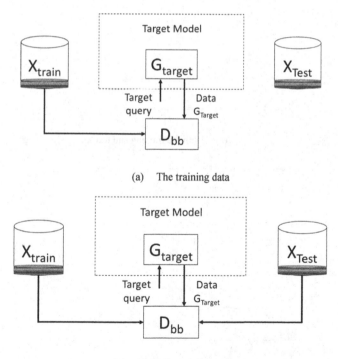

(a)   The training data

(b)   Both the training and test

Fig. 4. The Black-box attack with auxiliary knowledge

model labeled as real. If the target model overfitted the training dataset, D would learn to distinguish between the samples used for training and those generated from the target generator. In (b), D was provided for both the samples generated by the target model and the included auxiliary samples labeled as real, and the samples from the auxiliary test set labeled as fake. When the adversary knew some test samples (i.e. fake samples) in order to train a binary classification (this classification was divided into two parts: the sample contained in the set trained or not in the training sample set).

4.5 Experimental Results

The test environment was set up that the experiments were performed using PyTorch on google colab platform running in GPU. Dataset was generated that the attacks were performed on two machine learning datasets, that is LFW [2] and CIFAR-10 [3].

In white box attacks, measuring the accuracy of the member inference attack at epochs in successive run of training the target model, where one epoch corresponded to

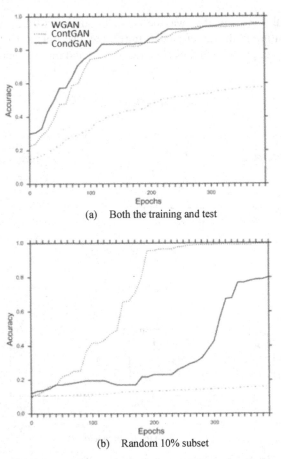

(a)   Both the training and test

(b)   Random 10% subset

Fig. 5. Accuracy of white-box attack with LFW dataset

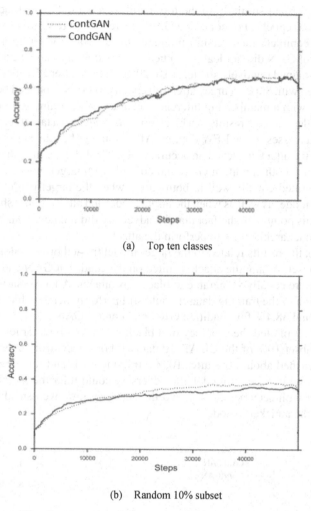

(a)   Top ten classes

(b)   Random 10% subset

Fig. 6. Accuracy of black-box attack on LFW datasets

one training round on all inputs to the training dataset. For black box attacks, the target model was fixed and the membership inference accuracy was measured at successive training steps of the attacker model, where a training step was defined as one iteration of training on mini-batch. The adversarial model was trained using pre-assigned labels and perturbation as suggested in [14]. In addition, the labels would also be swapped when the discriminator was trained. These modifications to the GAN had been shown to stabilize training in practice.

In Fig. 5(b), the results of white box attacks against a target model trained on random 10% subset of the LFW dataset. Similar to Fig. 5(a), both condition GAN (CondGAN) and control GAN (ContGAN) were vulnerable: when they were trained for 250 epochs, an adversary could completely infer whether a data was a member data of the training set. WGAN performed similarly to the top ten classes white box test, achieving 58% accuracy after 400 epochs. For condition GAN, the results were similar to control GAN on LFW, with complete membership inference on the training set after 400 epochs. However, control GAN did not leak too much information until about 250 epochs, and the accuracy remained relatively stable, at 10–20%. Instead, after 250 epochs, the model performed better, with 80% accuracy after 400 epochs. WGAN produced better samples, without overfit, with a membership inference accuracy of the training set of 19%.

Figure 6(a) depicted the results of a black box attack against a target model trained on ten labeled data classes of the LFW dataset. After training the attacker model on target model queries, the attack achieved an accuracy of 62% of the membership inference of the training set for both condition GAN and control GAN target models. To a surprise, the attack performed equally well in both cases: when the target model was different from the attack model as well as when the target model and attack model shared the same architecture. This pointed to the fact that the adversary did not need to have knowledge of the target model architecture to perform the attack.

In Fig. 6(b), the results related to the target model trained on a random subset 10% of the LFW dataset. Again, the attack resulted on the condition GAN and control GAN target models were equally vulnerable to black box attacks. An adversary without auxiliary information of the training dataset could still perform membership inference with accuracy 38% and 38.4% for condition GAN and control GAN.

Finally, Fig. 7 plotted the accuracy of a black box attack on a target model trained on a random subset 10% of the CIFAR-10 dataset. For the control GAN target model, the accuracy reached about 20% after 10,000 training steps and remained unchanged. For the condition GAN target model, the adversary could infer the membership of the training dataset with accuracy 36.4%, and the accuracy improved steadily throughout the training of the attacker's model.

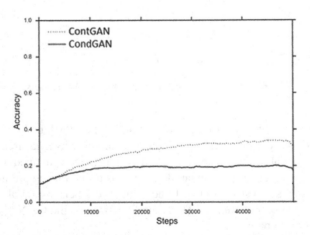

Fig. 7. Accuracy of black-box attack on CIFAR-10 datasets

In a black box attack with auxiliary knowledge, it supposed that the adversary had information about the training dataset and the number of training datasets were gradually increased to evaluate the leakage as well as the model's tolerance: (i) The adversary had 20% knowledge of the test data set; (ii) The adversary had 30% knowledge of both training and test data set.

Figure 8 showed the accuracy results for both setting, the attack failed with both datasets (LFW dataset and CIFAR dataset) when the adversary only had information about the test dataset. Whereas, if the adversary had both training data and test data; with LFW, the adversary achieved 50% accuracy, while for CIFAR-10 the accuracy reached 33%. These approaches did not improve the results of the black box attack methods on the CIFAR-10 dataset without supporting information, and only slightly improved the results on the LFW dataset. Therefore, the test methods also approached on (i) for black box attacks with supporting information about the adversary.

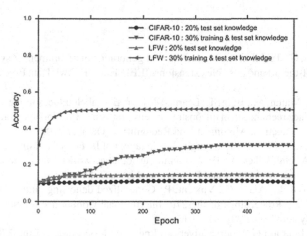

Fig. 8. Membership inference accuracy using a discriminator model of condition GAN

5 Conclusions

In this research, the leak evaluation of the generative model is presented, showing that many models are vulnerable to membership inference attacks, i.e. the presence of data in the training dataset. Attacks can be used to detect overfitting in generative models and help to choose an appropriate model that reduce leak information during the training process. Furthermore, GAN variant methods can be attacked to leak information members in the training dataset, and an adversary with limited supporting information about the data set can significantly improve their success.

Developments in machine learning have been enabled by the ability to analyze large amounts of data and refine model parameters to better encode information patterns in the data. On the one hand, generative models generate additional data samples to the training set so that more general models can be trained and processed new unseen data. On the other hand, attributes of the training data or dataset specific credentials should not be memorized during training and then revealed by membership inference attacks to avoid violating user privacy. But in the [18, 25], as well as the experiments in this study, it is shown that generative models can lead to information leakage of the user's training data. However, the close association between privacy violation and overfitting leads to the conclusion that security mechanisms such as differential privacy can work in conjunction with standardized training techniques to achieve the goals of protecting user privacy in generative models.

Acknowledgment. This work is supported by a project with the Department of Science and Technology, Ho Chi Minh City, Vietnam (contract with HCMUT No. 42/2019/HD-QPTKHCN, dated 11/7/2019). We also thank all members of AC Lab and D-STAR Lab for their great supports and comments during the preparation of this paper.

References

1. Adomavicius, G., Tuzhilin, A.: Toward the next generation of recommender systems: a survey of the state-of-the-art and possible extensions. IEEE Trans. Knowl. Data Eng. **17**(6), 734–749 (2005)
2. Huang, G.B., Mattar, M., Berg, T., Learned-Miller, E.: Labeled faces in the wild: a database for studying face recognition in unconstrained environments. In: Workshop on Faces in 'Real-Life' Images: Detection, Alignment, and Recognition, October 2008
3. Krizhevsky, A., Hinton, G.: Learning multiple layers of features from tiny images (2009)
4. Narayanan, A., Shmatikov, V.: De-anonymizing social networks. In: Proceedings of the 30th IEEE Symposium on Security and Privacy, pp. 173–187, May 2009
5. Bengio, Y., Yao, L., Alain, G., Vincent, P.: Generalized denoising auto-encoders as generative models. In: Proceedings of the 27th International Conference on Neural Information Processing Systems, pp. 899–907 (2013)
6. Goodfellow, I., et al.: Generative adversarial nets. In: Proceedings of the 27th International Conference on Neural Information Processing Systems, pp. 2672–2680 (2014)
7. Mirza, M., Osindero, S.: Conditional generative adversarial nets. arXiv preprint arXiv:1411.1784 (2014)
8. Ateniese, G., Mancini, L.V., Spognardi, A., Villani, A., Vitali, D., Felici, G.: Hacking smart machines with smarter ones: how to extract meaningful data from machine learning classifiers. Int. J. Secur. Netw. **10**(3), 137–150 (2015)
9. Ji, S., Li, W., Gong, N.Z., Mittal, P., Beyah, R.A.: On your social network de-anonymizablity: quantification and large-scale evaluation with seed knowledge. In: Proceedings of the 22nd Annual Network and Distributed System Security Symposium (NDSS), San Diego, California, USA, February 2015
10. Asghar, H.J., Melis, L., Soldani, C., De Cristofaro, E., Kaafar, M.A., Mathy, L.: Splitbox: toward efficient private network function virtualization. In: Proceedings of the ACM SIG-COMM Workshop on Hot Topics in Middleboxes and Network Function Virtualization, pp. 7–13, August 2016
11. Zhu, J.-Y., Krähenbühl, P., Shechtman, E., Efros, A.A.: Generative visual manipulation on the natural image manifold. In: Leibe, B., Matas, J., Sebe, N., Welling, M. (eds.) ECCV 2016. LNCS, vol. 9909, pp. 597–613. Springer, Cham (2016). https://doi.org/10.1007/978-3-319-46454-1_36
12. Oord, A.V.D., Kalchbrenner, N., Vinyals, O., Espeholt, L., Graves, A., Kavukcuoglu, K.: Conditional image generation with PixelCNN decoders. In: Proceedings of the 30th International Conference on Neural Information Processing Systems, pp. 4797–4805 (2016)

13. Qian, J., Li, X.Y., Zhang, C., Chen, L.: De-anonymizing social networks and inferring private attributes using knowledge graphs. In: Proceedings of the 35th Annual IEEE International Conference on Computer Communications, pp. 1–9, April 2016
14. Salimans, T., Goodfellow, I., Zaremba, W., Cheung, V., Radford, A., Chen, X.: Improved techniques for training GANs. In: Proceedings of the 30th International Conference on Neural Information Processing Systems, pp. 2234–2242, December 2016
15. Arjovsky, M., Chintala, S., Bottou, L.: Wasserstein generative adversarial networks. In: Proceedings of the 34th International Conference on Machine Learning, vol. 70, pp. 214–223, August 2017
16. Gulrajani, I., Ahmed, F., Arjovsky, M., Dumoulin, V., Courville, A.: Improved training of Wasserstein GANs. In: Proceedings of the 31st International Conference on Neural Information Processing Systems, pp. 5769–5779, December 2017
17. Ledig, C., et al.: Photo-realistic single image super-resolution using a generative adversarial network. In: Proceedings of the IEEE Conference on Computer Vision and Pattern Recognition, pp. 4681–4690 (2017)
18. Shokri, R., Stronati, M., Song, C., Shmatikov, V.: Membership inference attacks against machine learning models. In: Proceedings of the 38th IEEE Symposium on Security and Privacy (SP), pp. 3–18, May 2017
19. Wu, B., Duan, H., Liu, Z., Sun, G.: SRPGAN: perceptual generative adversarial network for single image super resolution. arXiv preprint arXiv:1712.05927 (2017)
20. Zhu, J.Y., Park, T., Isola, P., Efros, A.A.: Unpaired image-to-image translation using cycle-consistent adversarial networks. In: Proceedings of the IEEE International Conference on Computer Vision, pp. 2223–2232 (2017)
21. Chen, Y., Lai, Y.K., Liu, Y.J.: CartoonGAN: generative adversarial networks for photo cartoonization. In: Proceedings of the IEEE Conference on Computer Vision and Pattern Recognition, pp. 9465–9474 (2018)
22. Karras, T., Laine, S., Aila, T.: A style-based generator architecture for generative adversarial networks. In: Proceedings of the IEEE/CVF Conference on Computer Vision and Pattern Recognition, pp. 4401–4410 (2019)
23. Ha, T., Dang, T.K., Dang, T.T., Truong, T.A., Nguyen, M.T.: Differential privacy in deep learning: an overview. In: Proceedings of the 13th International Conference on Advanced Computing and Applications (ACOMP), pp. 97–102, November 2019
24. Melis, L., Song, C., De Cristofaro, E., Shmatikov, V.: Exploiting unintended feature leakage in collaborative learning. In: Proceedings of the 40th IEEE Symposium on Security and Privacy (SP), pp. 497–512, April 2019
25. Nasr, M., Shokri, R., Houmansadr, A.: Comprehensive privacy analysis of deep learning: passive and active white-box inference attacks against centralized and federated learning. In: Proceedings of the 40th IEEE Symposium on Security and Privacy (SP), pp. 1021–1035, April 2019
26. Shen, Y., Gu, J., Tang, X., Zhou, B.: Interpreting the latent space of GANs for semantic face editing. In: Proceedings of the IEEE/CVF Conference on Computer Vision and Pattern Recognition, pp. 9243–9252 (2020)
27. Ha, T., Dang, T.K., Le, H., Truong, T.A.: Security and privacy issues in deep learning: a brief review. SN Comput. Sci. 1(5), 1–15 (2020)

Face Recognition in the Wild for Secure Authentication with Open Set Approach

Hieu Dao[1,3], Dinh-Huan Nguyen[1,3], and Minh-Triet Tran[1,2,3](✉)

[1] University of Science, Ho Chi Minh City, Vietnam
[2] John von Neumann Institute, Ho Chi Minh City, Vietnam
[3] Vietnam National University, Ho Chi Minh City, Vietnam
{dhieu,ndhuan}@apcs.fitus.edu.vn, tmtriet@fit.hcmus.edu.vn

Abstract. In everyday life, authentication is an indispensable process of human activities. Bio-metric authentication system is one of the effective solutions, because it uses human-based features, instead of other traditional features, such as pin, password, etc. However, to apply a face authentication system in practical applications, we need to ensure that the system must not try to recognize the face of an unknown person into known categories, meaning we need to reject faces of unknown people in our application. In this paper, we present the limitations of recent Deep Learning based methods in Face Recognition tasks. We then propose two methods helping Face Recognition system have the ability to reject faces from unknown people by using Open-Set concepts. We conduct the experiments on a subset of CASIA-WebFace dataset, with a train set that includes 7000 images of 100 known people and a test set that includes both known and unknown people. Without rejecting unknown faces, the regular face recognition, i.e. the baseline method, yields the accuracy of only 45.9%, as the method tries to classify all face photos into known classes. Our proposed methods, which are combined deep network of Facenet system with recent Open Set methods, are called Learning Placeholder on Facenet (P-Facenet) and Facenet with OpenMax (O-Facenet). They achieve the accuracy of 83.6% and 88.5% respectively. This is a potential approach for authentication with face recognition to decrease the error rate of the model when recognizing faces of unknown people in the wild.

Keywords: Face recognition · Face authentication · Open set recognition

1 Introduction

User authentication is the process of verifying a person's identity when accessing a computer system. This is the most essential and initial stage in any user-based system. There are various methods for authentication, and some of famous methods are token, password, PIN, bio-metric, etc.

H. Dao and D.-H. Nguyen—These authors contributed equally.

© Springer Nature Switzerland AG 2021
T. K. Dang et al. (Eds.): FDSE 2021, LNCS 13076, pp. 338–355, 2021.
https://doi.org/10.1007/978-3-030-91387-8_22

Bio-metric authentication is a technique of identifying a person by using human-based features such as face, fingerprint, voice, and so on. As a result, we do not need to remember additional characteristics like passwords or PINs, and the system can still easily and securely identify us. Nowadays, Bio-metric authentication achieves a lot of success in user authentication and is extensively utilized in a variety of systems, from those that require acceptable-level security to those that require high-level security.

Face authentication uses face features to verify a person's identity. Therefore, with the rapid development of camera devices nowadays, this approach becomes increasingly popular and has an urgent demand. Face authentication is based on Face Recognition, which is a task trying to recognize faces of various people and is a long-standing research topic in CVPR community [24]. Deep learning-based methods applied to Face Recognition achieve many significant results.

Traditional deep learning approaches, on the other hand, have the drawback of attempting to categorize every input into the categories that the model has learned. This can result in incorrect predictions. In particular, in real-world face recognition applications, we must ensure that when encountering photos from unknown people, the face recognition system does not try to identify these objects as known people. As a result, these models must be able to reject faces from unknown people, which is a critical requirement in many real-world applications.

Open-Set Recognition describes a scenario where only known classes are available during the training phase and requires the models not only to achieve high performance in known samples but also to recognize unknown samples in the testing process. Open-Set Recognition helps the model to reject unknown and helps many real-world applications in reducing the error rate when facing unknown examples.

In this paper, our main contribution is that we proposed and successfully implemented two improved versions of the Facenet [17] model inspired by two Open-Set based methods to overcome this predicament in Face Recognition for secure authentication. The first method, based on OpenMax idea [1], is called Facenet with OpenMax extension (O-Facenet). O-Facenet replaces the SoftMax layer with OpenMax layer and extends the original network ability to reject unknown people without modification. Besides that, we also apply another state-of-the-art method in Open-Set Recognition, PROSER [29], in Face Recognition to propose a method, called Learning Placeholder on Facenet (P-Facenet). P-Facenet has the ability to mimic the novel people distribution and also reject the unknowns. We conduct the experiments on a subset of CASIA-WebFace dataset, with a train set that includes 7000 images of 100 known people and a test set that includes both known and unknown people with the ratio of 50:50. We evaluate the test set with a regular face recognition network, i.e. the baseline method, yielding the accuracy of 45.9%. Meanwhiles, our two proposed methods, P-Facenet, and O-Facenet, achieve the accuracy of 83.6% and 88.5% respectively.

The content of this paper is organized as follows. In Sect. 2, we present the related work on Face Recognition and Open-Set Recognition. In Sect. 3, we introduce our proposed methods for Face Recognition based on Open-Set approach. The experiments of our Face Recognition methods are discussed in Sect. 4. Finally, we conclude and present open problems for future work in Sect. 5.

2   Related Work

2.1   Face Recognition

Combining machine learning techniques, such as principal component analysis, linear discriminant analysis or support vector machines, with on hand-crafted features, such as edges and texture descriptors, are regarded as the traditional methods. However, different variations encountered in unconstrained environments made it difficult to engineer features that are robust to them, thus researchers had to focus on specialized methods for each type of variation such as age-invariant methods [9,13], pose-invariant methods [5], and illumination-invariant methods [21,31]. Recently, deep learning methods based on convolutional neural networks (CNNs) have taken over traditional face recognition methods. The main advantage of deep learning methods is that they can be trained with very large datasets to learn the best features to represent the data.

In order to tackle the face recognition problem by deep learning approach, a model usually needs to have 3 modules: face detection, feature extraction, and classification module. Face detection is tasked with extracting the face so that other modules can learn it better because an image normally has more features in it than just a human face. Feature extraction is generally the backbone of a deep convolutional network (DCNN), and is tasked with extracting the generic features of a human face. Module classifier is used to assign identity for an input face.

To train DCNNs for face recognition, there are generally two research approaches to it: the ones that train a multi-class classifier which can separate different identities in the training set, and the ones that directly learn an embedding. The former can be achieved by using a softmax classifier [2,20], and the triplet loss [17], Large Margin Cosine Loss (LMCL) [23] and ArcFace loss [3] can be used for the latter. Both the softmax-loss-based methods [2] and the triplet-loss-based methods in Facenet [17], the ArcFace-loss-based in ArcFace [3] and the LMCL based in CosFace [23] can produce outstanding performance on face recognition based on the detailed DCNN architectures and the large-scale training data.

2.2   Open-Set Recognition

In general, generative models and discriminative models are the two approaches to Open-Set Recognition (OSR). Discriminative models place a strong emphasis on determining the border between classes in a dataset. Its purpose is to determine the most effective decision boundary to distinguish one class from another. In contrast, generative models concentrate on the distribution of individual classes in a dataset, while learning algorithms tend to mimic the data points' underlying patterns. As a result, discriminative models are resistant to outliers and unable to generate new data instances, whereas generative models are the opposite. Figure 1 shows the difference between discriminative and generative models.

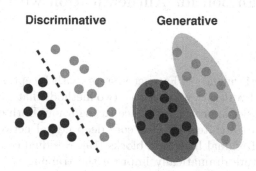

Fig. 1. Discriminative models and generative models Source: https://dataisutopia. com/blog/discremenet-generative-models

To approach the Open-Set problem using discriminative models, substantial progress has been made in both traditional machine learning methods and deep learning methods. Traditional machine learning methods including Support Vector Machines (SVMs), Nearest Neighbors, and Sparse Representation are commonly employed in a range of approaches where the training and testing data come from the same distribution. In OSR, however, such an assumption is no longer valid. Therefore, many changes to the existing models have been made to adapt these traditional methods to such situations [7,16,27]. Deep networks have been adopted in Open-Set Learning due to the substantial outcomes of Deep Learning in many Computer Vision problems and acquire more attractive results due to the powerful representation ability. [1], one of the early pioneers, proposes OpenMax, a new method that replaces the Softmax layer in the network by adjusting the output probability with the Weibull distribution. K J Joseph et al. [8] recently proposed an Energy Based Unknown Identifier in their models, building on [10]'s work. In addition, various efforts have been made to develop end-to-end models to deal with Open-Set situations by affecting the embedding space of known and unknown cases [4,6].

In contrast to discriminative models, generative approaches take advantage of Generative Adversarial Network (GAN) to create additional unknown samples from known data. As a result, generative methods may directly forecast the distribution of unknown cases, which could be beneficial for OSR. Geet al. [30] introduces the Generative OpenMax approach, which uses generative models to create unknown samples to train the network, which is inspired by OpenMax.

Besides GANs, Auto-Encoder based models such as Variational Auto Encoder (VAE) and Ladder Variational Autoencoders (L-VAE) are often used as generative models [12, 18, 26]. To improve their outcomes, numerous approaches have recently used generative and discriminative models. GANs or the Mixup [22] approach are frequently used in these methods to create pictures that lie between decision boundaries as counterfactual examples [11, 14, 29].

3   Face Recognition for Authentication with Open-Set Approach

3.1   Overview

Inspired by the performance of Facenet system, with Inception Resnet V1 as a core deep neural network [17], we proposed two methods that combined the above network and Open Set approaches in Face Recognition. Both our methods used Inception Resnet V1 architecture as a core deep neural network and inherited the advantages of Residual Inception blocks [19]. Residual connections help the Inception architecture dramatically improve the training speed, but still keep the significant performance. Then, we apply the ideas of Open-Set Recognition in our models, in order to help the model has the ability to reject unknown. Our first proposed method, which used OpenMax idea [1], is called Facenet with OpenMax (O-Facenet). And the second method, which inherited the Learning Placeholder idea in [29], is called Learning Placeholder with Facenet (P-Facenet).

3.2   Facenet with OpenMax

3.2.1   Overview

In order to tackle Face Recognition in the wild problem, we need to find an effective method that satisfies several essential qualities: firstly, it must not employ faces from unknown people during training phase; secondly, training models using that method must be fast and require less computing effort; and finally, it must be simple. While it is very difficult to satisfy all three conditions, OpenMax, one of the first deep network approaches to the Open-Set Recognition problem, is able to achieve such a feat. Despite being one of the pioneering methods, OpenMax has not at all become obsolete, and even managed to attain great results. This motivates us to utilize OpenMax as our first step to approach Face Recognition in the real-life scenario.

A. Bendale et al. [1] propose OpenMax method, and extended SoftMax function in unknown rejection. OpenMax is used as the final layer to compute the probability for both known and unknown classes. Therefor, we replace the SoftMax layer of Inception Resnet V1 in Facenet with OpenMax layer. The Facenet with OpenMax (O-Facenet) architecture is illustrated in Fig. 2.

OpenMax layer uses the distance from known training data in decision function of the network. The process of OpenMax can separate into two steps:

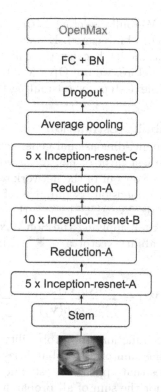

Fig. 2. Architecture of Facenet with OpenMax (O-Facenet)

- The first step, which is called Meta-recognition process, tries to create Weibull model p on training examples for estimating the outlier probability.
- The second step estimates the OpenMax probabilities for both known and unknown categories by continuously calibrating the threshold of Weibull models p. Therefore, the deep network has the ability to reject the faces of unknown people.

3.2.2 Meta-recognition Process

Meta-recognition algorithm allows deep networks to analyze scores and recognize when the input is not from a known class [1]. In [15], authors apply Extreme Value Theory on meta-recognition and claim the distribution of final scores in a deep network follow Weibull distribution. Therefore, OpenMax inherits this idea and creates Weibull model p for estimating the outlier probability of an input [1].

A. Bendale et al. [1] adapt the concept of Nearest Class Mean in the activation vector and results in a single mean, which is called mean activation vector (MAV). This mean μ is used for representing a class as a point. Then, for each class i, η largest distances between all positive training examples and their corresponding mean u_i are fitted into a Weibull distribution p_i by FitHigh function from libMR library [1]. The Weibull model p has the ability to define a threshold that decides if an input should be rejected. However, it is difficult to calibrate an absolute threshold for this model, because the calibration process depends on the unknown examples. Therefore, we use OpenMax algorithm to continuously adjust this threshold in the next step, for estimating the OpenMax probability.

3.2.3   OpenMax Probability Estimation

The SoftMax function is a multi-dimensional generalization of the logistic function. It is often used in the last layer of neural network as an activation function, in order to normalize the output of the network to a categorical probability distribution. Given d data dimension, N number of classes, a sample vector $\mathbf{x} \in \mathbb{R}^{d+1}$ and weight matrix $\mathbf{W} \in \mathbb{R}^{(d+1) \times N}$, the output of the network is $\mathbf{v} = f(\mathbf{x}) = \mathbf{W}^T \mathbf{x} \in \mathbb{R}^N$. After applying the SoftMax function for this output vector $\mathbf{v}$, we have the probability vector $\mathbf{a} \in \mathbb{R}^N$. The probability $\mathbf{a}_i$ for each class i is computed by:

$$\mathbf{a}_i = P(y = i | \mathbf{x}; \mathbf{W}) = \frac{\exp(\mathbf{v}_i)}{\sum_{j=1}^{N} \exp(\mathbf{v}_j)} = \frac{\exp(\mathbf{W}_i^T \mathbf{x})}{\sum_{j=1}^{N} \exp(\mathbf{W}_i^T \mathbf{x})} \tag{1}$$

After using the SoftMax function, the probability $\mathbf{a}_i$ of each class i will be in the interval $[0, 1]$, and the sum of all probabilities are equal to 1. However, there are unknown examples that appear at test time in Open-Set Recognition, hence, it is not acceptable for the sum of all probabilities to be 1 [1].

OpenMax extends SoftMax in Open-Set Recognition by using Meta-recognition Process to estimate both known and unknown probabilities. For convenience, we define unknown class as index 0. Suppose we have Weibull model p, and mean activation vector μ, computed from Sect. 3.2.2. The algorithm of OpenMax probability estimation is applied. First of all, we use α largest activation classes to compute weight ω for scaling the Weibull CDF probability. After that, we compute revised activation vector $\hat{v}(x) = v(x) \circ \omega(x)$ and a pseudo-activation for unknown class $\hat{v}_0(x)$. Combining the revised activation vector $\hat{v}$ and unknown activation $\hat{v}_0$, we can compute the OpenMax probabilities as below:

$$\hat{P}(y = i | \mathbf{x}) = \frac{\exp(\hat{v}_i)}{\sum_{j=1}^{N} \exp(\hat{v}_j)} \tag{2}$$

Finally, OpenMax can reject the unknown when the unknown probability is largest. Moreover, OpenMax also rejects the uncertain example as the unknown category when the probability of this example is smaller than a threshold ε.

3.3   Learning Placeholder with Facenet

While OpenMax is able to produce substantial results and fulfill our requirements, it is not trainable because it is created after the training process. Thus, it won't make the model smarter in terms of Open-Set Recognition, but merely uses its predictions in a more clever manner. Specifically, the model predicts the faces of unknown people by rejecting a face if it doesn't belong to the distribution of known faces or if it belongs to the distribution of unknown faces. Yet, without information for unknown faces, creating their distribution can be inaccurate. As such, we need to find a way to generate data for unknowns using existing data from known classes. Fortunately, there is a recent method that can meet the requirement of creating unknown samples from known samples called Learning Placeholder for Open-Set Recognition.

We apply Learning Placeholder idea on Facenet model architecture, which is called Learning Placeholder on Facenet (P-Facenet). This model can mimic the distribution of faces of unknown people from known people in training and have the ability to reject unknown ones. The architecture of P-Facenet is illustrated in Fig. 3.

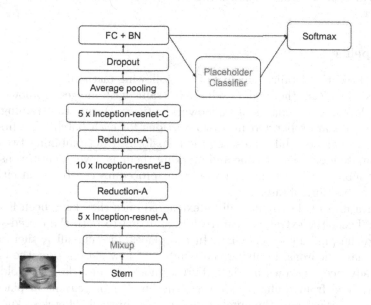

Fig. 3. Architecture of Learning Placeholder on Facenet (P-Facenet)

3.3.1   Overview Methods
Wei Zhou et al. [29] created Learning Placeholder for Open-Set Recogntion (PROSER) as a result of combining generative and discriminative models. PROSER aims to convert current closed-set to open-set models by adding two supplements. In detail, an extra **classifier placeholder** is introduced to a closed-set classifier instead of naively applying a specific threshold disregarding the classes applied to softmax score as a boundary between known and

unknown. The extra classifier can learn the dynamic thresholds for each learnt class by leveraging this. Furthermore, the persistent knowledge of current classes can be retained while learning new information for new unknown classes.

Additionally, **data placeholders** are utilized as an alternate approach for GANs to approximate open-set categories while maintaining a low complexity cost. Therefore, the model has the generative model property of directly estimating the distribution of novel classes. As a consequence, many state-of-the-art (SOTA) image classifiers can use this approach to improve their classification skill for instances beyond their understanding.

However, as stated in the original work, obtaining a well-trained closed-classifier is required before successfully expanding its capacity to determine novel instances. Since data placeholders are responsible for producing confusing examples as novel instances that sit between the decision boundaries of each class, if the boundaries are not adequately established, many known examples lying nearby will be treated as unknown. Inevitably, the present model's performance will suffer.

3.4   Pipeline

Figure 4 shows the detailed pipeline of our following method.

As seen in Fig. 4, the model exclusively trains with images of known identities and does not use images of unknown identities. During the training phase, the input data are separated into two portions for each batch. The first component is left alone, while the second is transformed by combining two images from separate identities into one and marking them as novel identities using the manifold mixup [22] technique. The mixing procedure can occur in either the input or embedding phrase.

The features of the inputs will be extracted once they have been fed to the network. Then, the extracted features are assigned to both the closed-set classifier and the placeholder classifier. In the placeholder classifier, there are not one but many neurons, implying the number of new classes that the classifier takes in advance, as shown in Fig. 4. Furthermore, because the placeholder classifier is trained from confusing pictures, the number of neurons in placeholder classifiers also represents the number of dynamic thresholds between known and unknown identities. Finally, the largest element in the placeholder classifier is selected and appended to the closed-set classifier. As a consequence, the model can distinguish between learned and novel identities.

3.4.1   Classifier Placeholders

An open-set model must fulfill certain characteristics, including the capacity to detect unexpected cases while maintaining acceptable performance on previously learned ones. As a result, introducing another classifier is logical since it can acquire the knowledge of new classes without engaging the present classifier.

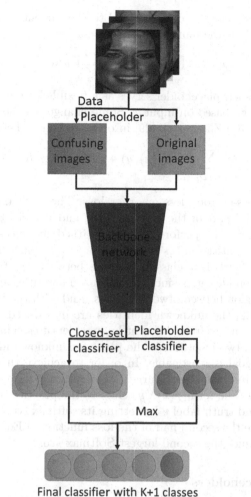

Fig. 4. Pipeline of learning placeholder for open-set recognition

Based on this concept, an additional classifier is generated as a classifier placeholder to supplement the current classifier. The output layer can be presented as below:

$$\hat{f}(x) = [W^T h(x), \hat{w}^T h(x)] \tag{3}$$

where W and $\hat{w}$ are used to denote a closed-set and open-set classifier, respectively. As presented in Fig. 4 and Eq. 3, the extra classifier $\hat{w}$ is embedded by $h(\cdot)$ in the same way as the current classifier. The placeholder classifier is built as a linear layer that can hold up to C classes in advance rather than just one. However, only the largest result of $\hat{w}_k^T h(\cdot)$, $k \in \{1, ..., C\}$, will be accepted as the unknown class's score and concatenated to the current classifier as the class K+1 result, where K is the number of known classes.

$$\hat{f}(x) = [W^T h(x), \max_{k=1,\ldots,C} \hat{w}_k^T h(x)] \tag{4}$$

Intuitively, the classifier placeholder serves as a flexible border that distinguishes unknown from known based on input x. Some changes to the classification loss have been done by Wei Zhou et al. [29] in order to comprehend this intuition:

$$l_1 = \sum_{(x,y) \in D_{train}} l(\hat{f}(x), y) + \beta l(\hat{f}(x) \setminus y, K+1) \tag{5}$$

where l can be cross-entropy loss or other loss. The normal classification loss function is the initial part of the loss function, and it seeks to at least reserve and improves the model's performance on learned instances. The process of learning unknown identities takes place in the second part. The second part's goal is to set the placeholder classifier's score between the true identities and neighboring mistaken identities since the unknown identities are likely to appear at the uncertain region between two identities. And with the help of data placeholders in Sect. 3.4.2, the unknown identities are introduced to the model as a combination of two images from different known identities. Therefore, the uncertain region between two identities is the place for unknown identities, which to the second part's goal is reasonable. In order to achieve this, the placeholder classifier must produce the second-largest score of known classes, as implied by the phrase $\hat{f}(x) \setminus y$. The term $\hat{f}(x) \setminus y$ refers to the practice of disregarding the output of the ground truth label y by setting its softmax score to 0. As a result, in order to optimize the second half of the loss function in Eq. 5, the placeholder classifier must produce the second-largest Softmax score.

3.4.2   Data Placeholders

The goal of learning data placeholders, as mentioned in the study, is to extend the closed-set model training method to open-set model training. In order to contribute to the learning of novel cases, data placeholders must fulfill two criteria. The first requirement is the ability to approximately generate distributions of real novel instances. The second requirement is that the generating process should be quick and simple. Despite being powerful in terms of perspective generation, the traditional method of using GANs or Auto-Encoder models as generative-based open-set models frequently necessitates more time to estimate the distribution of novel instances before effectively applying the result for training the classifier, resulting in a nearly doubled training time. Taking the aforementioned requirements into account, Wei Zhou et al. developed a simple yet efficient approach for imitating new patterns using manifold mixup.

The basic idea behind manifold mixup is that instead of blending two images together only at the input phase and then feeding them to a backbone network, the blending process can occur at any embedding stage within the network at random. Consider a backbone network to be a function $f(x)$ that learns to map an input picture x to its associated label y. Before reaching the final embedding layer, an image goes through many intermediate layers $h_m(x)$, which transform the input to the hidden representation. The final embedding layer can be denoted as $h_{post}(x)$. Therefore, the overall deep neural network can be described as $f(x) = h_{post}(h_m(x))$. As proposed in the original paper, the random layer m from a set of eligible layers S in the neural network is selected to perform the manifold mixup process. However, from the experiments we have conducted in Sect. 4, we discovered that the manifold mixup process should be performed after the STEAM layer of Facenet if we wished to improve Facenet and help it achieve better results. As such, we propose P-Facenet, a method to extend Facenet by applying PROSER method to it. The detailed architecture of P-Facenet is demonstrated in Fig. 3. Instead of choosing any two pictures to linearly merge in the hidden representation, Wei Zhou et al. add a new requirement: the two instances must belong to distinct classes. The mixup formulation is:

$$x_{mix} = \lambda \cdot h_{STEAM}(x_i) + (1 - \lambda) \cdot h_{STEAM}(x_j), y_i \neq y_j \qquad (6)$$

The mixing coefficient λ has the value from the intervals [0,1] and is generated from the Beta distribution. The x_{mix} then travel through later layers to obtain $h_{post}(x_{mix})$.

By doing this, the distribution of each known class is pushed away from each other during training process. As a result, the interpolation between two different clusters becomes the ideal places of unknown classes. Therefore, the embedding of x_{mix} can be treated as the embedding of open-set classes, and train them as novel ones:

$$l_2 = \sum_{(x_i,x_j) \in D_{train}} ([W, \hat{w}]^T h_{post}(x_{mix}), K + 1) \qquad (7)$$

$$x_{mix} = \lambda \cdot h_m(x_i) + (1 - \lambda) \cdot h_m(x_j), y_i \neq y_j$$

The mixup process occurs just inside each mini-batch throughout the training phase, as shown in Fig. 4. As a result, there is no need to preprocess the whole dataset to join all potential pairs of (x_i, x_j).

In summary, the training process is guided with these loss functions:

$$l_1 = \sum_{(x,y) \in D_{train}} l(\hat{f}(x), y) + \beta l(\hat{f}(x) \setminus y, K + 1)$$

$$l_2 = \sum_{(x_i,x_j) \in D_{train}} ([W, \hat{w}]^T h_{post}(x_{mix}), K + 1) \qquad (8)$$

$$l_{total} = l_1 + \gamma \cdot l_2$$

4   Experiments

We conduct the experiments for our proposed method on the CASIA-WebFace dataset [25], which is among one of the bigger datasets for face recognition, and

a) original images

b) images cropped
by using MTCNN

Fig. 5. Examples from CASIA-WebFace dataset.

is used for face verification and face identification tasks. The dataset contains 494,414 face images of 10,575 real identities collected from the web. This dataset contains not only frontal face images but also faces of people in the real environment, where there are many variations such as head poses, aging, occlusions, illumination conditions, and facial expressions. Figure 5 shows some examples from the CASIA-WebFace dataset.

From this dataset, we sample a subset of 100 people who were already registered into the system to act as known identities, each of which has 70 face images used for training and 10 for testing. To achieve a good diversity for unknown identities, we sample 1000 identities from CASIA-WebFace that are not identical to the aforementioned 100 known identities, and each unknown identity will contain 1 image. Thus, the test set will include 1000 images from known identities and 1000 images of unknown identities, yielding a known-unknown ratio of 1:1.

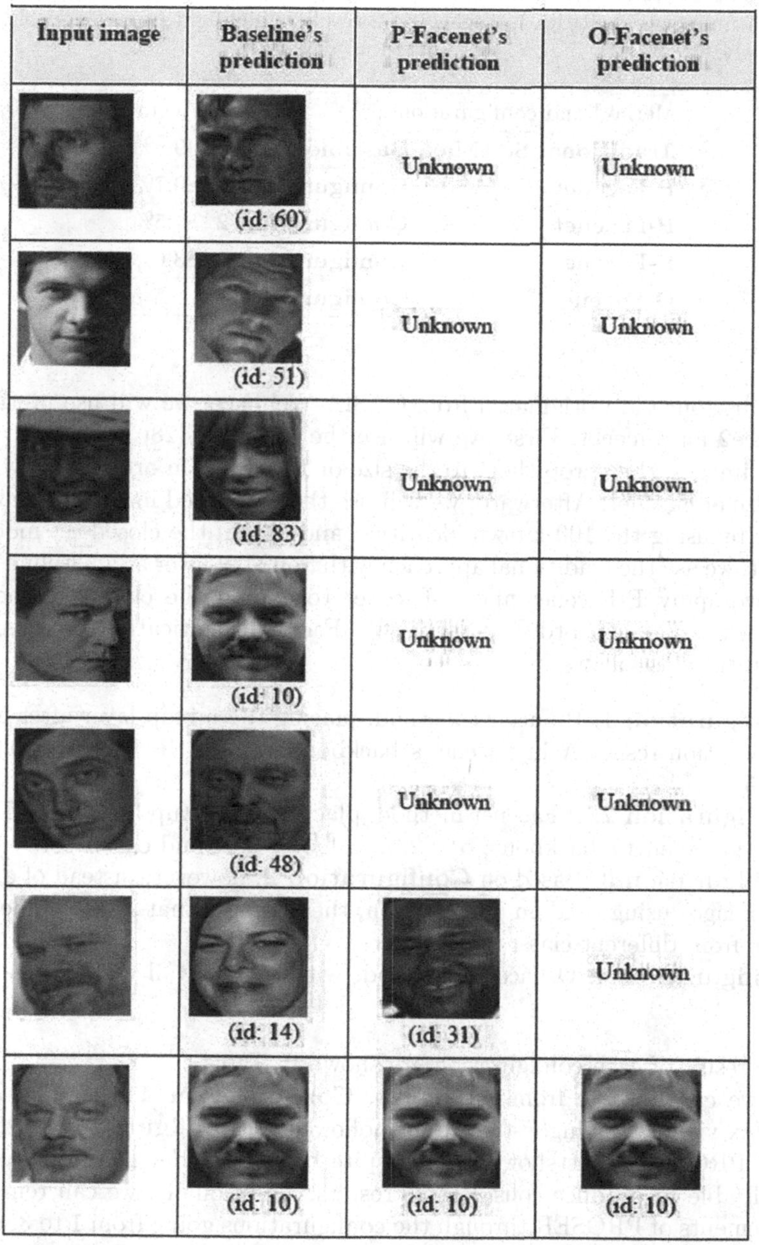

Input image	Baseline's prediction	P-Facenet's prediction	O-Facenet's prediction
	 (id: 60)	Unknown	Unknown
	 (id: 51)	Unknown	Unknown
	 (id: 83)	Unknown	Unknown
	 (id: 10)	Unknown	Unknown
	 (id: 48)	Unknown	Unknown
	 (id: 14)	 (id: 31)	Unknown
	 (id: 10)	 (id: 10)	 (id: 10)

Fig. 6. Example results for face recognition with unknown identities. The first column is the input image of an unknown person. Column 2, 3, and 4 represent the results of recognizing the input image with baseline approach (without rejecting unknown person), P-Facenet and O-Facenet, respectively

352   H. Dao et al.

Table 1. The result of baseline and 4 configurations on subset of CASIA-WebFace dataset. The performance is evaluated by accuracy in 101 identities (100 known identities and unknown identities). Top-1, top-2 result is highlight by red and blue color, respectively.

Method and configuration		Accuracy
Traditional SoftMax	**Baseline**	49.5%
P-Facenet	**Configuration 1**	80.7%
P-Facenet	**Configuration 2**	81.5%
P-Facenet	**Configuration 3**	83.6%
O-Facenet	**Configuration 4**	88.5%

Because our data originated from CASIA-WebFace, we will use pretrain on VGGFace2 for Facenet. First, we will use the MTCNN [28] network to detect faces in images, then crop them to the size of 160×160 in order to fit with the input size of Facenet. Afterward, we will use those cropped images to train with Facenet by using the 100 known identities, and obtain the closed-set model.

First, we use the traditional approach with SoftMax layer as a baseline model. Then, we apply P-Facenet and O-Facenet to extend the closed-set model to an open-set model in order to tackle the Face Authentication problem, with configurations as follows:

- **Configuration 1**: P-Facenet method, placing the mixup layer after 5 layers of Inception-resnet-A in Facenet's backbone, $\alpha = 2$, $\beta = 1$, $\lambda = 0.01$ and $\mathcal{C} = 5$.
- **Configuration 2**: P-Facenet method, placing the mixup layer after STEAM layer in Facenet's backbone, $\alpha = 2$, $\beta = 1$, $\lambda = 0.01$ and $\mathcal{C} = 5$.
- **Configuration 3**: Based on **Configuration 2**, however, instead of a mixup of 2 images using random permutation, the 2 images that are combined must come from different classes.
- **Configuration 4**: O-Facenet method, with $\alpha = 2$, tail = 10, threshold = 0.7.

The result of each configuration are shown in Table 1.

As we can observe from the results, **Configuration 4** where we applied OpenMax yielded the highest accuracy, followed by **Configuration 3** where we applied PROSER. This shows that, despite being an older method, OpenMax was still able to produce considerable results. Additionally, we can remark the improvements of PROSER through the configurations going from 1 to 3, demonstrating how PROSER's results can become better by applying the appropriate parameters. Overall, all configurations were able to generate greater results than using the traditional approach with SoftMax layer.

Figure 6 shows some successful and unsuccessful cases of P-Facenet and o-Facenet on examples with faces that come from unknown people, as well as the prediction of **Baseline** on the same examples.

The first column contains input images of unknown people where their expected classification should be unknown (ground truth). The second column illustrates incorrect results that were classified by **Baseline** method, because **Baseline** method will always try to classify any given face image into a known category even if they are unknown ones. The third and the fourth column show the result of our two proposed models which are P-Facenet and O-Facenet corresponding to **Configuration 3** and **Configuration 4**, respectively. If the result is unknown, which is the same compared to the ground truth, then it is the correct answer. From this, it can be seen that our proposed Facenet models still suffer in some cases where it provided the wrong classification, in which case we visualize one of the examples of the true person that belongs to the corresponding wrong predicted category.

5   Conclusion

Face authentication, a user authentication approach with human-based features, is more and more popular in many real-life applications with the help of newer technologies, such as Deep Learning. However, the recent method with recognition process have a limit, that is the model tries to classify the face of an unknown person into known categories. This leads to the wrong prediction, causes many errors, and unsecured authentication in real-life scenarios. Therefore, in this paper, we introduce two methods, O-Facenet and P-Facenet, based on Open-Set idea to tackle this problem, by extends the unknown face rejection ability to the Face Recognition model.

The experiment results on a sub-set of CASIA-WebFace dataset [25], which consists of 7000 face images of 100 known people for training process. Besides that, the test set includes face images of both known and unknown people. Without rejecting unknown faces, the regular face recognition, i.e. the baseline method, yields the accuracy of only 45.9%, as the method tries to classify all face photos into known classes. Our methods, O-Facenet and P-Facenet, achieve the accuracy of 88.5% and 83.6%, respectively. This is a potential approach in reducing the error rate of the model in real-life scenarios, which contains faces of various unknown people.

Currently, we are focusing our effort to improve our solution for Face Recognition for Secure Authentication in the Wild, and we also are trying to apply our methods on a larger subset of dataset with 1000, and 10000 categories from CASIA-WebFace. In the future, we consider other Open-Set learning methods for this problem.

Acknowledgements. This research is supported by research funding from Faculty of Information Technology, University of Science, Vietnam National University - Ho Chi Minh City.

References

1. Bendale, A., Boult, T.: Towards open set deep networks. In: Computer Vision and Pattern Recognition (CVPR), 2016 IEEE Conference on. IEEE (2016)

2. Cao, Q., Shen, L., Xie, W., Parkhi, O.M., Zisserman, A.: Vggface2: a dataset for recognising faces across pose and age. In: 2018 13th IEEE International Conference on Automatic Face Gesture Recognition (FG 2018), pp. 67–74 (2018). https://doi.org/10.1109/FG.2018.00020
3. Deng, J., Guo, J., Xue, N., Zafeiriou, S.: Arcface: additive angular margin loss for deep face recognition. In: 2019 IEEE/CVF Conference on Computer Vision and Pattern Recognition (CVPR), pp. 4685–4694 (2019). https://doi.org/10.1109/CVPR.2019.00482
4. Dhamija, A., Günther, M., Boult, T.: Reducing network agnostophobia. In: NeurIPS (2018)
5. Ding, C., Tao, D.: A comprehensive survey on pose-invariant face recognition. ACM Trans. Intell. Syst. Technol. (TIST) **7**, 1–42 (2016)
6. Hassen, M., Chan, P.K.: Learning a neural-network-based representation for open set recognition. In: Proceedings of the 2020 SIAM International Conference on Data Mining, pp. 154–162. SIAM (2020)
7. Jain, L.P., Scheirer, W.J., Boult, T.E.: Multi-class open set recognition using probability of inclusion. In: Fleet, D., Pajdla, T., Schiele, B., Tuytelaars, T. (eds.) ECCV 2014. LNCS, vol. 8691, pp. 393–409. Springer, Cham (2014). https://doi.org/10.1007/978-3-319-10578-9_26
8. Joseph, K.J., Khan, S., Khan, F.S., Balasubramanian, V.N.: Towards open world object detection. In: Proceedings of the IEEE/CVF Conference on Computer Vision and Pattern Recognition (CVPR 2021) (2021)
9. Li, Z., Park, U., Jain, A.K.: A discriminative model for age invariant face recognition. IEEE Trans. Inf. Forensics Secur. **6**(3), 1028–1037 (2011). https://doi.org/10.1109/TIFS.2011.2156787
10. Liu, W., Wang, X., Owens, J., Li, Y.: Energy-based out-of-distribution detection. Advances in Neural Information Processing Systems (2020)
11. Neal, L., Olson, M., Fern, X., Wong, W.K., Li, F.: Open set learning with counterfactual images. In: Proceedings of the European Conference on Computer Vision (ECCV) (September 2018)
12. Oza, P., Patel, V.M.: C2ae: class conditioned auto-encoder for open-set recognition. In: 2019 IEEE/CVF Conference on Computer Vision and Pattern Recognition (CVPR), pp. 2302–2311 (2019). https://doi.org/10.1109/CVPR.2019.00241
13. Park, U., Tong, Y., Jain, A.K.: Age-invariant face recognition. IEEE Trans. Pattern Anal. Mach. Intell. **32**(5), 947–954 (2010). https://doi.org/10.1109/TPAMI.2010.14
14. Perera, P., et al.: Generative-discriminative feature representations for open-set recognition. In: Proceedings of the IEEE/CVF Conference on Computer Vision and Pattern Recognition (CVPR) (June 2020)
15. Scheirer, W., Rocha, A., Micheals, R., Boult, T.: Meta-recognition: the theory and practice of recognition score analysis. IEEE Trans. Pattern Anal. Mach. Intell. **33**(8), 1689–1695 (2011). https://doi.org/10.1109/TPAMI.2011.54
16. Scheirer, W.J., de Rezende Rocha, A., Sapkota, A., Boult, T.E.: Toward open set recognition. IEEE Trans. Pattern Anal. Mach. Intell. **35**(7), 1757–1772 (2013). https://doi.org/10.1109/TPAMI.2012.256
17. Schroff, F., Kalenichenko, D., Philbin, J.: Facenet: a unified embedding for face recognition and clustering. In: 2015 IEEE Conference on Computer Vision and Pattern Recognition (CVPR) (June 2015). https://doi.org/10.1109/cvpr.2015.7298682, http://dx.doi.org/10.1109/CVPR.2015.7298682
18. Sun, X., Yang, Z., Zhang, C., Peng, G., Ling, K.V.: Conditional gaussian distribution learning for open set recognition (2021)

19. Szegedy, C., Ioffe, S., Vanhoucke, V., Alemi, A.A.: Inception-v4, inception-resnet and the impact of residual connections on learning. In: Proceedings of the Thirty-First AAAI Conference on Artificial Intelligence, pp. 4278–4284. AAAI 2017, AAAI Press (2017)

20. Taigman, Y., Yang, M., Ranzato, M., Wolf, L.: Deepface: closing the gap to human-level performance in face verification. In: Proceedings of the IEEE Conference on Computer Vision and Pattern Recognition, pp. 1701–1708 (2014)

21. Tan, X., Triggs, B.: Enhanced local texture feature sets for face recognition under difficult lighting conditions. IEEE Trans. Image Process. 19(6), 1635–1650 (2010). https://doi.org/10.1109/TIP.2010.2042645

22. Verma, V., et al.: Manifold mixup: better representations by interpolating hidden states. In: Chaudhuri, K., Salakhutdinov, R. (eds.) Proceedings of the 36th International Conference on Machine Learning. Proceedings of Machine Learning Research, vol. 97, pp. 6438–6447. PMLR, Long Beach, California, USA, 09–15 June 2019. http://proceedings.mlr.press/v97/verma19a.html

23. Wang, H., et al.: Cosface: large margin cosine loss for deep face recognition. In: 2018 IEEE/CVF Conference on Computer Vision and Pattern Recognition, pp. 5265–5274 (2018). https://doi.org/10.1109/CVPR.2018.00552

24. Wang, M., Deng, W.: Deep face recognition: a survey. Neurocomputing 429, 215–244 (2021)

25. Yi, D., Lei, Z., Liao, S., Li, S.Z.: Learning face representation from scratch. arXiv preprint arXiv:1411.7923 (2014)

26. Yue, Z., Wang, T., Zhang, H., Sun, Q., Hua, X.: Counterfactual zero-shot and open-set visual recognition. In: CVPR (2021)

27. Zhang, H., Patel, V.M.: Sparse representation-based open set recognition. IEEE Trans. Pattern Anal. Mach. Intell. 39(8), 1690–1696 (2017). https://doi.org/10.1109/TPAMI.2016.2613924

28. Zhang, K., Zhang, Z., Li, Z., Qiao, Y.: Joint face detection and alignment using multitask cascaded convolutional networks. IEEE Signal Process. Lett. 23(10), 1499–1503 (2016)

29. Zhou, D.W., Ye, H.J., Zhan, D.C.: Learning placeholders for open-set recognition. In: CVPR (2021)

30. Zongyuan Ge, S.D., Garnavi, R.: Generative openmax for multi-class open set classification. In: Kim, T.-K., Stefanos Zafeiriou, G.B., Mikolajczyk, K. (eds.) Proceedings of the British Machine Vision Conference (BMVC), pp. 42.1-42.12. BMVA Press (September 2017). https://doi.org/10.5244/C.31.42, https://dx.doi.org/10.5244/C.31.42

31. Zou, X., Kittler, J., Messer, K.: Illumination invariant face recognition: a survey. In: 2007 First IEEE International Conference on Biometrics: Theory, Applications, and Systems, pp. 1–8 (2007). https://doi.org/10.1109/BTAS.2007.4401921

Intrusion Detection in Software-Defined Networks

Quang-Vinh Dang[✉]

Industrial University of Ho Chi Minh City, Ho Chi Minh City, Vietnam
dangquangvinh@iuh.edu.vn

Abstract. Software-Defined Network (SDN) is being implemented in many data centers to reduce the complexity in controlling and managing the network. As the SDN gains popularity in practice, it attracts more attention from the attackers. An intrusion detection system (IDS) is a determining component of a cybersecurity system in dealing with outsider threats. Fortunately, the centralization of the SDN makes the training of a machine learning-based IDS easier. On the other hand, there are not many yet research studies that address the intrusion detection problem, particularly to the SDN. The majority of published literature considers the intrusion detection problem of SDN similar to a general computer system. Many of them rely on the intrusion datasets that are generated for a normal computer system. In this paper, we study the problem of intrusion detection for the SDN using the latest public dataset. We present novel evaluation scenarios that have not been studied in the literature. We showed that our chosen algorithm can perform very well in some particular settings, such as classifying a single attack type or classifying multiple attack types coming from the same data source with the training dataset. We pointed out the limitation of the current approach and draw some potential research ideas to overcome these limitations.

Keywords: Intrusion detection system · Machine learning · Classification · Software-defined network

1 Introduction

In conventional computer networks, the functions of the network are performed in a distributed manner. The core functions of the networks, including decision making and transferring data, are implemented at the network devices such as routers and switches [22]. These devices usually perform their tasks without knowing the state of the entire network.

The idea of the software-defined network (SDN) [23] is to provide the programmability and the centralization of the network control [17]. The SDN applies the loose coupling principle between the control plane and the data plane, hence the entire system can be controlled from a controller which might be a remote device [22]. The network is more flexible and robust to update, and the developers can implement new algorithms and services on top of the SDN for additional network functions.

© Springer Nature Switzerland AG 2021
T. K. Dang et al. (Eds.): FDSE 2021, LNCS 13076, pp. 356–371, 2021.
https://doi.org/10.1007/978-3-030-91387-8_23

As the result of the above-mentioned benefits of the SDN, the market size of SDN is reported to reach 13,8 billion USD in 2021 [45] as demonstrated in Fig. 1 and it might reach the value of more than 72 billion USD in 2027 [52].

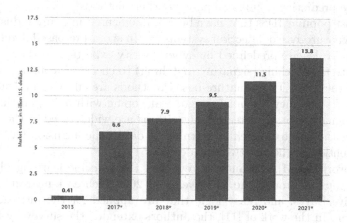

Fig. 1. SDN market size in billion USD [45].

Indeed, new architectures will attract new attacks, and the SDN is no exception. Particularly, as the entire SDN can be controlled from the controller, it might become a single point of failure [49]. In order to secure the SDN, the intrusion detection system (IDS) [6] plays a crucial role. The IDS would stop all the malicious traffic coming from outside the system. Despite the recent success of IDS [15], there is not yet much research attention focusing on SDN threats [22]. Most of the time the IDS are designed for conventional computer networks [13]. Furthermore, many published studies only use one single data source for both training and testing purposes, hence the reported performance does not reflect the behavior of the algorithm in dealing with similar attacks but from various sources.

In this paper, we study the problem of intrusion detection for the SDN. We utilize the most recent intrusion dataset generated for the SDN context [22]. We introduce new evaluation settings that have not been studied in the previous literature. We review the related works in terms of methodology and datasets in Sect. 2. We describe in detail the dataset InSDN in Sect. 3. We discuss our methodology and the results in Sect. 4. We conclude our paper and discuss further research in Sect. 5.

2 Related Works

In this section we review the related published studies in two criteria: i) the methods to detect the intrusion, and ii) the intrusion detection public datasets.

2.1 Methods

Intrusion detection systems and their techniques to monitor and detect malicious traffic have been discussed in several reviews [1,43]. Recently, intrusion detection system research has been dominated by machine learning methods [44] due to the effective predictive results compare to other methods [11].

The most popular IDSs that have been implemented in many industrial systems are signature-based detection systems [3]. In signature-based detection systems, several rules are predefined by cybersecurity experts, and the system can perform pattern matching (signature matching) to check if incoming traffic falls into one of the rules. The signature-based methods are effective in dealing with known attacks but they are not very good at coping with unknown threats [6].

While signature-based detection algorithms are widely used in the industry, in literature the usage of machine learning to detect the intrusion has attracted a lot of attention in the last two decades [35,51].

In the work of [51], the authors reviewed the machine learning algorithms which are popular around the early years of 2000 such as k-nearest neighbors (kNN), naive Bayes, and fuzzy logic that are applied for the intrusion detection problem. In the work of [11], the authors extended the survey by including more recent algorithms, including some tree-based ensemble methods [41] such as xgboost [9]. The authors claimed the best performance belongs to the xgboost, which achieved nearly perfect predictive results on the testing dataset. The result is confirmed in the research work of [24]. The line of work is continued in the study of [32] where the authors studied and evaluated several algorithms, including random forest, multi-layer perceptron (MLP), SVM, decision tree, and kNN using the dataset CICIDS'2017 [39]. However, the authors do not report the Are-Under-the-Curve (AUC) metric which is very important in an imbalanced classification problem [25].

The ensemble supervised machine learning usually achieved the highest predictive performance, but they require a huge labelled dataset which is quite costly to acquire [39]. Furthermore, they also require a high computational power that might not be available at the IoT devices [24,41]. Therefore, different approaches have been studied as well in the literature.

The authors of [10] studied the intrusion detection problem from the outlier analysis point of view. Based on an assumption that the benign network flows are the major part of the traffic and the attacks are outliers, the authors studied the outlier detection algorithm like Isolation Forest [29] to distinguish the outliers from the normal traffic. The outlier detection algorithm is different from the classification algorithm [11] as in the outlier detection problem, the malicious traffic does not form a separated class, but rather a set of data points that are different from the majority. Hence, the outlier detection algorithms are considered unsupervised machine learning algorithms.

Among the basis of the unsupervised machine learning algorithms, clustering algorithms are probably the most well-known. The authors of [27] studied the popular k-means algorithm for the problem of intrusion detection. More recently, the authors of [16] revisited the fuzzy clustering algorithms [33] for the problem and get some promising results. The difference between fuzzy clustering and hard clustering algorithms like k-means is that the fuzzy clustering algorithms allow a single data point can belong to multiple clusters with different weights or *belongingness*.

As we mentioned above, one of the problems with supervised learning is that they require a huge number of training instances and computational power. Active learning [34] is a technique to reduce the requirement on both training data and the training time of the algorithms. The authors of [2,12] studied active learning technique based on the top of the established IDS. The authors claimed that active learning can significantly reduce the time and the required training data while maintaining comparable performance to the previous works.

The researchers have studied deep learning techniques for the intrusion detection problem. In [21], the authors designed an LSTM-based Autoencoder combined with a One-Class SVM algorithm to detect the network anomaly. The convolution neural networks are studied in other studies [28,53,54]. However, according to [31], the deep neural networks did not gain a significant improvement compared to the ensemble methods. The reason might be that the ensemble methods have already achieved very good performance.

Reinforcement learning is another approach in dealing with attacks. In the reinforcement learning context, the IDS learn by try some actions and observe the response from the protected computer system. The reinforcement learning for IDS has been studied since 2000 [8] but somehow they disappear in the literature, probably due to the fact that the computing system at that time is not ready to process the big dataset. In recent years, there are several research studies study reinforcement learning again [18,19,26,30]. The reinforcement learning algorithms did not perform as well as the supervised machine learning algorithms, but they do have a lot of potentials as they do not require a huge training dataset at the beginning.

2.2   Datasets

In this section, we review some popular public datasets that have been used a lot in literature. We focus on only some widely used datasets and their relationship to the SDN environment. The detailed analysis of the public datasets for network intrusion detection can be found in the work of [38].

KDD'99 and NSL-KDD. The dataset KDD'99 [20,35] and its revised version NSL-KDD [48] are probably the most popular intrusion detection datasets [11] using in the literature. Recent published studies still keep using the dataset for the evaluation [5,42].

The KDD'99 dataset is presented with 41 features derived from DARPA packet traces. The features are divided into three groups: basic, traffic, and content features[1]. There are four attack classes presented in the KDD'99 dataset:

---

[1] The dataset is available at https://kdd.ics.uci.edu/databases/kddcup99/task.html.

Denial of Service (DoS), Remote to Local (R2L), User to Root (U2R), and probe attacks.

NSL-KDD is a revised version of the KDD'99 which is introduced to solve some problems of KDD'99 such as the duplicate records. Some interesting properties of NSL-KDD somewhat are missing in the more recent datasets: i) the dataset is divided beforehand to the training and testing set, so everyone can make compare their algorithms easily; and ii) some attacks are presented only in the testing set.

Due to the fact that the two datasets are released more than twenty years ago, they do not reflect the current Internet standards, let beside the SDN context [11,22]. The authors of [46,47] analyzed the KDD'99 and the NSL-KDD datasets in the SDN settings and they found that there are only six over 41 features of the KDD datasets are available for the SDN environment.

ISCX2012. The dataset ISCX2012 [40] is introduced by the Canadian Institute of Cybersecurity[2]. The dataset contains only two types of attack: DoS and brute force. Furthermore, only HTTP traffic is included, which is not a current standard on the Internet today. For instance, many web browsers are forcing users to use the HTTPS version of the websites [7]. Hence the dataset is not suitable for modern evaluation.

CICIDS 2017. The dataset CICIDS 2017[3] is introduced as the updated version of the dataset ISCX 2012 [39]. The dataset has been used by many recent research studies [11]. However, the dataset has the same issues as the previous one ISCX 2012 [36] and is not suitable for the intrusion detection problem in the SDN context [22].

CICIDS 2018. The dataset CICIDS 2018[4] is introduced to cover some problems of the dataset CICIDS 2017. However, all three datasets ISCX 2012, CICIDS 2017, and CICIDS 2018 are generated using synthetic traffic. Nevertheless, the datasets are being used extensively in recent published literature [50].

Above we review some generic intrusion datasets. Other datasets focusing particularly on some domains such as Port-scan [37] or attacks on DNS [14]. We do not consider them in this section.

InSDN. The InSDN dataset [22] is the dataset we used in this research. The dataset is the result of an effort to produce a high-quality dataset for the SDN research. We will review the dataset in detail in Sect. 3.

---

[2] https://www.unb.ca/cic/datasets/ids.html.
[3] https://www.unb.ca/cic/datasets/ids-2017.html.
[4] https://www.unb.ca/cic/datasets/ids-2018.html.

3   The InSDN Dataset

The InSDN dataset[5] is released in 2020 as an open-source dataset for training and evaluating the IDS in the SDN context [22]. The dataset is generated by using a network of four virtual machines as visualized in Fig. 2.

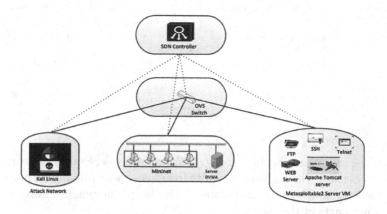

Fig. 2. The InSDN network architecture [22].

The dataset includes three groups.

- The normal traffic includes 68, 424 instances.
- The attack traffics targeting the mealsplotable 2 servers that include 136, 743 instances with the following attacks:
  - DoS: 1145 instances
  - DDoS: 73529 instances
  - Probe: 61757 instances
  - BFA: 295 instances
  - U2R: 17 instances
- The attack traffics targeting the Open vSwitch (OVS) machine that includes 138, 722 instances with the following attacks.
  - DoS: 52471 instances
  - DDoS: 48413 instances
  - Probe: 36372 instances
  - BFA: 1110 instances
  - Web-Attack: 192 instances
  - BOTNET: 164 instances

Due to the limited number of training instances of the attack types U2R, Web-Attack, and BOTNET, we exclude them from the analysis.

The dataset is presented with 83 features and one label column. The first part of the normal traffic data is displayed in Table 1. We presented the correlation

---

[5] http://aseados.ucd.ie/datasets/SDN/.

Table 1. A first few rows and columns of the normal traffic data.

	Flow ID	Src IP	Src Port	Dst IP	Dst Port	Protocol	Timestamp	Flow Duration	Tot Fwd Pkts	Tot Bwd Pkts	TotLen Fwd Pkts	TotLen Bwd Pkts	Fwd Pkt Len Max
0	185.127.17.56-192.168.20.133-443-53648-6	185.127.17.56	443	192.168.20.133	53648	6	5/2/2020 13:58	245230	44	40	124937.0	1071.0	9100
1	185.127.17.56-192.168.20.133-443-53650-6	192.168.20.133	53650	185.127.17.56	443	6	5/2/2020 13:58	1605449	107	149	1071.0	439537.0	517
2	192.168.20.133-192.168.20.2-35108-53-6	192.168.20.133	35108	192.168.20.2	53	6	5/2/2020 13:58	53078	5	5	66.0	758.0	66
3	192.168.20.133-192.168.20.2-35108-53-6	192.168.20.2	53	192.168.20.133	35108	6	5/2/2020 13:58	6975	1	1	0.0	0.0	0
4	154.59.122.74-192.168.20.133-443-60900-6	192.168.20.133	60900	154.59.122.74	443	6	5/2/2020 13:58	190141	13	16	780.0	11085.0	427

matrix of the normal traffic data in Fig. 3. We might observe that the correlation coefficients between the features are quite low, hence we probably do not need to worry about the multi-collinearity issue. A similar phenomenon can be observed on two other parts of the dataset.

Furthermore, we see that the InSDN dataset contains the same attack types from different sources, which is quite different from other datasets. In this study, we will pay attention to this property. We note that this property is usually ignored in other studies, such as the work that introduced the InSDN dataset itself [22].

4   Methods and Experimental Results

4.1   Methods

In this study, we focus on two main research questions:

- Can we effectively classify the attacks that come from the same source that we learned before?
- Can we effectively classify the attacks that come from a different source that we learned before?

The first research question is similar to previous studies [11] but the second research question is relatively new.

We keep using the algorithm *xgboost* [9] as it is proved as the best intrusion detector [11,14,24] compared to other classification algorithms. The xgboost model is visualized in Fig. 4. The core idea of the xgboost model is to build multiple sequential decision trees, in which the following tree will try to recover the prediction error of the predecessor.

We divide the training/testing set by the ratio of 50:50.

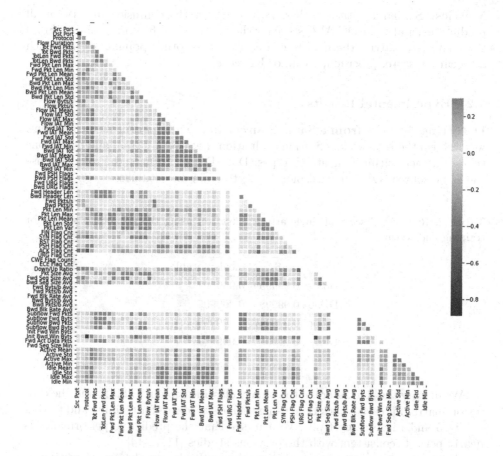

Fig. 3. Correlation matrix of the normal traffic data

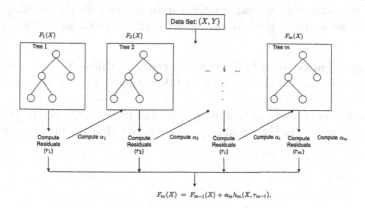

Fig. 4. xgboost model [4]

Metrics. Similar to other studies, we will report the confusion matrix for all predictions, and the ROC AUC score for the binary classification problem. From the confusion matrix, the readers can easily derive other popular metrics such as accuracy score, precision, recall, or F1-score.

4.2   Experimental Results

Detecting Attacks from a Same Source. In the first part of the experiment, we perform the binary classification evaluation, i.e. we only train and detect one type of attack within four attack types: DoS, DDoS, BFA, and Probe. We use the same source (OVS or mealsplotable 2) to train and evaluate.

Table 2. ROC AUC score of single attack detection using the same data source for training and testing.

	OVS	mealsplotable
DoS	0.999999	0.999997
DDoS	0.999985	0.999999
Probe	0.999997	0.999999
BFA	0.99998	0.9999

We display the AUC score of the binary classification in Table 2, and the confusion matrix for each attack type using each data source for training and testing in Fig. 5 and Fig. 6 respectively. We can see that the predictive performance is nearly perfect, consistent with the previous studies [11,12,14].

We display the confusion matrix of the multiple attack type detection in Fig. 7. The accuracy and other meta-metrics such as meta-precision and recall are near perfect. We conclude that the xgboost model can effectively classify the attacks if the attacks come from a similar data source with the training dataset. The results we achieved here are consistent with the previous studies [22].

Detecting Attacks from Different Sources. In this section, we try to train the model using a data source, such as OVS, and to predict the attacks that come from another data source.

Similar to the previous section, we display the ROC AUC score in Table 3 and the confusion matrix in Fig. 8 and Fig. 9. We can see that the performance is still

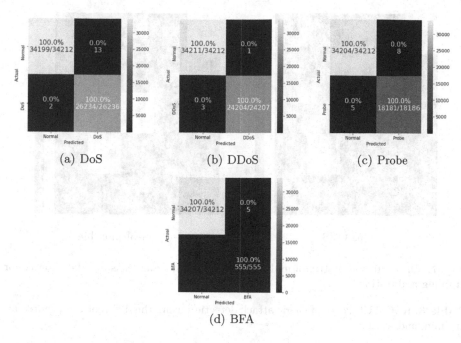

Fig. 5. The confusion matrix of single attack detection using the same data source OVS for training and testing.

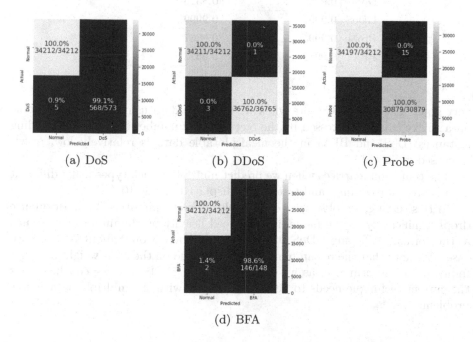

Fig. 6. The confusion matrix of single attack detection using the same data source mealsplotable for training and testing.

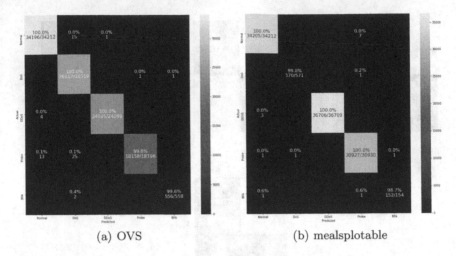

(a) OVS   (b) mealsplotable

Fig. 7. The confusion matrix of multi attack detection using the same data source for training and testing.

Table 3. ROC AUC score of single attack detection using the different data source for training and testing.

	OVS - mealsplotable	mealsplotable - OVS
DoS	0.9997	0.8326
DDoS	0.9997	0.9999
Probe	0.9995	0.9987
BFA	0.9986	0.6522

very good. The low AUC score when we train the data using the mealsplotable data source can be addressed by the fact that the number of the corresponding instances (DoS and BFA) in the mealsplotable data is relatively low, as we discussed in Sect. 3.

The confusion matrices when we predict multiple attack types using different data sources for training and testing are displayed in Fig. 10.

In this setting, we observe a completely different picture. The performance drops significantly as the model is influenced heavily by the major classes such as the normal traffic and DDoS attacks. The accuracy is only about 70% in both cases. We note that the re-sampling technique based on the class weight does not improve the performance, as we displayed in Fig. 11. Hence, we conclude that the current technique needs to be revised to deal with the multiple data source problem.

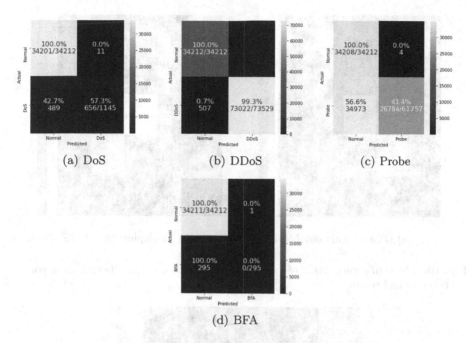

Fig. 8. The confusion matrix of single attack detection using the data source OVS for training and the data source mealsplotable for testing.

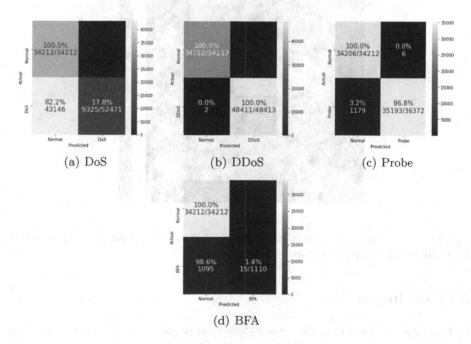

Fig. 9. The confusion matrix of single attack detection using the data source mealsplotable for training and the data source OVS for testing.

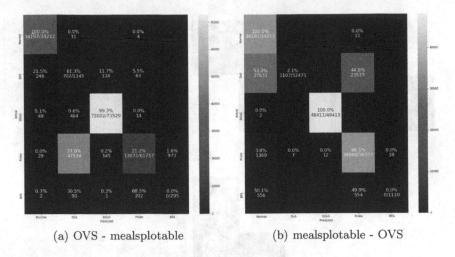

(a) OVS - mealsplotable   (b) mealsplotable - OVS

Fig. 10. The confusion matrix of multi attack detection using different data sources for training and testing.

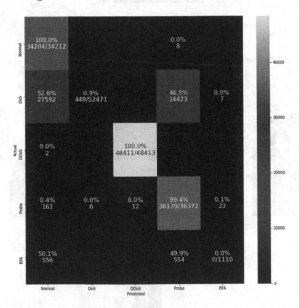

Fig. 11. The confusion matrix of multi attack detection using different data sources for training and testing with class weights

5 Conclusions

In this paper, we study the intrusion detection problem in the software-defined network environment. We analyze multiple settings in detecting the intrusion in the SDN environment. We show that the current classifier can effectively classify the attacks in some particular scenarios, but in the general scenario more research studies are required. In the future, we will focus on the work of transfer learning between data sources to improve cross-dataset predictive performance.

References

1. Agrawal, D., Agrawal, C.: A review on various methods of intrusion detection system. Comput. Eng. Intell. Syst. **11**(1), 7–15 (2020)
2. Almgren, M., Jonsson, E.: Using active learning in intrusion detection. In: CSFW, pp. 88–98. IEEE (2004)
3. Amanowicz, M., Jankowski, D.: Detection and classification of malicious flows in software-defined networks using data mining techniques. Sensors **21**(9), 2972 (2021)
4. Amazon: How xgboost works. https://docs.aws.amazon.com/sagemaker/latest/dg/xgboost-HowItWorks.html. Accessed 21 June 2021
5. Awujoola, O.J., Ogwueleka, F.N., Irhebhude, M.E., Misra, S.: Wrapper based approach for network intrusion detection model with combination of dual filtering technique of resample and SMOTE. In: Misra, S., Kumar Tyagi, A. (eds.) Artificial Intelligence for Cyber Security: Methods, Issues and Possible Horizons or Opportunities. SCI, vol. 972, pp. 139–167. Springer, Cham (2021). https://doi.org/10.1007/978-3-030-72236-4_6
6. Bansal, P., Ahmad, T., et al.: Methods and techniques of intrusion detection: a review. In: Unal, A., Nayak, M., Mishra, D.K., Singh, D., Joshi, A. (eds.) Smart Trends in Information Technology and Computer Communications. SmartCom 2016. Communications in Computer and Information Science, vol. 628, pp. 518–529. Springer, Singapore (2016). https://doi.org/10.1007/978-981-10-3433-6_62
7. Boulevard, S.: Chrome to enforce https web protocol (like it or not), 25 March 2021. https://securityboulevard.com/2021/03/chrome-to-enforce-https-web-protocol-like-it-or-not
8. Cannady, J.: Next generation intrusion detection: autonomous reinforcement learning of network attacks. In: Proceedings of the 23rd National Information Systems Security Conference, pp. 1–12 (2000)
9. Chen, T., Guestrin, C.: Xgboost: a scalable tree boosting system. In: KDD, pp. 785–794. ACM (2016)
10. Dang, Q.V.: Outlier detection in network flow analysis. arXiv:1808.02024 (2018)
11. Dang, Q.-V.: Studying machine learning techniques for intrusion detection systems. In: Dang, T.K., Küng, J., Takizawa, M., Bui, S.H. (eds.) FDSE 2019. LNCS, vol. 11814, pp. 411–426. Springer, Cham (2019). https://doi.org/10.1007/978-3-030-35653-8_28
12. Dang, Q.V.: Active learning for intrusion detection systems. In: IEEE RIVF (2020)
13. Dang, Q.-V.: Understanding the decision of machine learning based intrusion detection systems. In: Dang, T.K., Küng, J., Takizawa, M., Chung, T.M. (eds.) FDSE 2020. LNCS, vol. 12466, pp. 379–396. Springer, Cham (2020). https://doi.org/10.1007/978-3-030-63924-2_22
14. Dang, Q.-V.: Detecting the attacks to DNS. In: Antipova, T. (ed.) ICCS 2021. LNNS, vol. 315, pp. 173–179. Springer, Cham (2022). https://doi.org/10.1007/978-3-030-85799-8_15
15. Dang, Q.V.: Improving the performance of the intrusion detection systems by the machine learning explainability. Int. J. Web Inf. Syst. **17**(5), 537–555 (2021)

16. Dang, Q.V.: Studying the fuzzy clustering algorithm for intrusion detection on the attacks to the domain name system. In: WorldS4. IEEE (2021)
17. Dang, Q., François, J.: Utilizing attack enumerations to study SDN/NFV vulnerabilities. In: NetSoft, pp. 356–361. IEEE (2018)
18. Dang, Q.V., Vo, T.H.: Reinforcement learning for the problem of detecting intrusion in a computer system. In: Proceedings of ICICT (2021)
19. Dang, Q.V., Vo, T.H.: Studying the reinforcement learning techniques for the problem of intrusion detection. In: ICAIBD. IEEE (2021)
20. Dhanabal, L., Shantharajah, S.: A study on NSL-KDD dataset for intrusion detection system based on classification algorithms. Int. J. Adv. Res. Comput. Commun. Eng. **4**(6), 446–452 (2015)
21. Elsayed, M.S., Le-Khac, N., Dev, S., Jurcut, A.D.: Network anomaly detection using LSTM based autoencoder. In: Q2SWinet, pp. 37–45. ACM (2020)
22. Elsayed, M.S., Le-Khac, N.A., Jurcut, A.D.: InSDN: a novel SDN intrusion dataset. IEEE Access **8**, 165263–165284 (2020)
23. Goransson, P., Black, C., Culver, T.: Software Defined Networks: A Comprehensive Approach. Morgan Kaufmann, Burlington (2016)
24. Gouveia, A., Correia, M.: Network intrusion detection with XGBoost. In: Recent Advances in Security, Privacy, and Trust for Internet of Things (IoT) and Cyber-Physical Systems (CPS), p. 137 (2020)
25. Hand, D.J., Till, R.J.: A simple generalisation of the area under the roc curve for multiple class classification problems. Mach. Learn. **45**(2), 171–186 (2001)
26. Hsu, Y.F., Matsuoka, M.: A deep reinforcement learning approach for anomaly network intrusion detection system. In: CloudNet, pp. 1–6. IEEE (2020)
27. Jianliang, M., Haikun, S., Ling, B.: The application on intrusion detection based on k-means cluster algorithm. In: IFITA, vol. 1, pp. 150–152. IEEE (2009)
28. Khan, R.U., Zhang, X., Alazab, M., Kumar, R.: An improved convolutional neural network model for intrusion detection in networks. In: CCC, pp. 74–77. IEEE (2019)
29. Liu, F.T., Ting, K.M., Zhou, Z.: Isolation forest. In: ICDM, pp. 413–422. IEEE Computer Society (2008)
30. Lopez-Martin, M., Carro, B., Sanchez-Esguevillas, A.: Application of deep reinforcement learning to intrusion detection for supervised problems. Expert Syst. Appl. **141**, 112963 (2020)
31. Mandru, D.B., Aruna Safali, M., Raghavendra Sai, N., Sai Chaitanya Kumar, G.: Assessing deep neural network and shallow for network intrusion detection systems in cyber security. In: Smys, S., Bestak, R., Palanisamy, R., Kotuliak, I. (eds.) Computer Networks and Inventive Communication Technologies. LNDECT, vol. 75, pp. 703–713. Springer, Singapore (2022). https://doi.org/10.1007/978-981-16-3728-5_52
32. Maseer, Z.K., Yusof, R., Bahaman, N., Mostafa, S.A., Foozy, C.F.M.: Benchmarking of machine learning for anomaly based intrusion detection systems in the cicids2017 dataset. IEEE Access **9**, 22351–22370 (2021)
33. Miyamoto, S., Ichihashi, H., Honda, K., Ichihashi, H.: Algorithms for Fuzzy Clustering. Springer, Heidelberg (2008)
34. Monarch, R.: Human-in-the-Loop Machine Learning: Active Learning and Annotation for Human-centered AI. Manning Publications, New York, NY, USA (2021)
35. Özgür, A., Erdem, H.: A review of kdd99 dataset usage in intrusion detection and machine learning between 2010 and 2015. PeerJ Prepr. **4**, e1954v1 (2016)
36. Panigrahi, R., Borah, S.: A detailed analysis of cicids2017 dataset for designing intrusion detection systems. Int. J. Eng. Technol. **7**(3.24), 479–482 (2018)

37. Ring, M., Landes, D., Hotho, A.: Detection of slow port scans in flow-based network traffic. PloS one **13**(9), e0204507 (2018)
38. Ring, M., Wunderlich, S., Scheuring, D., Landes, D., Hotho, A.: A survey of network-based intrusion detection data sets. Comput. Secur. **86**, 147–167 (2019)
39. Sharafaldin, I., Lashkari, A.H., Ghorbani, A.A.: Toward generating a new intrusion detection dataset and intrusion traffic characterization. In: ICISSP, pp. 108–116 (2018)
40. Shiravi, A., Shiravi, H., Tavallaee, M., Ghorbani, A.A.: Toward developing a systematic approach to generate benchmark datasets for intrusion detection. Comput. Secur. **31**(3), 357–374 (2012)
41. Sindhu, S.S.S., Geetha, S., Kannan, A.: Decision tree based light weight intrusion detection using a wrapper approach. Expert Syst. Appl. **39**(1), 129–141 (2012)
42. Singh, K., Kaur, L., Maini, R.: Comparison of principle component analysis and stacked autoencoder on NSL-KDD dataset. In: Singh, V., Asari, V.K., Kumar, S., Patel, R.B. (eds.) Computational Methods and Data Engineering. AISC, vol. 1227, pp. 223–241. Springer, Singapore (2021). https://doi.org/10.1007/978-981-15-6876-3_17
43. Singh, R., Kumar, H., Singla, R.K., Ketti, R.R.: Internet attacks and intrusion detection system: a review of the literature. Online Information Review (2017)
44. Singh, S., Banerjee, S.: Machine learning mechanisms for network anomaly detection system: A review. In: ICCSP, pp. 0976–0980. IEEE (2020)
45. Statista: Software-defined networking (SDN) market size worldwide from 2013 to 2021 (in billion u.s. dollars). https://www.statista.com/statistics/468636/global-sdn-market-size/. Accessed 21 June 2021
46. Tang, T.A., Mhamdi, L., McLernon, D., Zaidi, S.A.R., Ghogho, M.: Deep learning approach for network intrusion detection in software defined networking. In: 2016 international conference on wireless networks and mobile communications (WINCOM), pp. 258–263. IEEE (2016)
47. Tang, T.A., Mhamdi, L., McLernon, D., Zaidi, S.A.R., Ghogho, M.: Deep recurrent neural network for intrusion detection in sdn-based networks. In: 2018 4th IEEE Conference on Network Softwarization and Workshops (NetSoft), pp. 202–206. IEEE (2018)
48. Tavallaee, M., Bagheri, E., Lu, W., Ghorbani, A.A.: A detailed analysis of the KDD CUP 99 data set. In: CISDA, pp. 1–6. IEEE (2009)
49. Tayfour, O.E., Marsono, M.N.: Collaborative detection and mitigation of DDoS in software-defined networks. J. Supercomput. **77**(11), 13166–13190 (2021)
50. Thakkar, A., Lohiya, R.: A review of the advancement in intrusion detection datasets. Procedia Comput. Sci. **167**, 636–645 (2020)
51. Tsai, C., Hsu, Y., Lin, C., Lin, W.: Intrusion detection by machine learning: a review. Expert Syst. Appl. **36**(10), 11994–12000 (2009)
52. Valuates: SDN market size is projected to reach usd 72,630 million by 2027, 07 October 2020. https://www.prnewswire.com/in/news-releases/sdn-market-size-is-projected-to-reach-usd-72-630-million-by-2027-valuates-reports-815582808.html
53. Vinayakumar, R., Soman, K., Poornachandran, P.: Applying convolutional neural network for network intrusion detection. In: ICACCI, pp. 1222–1228. IEEE (2017)
54. Wang, H., Cao, Z., Hong, B.: A network intrusion detection system based on convolutional neural network. J. Intell. Fuzzy Syst. **38**(6), 7623–7637 (2020)

Emerging Data Management Systems
and Applications

Clustering Analyses of Two-Dimensional Space-Filling Curves

H. K. Dai[1(✉)] and H. C. Su[2]

[1] Computer Science Department, Oklahoma State University,
Stillwater, OK 74078, USA
dai@cs.okstate.edu
[2] Department of Computer Science, Arkansas State University,
Jonesboro, AR 72401, USA
suh@astate.edu

Abstract. A discrete space-filling curve provides a linear traversal or indexing of a multi-dimensional grid space. This paper presents two analytical studies on clustering analyses of the 2-dimensional Hilbert and z-order curve families. The underlying measure is the mean number of cluster over all identically shaped subgrids. We derive the exact formulas for the clustering statistics for the 2-dimensional Hilbert and z-order curve families. The exact results allow us to compare their relative performances with respect to this measure: when the grid-order is sufficiently larger than the subgrid-order (typical scenario for most applications), Hilbert curve family performs significantly better than z-order curve family.

Keywords: Space-filling curve · Hilbert curve · z-order curve · Clustering

1 Preliminaries

Discrete space-filling curves have a wide range of applications in databases, parallel computation, algorithms, in which linearization techniques of multi-dimensional arrays or computational grids are needed. Sample applications include heuristics for combinatorial algorithms and data structures: traveling salesperson algorithm [24] and nearest-neighbor finding [6], multi-dimensional space-filling indexing methods [2,5,12,18], image compression [20], dynamic unstructured mesh partitioning [16], and linearization and traversal of sensor networks [4,28]. Some recent diverse applications of space-filling curves extend to statistical sampling [14] and bioinformatics [17]. For a comprehensive historical development of classical space-filling curves, see [25] and [3].

For positive integer n, denote $[n] = \{1, 2, \ldots, n\}$. An m-dimensional (discrete) space-filling curve of length n^m is a bijective mapping $C : [n^m] \to [n]^m$, thus providing a linear indexing/traversal or total ordering of the grid points in $[n]^m$. An m-dimensional grid is said to be of order k if it has side-length $n = 2^k$;

T. K. Dang et al. (Eds.): FDSE 2021, LNCS 13076, pp. 375–391, 2021.
https://doi.org/10.1007/978-3-030-91387-8_24

a space-filling curve has order k if its codomain is a grid of order k. An m-dimensional space-filling curve C is continuous if the Euclidean distance between $C(i)$ and $C(i+1)$ is 1 for all $i \in [n^m - 1]$. The generation of a sequence of multi-dimensional space-filling curves of successive orders usually follows a recursive framework (on the dimensionality and order), which results in a few classical families, such as Gray-coded curves, Hilbert curves, Peano curves, and z-order curves (see, for examples, [1] and [22]).

Denote by H_k^m and Z_k^m an m-dimensional Hilbert and z-order, respectively, space-filling curve of order k. Figure 1 illustrates the recursive constructions of H_k^m and Z_k^m for $m = 2$ and $k = 1, 2$, and $m = 3$ and $k = 1$.

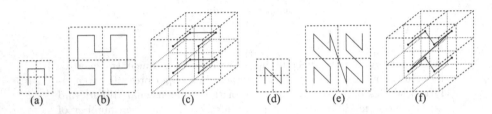

Fig. 1. Recursive self-similar generations of Hilbert and z-order curves of higher order (respectively, H_k^m and Z_k^m) by interconnecting symmetric subcurves, via reflection and/or rotation, of lower order (respectively, H_{k-1}^m and Z_{k-1}^m) along an order-1 subcurve (respectively, H_1^m and Z_1^m): (a) H_1^2; (b) H_2^2; (c) H_1^3; (d) Z_1^2; (e) Z_2^2; (f) Z_1^3.

We measure the applicability of a family of space-filling curves based on their common structural characteristics, which are informally described as follows. Clustering performance measures the distribution of continuous runs of grid points (clusters) over identically shaped subspaces of $[n]^m$, which can be characterized by the mean number of clusters and the average inter-cluster distance (in $[n^m]$) within a subspace. Locality preservation reflects proximity between the grid points of $[n]^m$, that is, close-by points in $[n]^m$ are mapped to close-by indices/numbers in $[n^m]$, or vice versa.

For an m-dimensional space-filling curve $C : [n^m] \to [n]^m$ and a subgrid G of $[n]^m$, a cluster of G induced by C is a maximal (contiguous) subinterval I of $[n^m]$ such that $C(I) \subseteq G$. We can partition and order $C^{-1}(G)$ into disjoint union of clusters. An inter-cluster gap of G is a subinterval of $[n^m]$ delimited by two consecutive clusters of G, and the corresponding inter-cluster distance is the length of the inter-cluster gap. Thus, the space-filling curve C induces the following statistics: (1) the mean number of clusters of $C^{-1}(G)$ over all identically shaped subgrids G of $[n]^m$, (2) the (universe) mean inter-cluster distance over all inter-cluster gaps from all identically shaped subgrids G of $[n]^m$, and (3) the mean total inter-cluster distance (in a subgrid) over all identically shaped subgrids G of $[n]^m$.

Empirical and analytical studies of clustering and inter-clustering performances of various low-dimensional space-filling curves have been reported in the literature (see [22] for details). Generally, the Hilbert and z-order curve families exhibit good performance in this respect.

Jagadish [15] derives exact formulas for the mean numbers of clusters over all rectangular 2×2 and 3×3 subgrids of an H_k^2-structural grid space. Moon, Jagadish, Faloutsos, and Saltz [22] prove that in a sufficiently large m-dimensional H_k^m-structural grid space, the asymptotic mean number of clusters over all rectilinear polyhedral queries with common surface area $S_{m,k}$ approaches $\frac{1}{2}\frac{S_{m,k}}{m}$ as k approaches ∞. They also extend the work in [Jag97] to obtain the exact formula for the mean number of clusters over all rectangular $2^q \times 2^q$ subgrids of an H_k^2-structural grid space.

Xu and Tirthapura [29] generalize the above asymptotic mean number of clusters over all rectilinear polyhedral queries with common surface area from m-dimensional Hilbert curves to arbitrary continuous space-filling curves. Note that rectangular queries with common volume yield the optimal asymptotic mean number of clusters for a continuous space-filling curve.

Dai and Su [7] obtain the exact formulas for the following three statistics for H_k^2 and Z_k^2: (1) the summation of all inter-cluster distances over all $2^q \times 2^q$ query subgrids, (2) the universe mean inter-cluster distance over all inter-cluster gaps from all $2^q \times 2^q$ subgrids, and (3) the mean total inter-cluster distance over all $2^q \times 2^q$ subgrids. Based on the analytical results, the asymptotic comparisons indicate that z-order curve family performs better than Hilbert curve family with respect to the statistics.

A few locality measures have been proposed and analyzed for space-filling curves in the literature. Denote by d and d_p the Euclidean metric and p-normed metric (Manhattan ($p = 1$) and maximum metric ($p = \infty$)), respectively.

Let $\mathcal{C}$ denote a family of m-dimensional curves of successive orders. For quantifying the proximity preservation of close-by points in the m-dimensional space $[n]^m$, Pérez, Kamata, and Kawaguchi [23] employ an average locality measure:

$$L_{\text{PKK}}(C) = \sum_{i,j \in [n^m] | i < j} \frac{|i - j|}{d(C(i), C(j))} \text{ for } C \in \mathcal{C},$$

and provide a hierarchical construction for a 2-dimensional $\mathcal{C}$ with good but suboptimal locality with respect to this measure.

Mitchison and Durbin [21] use a more restrictive locality measure parameterized by q:

$$L_{\text{MD},q}(C) = \sum_{i,j \in [n^m] | i < j \text{ and } d(C(i), C(j)) = 1} |i - j|^q \text{ for } C \in \mathcal{C}$$

to study optimal 2-dimensional mappings for $q \in [0, 1]$. For the case $q = 1$, the optimal mapping with respect to $L_{\text{MD},1}$ is very different from that in [23].

Dai and Su [8] consider a locality measure similar to $L_{\text{MD},1}$ conditional on a 1-normed distance of δ between points in $[n]^m$:

$$L_\delta(C) = \sum_{i,j \in [n^m] | i < j \text{ and } d_1(C(i), C(j)) = \delta} |i - j| \text{ for } C \in \mathcal{C}.$$

They derive exact formulas for L_δ for the Hilbert curve family $\{H_k^m \mid k = 1, 2, \ldots\}$ and z-order curve family $\{Z_k^m \mid k = 1, 2, \ldots\}$ for $m = 2$ and arbitrary δ that is an integral power of 2, and $m = 3$ and $\delta = 1$. With respect to the locality measure L_δ and for sufficiently large k and $\delta \ll 2^k$, the z-order curve family performs better than the Hilbert curve family for $m = 2$ and over the δ-spectrum of integral powers of 2. When $\delta = 2^k$, the domination reverses. The superiority of the z-order curve family persists but declines for $m = 3$ with unit 1-normed distance for L_δ.

For measuring the proximity preservation of close-by points in the indexing space $[n^m]$, Gotsman and Lindenbaum [13] consider the following measures for $C \in \mathcal{C}$:

$$L_{\mathrm{GL,min}}(C) = \min_{i,j \in [n^m] \mid i < j} \frac{d(C(i), C(j))^m}{|i - j|}, \text{ and}$$

$$L_{\mathrm{GL,max}}(C) = \max_{i,j \in [n^m] \mid i < j} \frac{d(C(i), C(j))^m}{|i - j|}.$$

They show that for arbitrary m-dimensional curve C,

$$L_{\mathrm{GL,min}}(C) = O(n^{1-m}), \text{ and}$$

$$L_{\mathrm{GL,max}}(C) > (2^m - 1)(1 - \frac{1}{n})^m.$$

For the m-dimensional Hilbert curve family $\{H_k^m \mid k = 1, 2, \ldots\}$, they prove that:

$$L_{\mathrm{GL,max}}(H_k^m) \leq 2^m (m + 3)^{\frac{m}{2}}.$$

Alber and Niedermeier [1] generalize $L_{\mathrm{GL,max}}$ to $L_{\mathrm{AN},p}$ by employing the p-normed metric d_p in place of the Euclidean metric d. They improve and extend the above tight bounds for the 2-dimensional Hilbert curve family to:

$$L_{\mathrm{AN},1}(H_k^2) \leq 9\frac{3}{5},$$

$$6(1 - O(2^{-k})) \leq L_{\mathrm{AN},2}(H_k^2) \leq 6\frac{1}{2}, \text{ and}$$

$$6(1 - O(2^{-k})) \leq L_{\mathrm{AN},\infty}(H_k^2) \leq 6\frac{2}{5}.$$

Dai and Su [9,10] provide analytical studies on the locality measure $L_{\mathrm{AN},p}$ for the 2-dimensional Hilbert curve family, and obtain exact formulas for $L_{\mathrm{AN},p}(H_k^2)$ for $p = 1$ and all reals $p \geq 2$. In addition, they identify all the representative grid-point pairs (which realize $L_{\mathrm{AN},p}(H_k^2)$) for $p = 1$ and all reals $p \geq 2$. A practical implication of their results on $L_{\mathrm{AN},p}(H_k^2)$ is that the exact formulas provide good bounds on measuring the loss in data locality in the index space, while spatial correlation exists in the 2-dimensional grid space.

The studies of clustering and inter-clustering performances for space-filling curves are motivated by the applicability of multi-dimensional space-filling indexing methods, in which an m-dimensional data space is mapped onto a 1-dimensional data space (external storage structure) by adopting a 1-dimensional indexing method based on an m-dimensional space-filling curve.

The space-filling index structure can support efficient query processing (such as range queries) provided that we minimize the average number of external fetch/seek operations, which is related to the clustering statistics. Asano, Ranjan, Roos, Welzl, and Widmayer [2] study the optimization of range queries over space-filling index structures, which aims at minimizing the number of seek operations (not the number of block accesses)—tradeoff between seek time to proper block (cluster) and latency/transfer time for unnecessary blocks (inter-cluster gap). Good bounds on the two inter-clustering statistics translate into good bounds on the average tolerance of unnecessary block transfers.

Figure 2 illustrates a range-query example that is mapped to subspaces of Z_k^2- and H_k^2-structural grid spaces (where $k = 4$) with different runs/numbers of clusters.

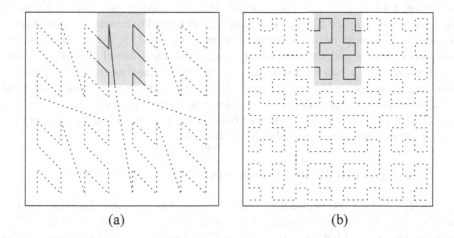

(a)   (b)

Fig. 2. Clusters within a subspace of (a) z-order curve and (b) Hilbert curve.

The transformation of a database of spatially extended objects (original space) into a database of higher-dimensional points (transform space), when coupling with a point access method for the transform space, produces a spatial access method for the original space. Formal analyses of locality preservation of discrete space-filling curves supply a goodness measure for the transformation-based multi-dimensional space-filling indexing. Dai, Whang, and Su [11] employ an appropriate distance metric on the space of multi-dimensional polytopes to justify the effectiveness of the transformation-based multi-dimensional space-filling indexing in preserving spatial locality in [19,26,27], and [19].

This paper presents two analytical studies on the clustering analyses of two representative 2-dimensional space-filling curve families: Hilbert and z-order. For each studied space-filling curve family, we derive the exact formulas for its clustering statistics that includes the mean number of clusters for the underlying curve family over all identically shaped rectangular subgrids of $[n]^2$. Note that our analytical and combinatorial approach in deriving the clustering statistics, uniform for both curve-families, are three-fold: (1) extending the work for the 2-dimensional Hilbert curve family in [22] while completing the relative-performance study of the 2-dimensional Hilbert and z-order curve families with respect to the clustering statistics, (2) complementing our earlier analytical and approximation studies [7] of inter-clustering performances of the 2-dimensional Hilbert and z-order curve families, and (3) more importantly, this common analytical approach provides us a natural geometric transition in computing the clustering statistics for their 3- and higher-dimensional curve families.

Note that we present the skeletons for proving the main results for the Hilbert and z-order curve families without lengthy details in the abstract. The computation of the mean-clustering statistics proceeds to establishing many systems of recurrences and their closed-form solutions over two geometric regions of 2-dimensional Hilbert and z-order curves (boundary edge- and vertex-regions). For each studied space-filling curve family, we provide the overall and successive stepwise ideas in developing the recurrence systems for the boundary edge- and vertex-regions and the final exact formula for the desired clustering statistics. Closed-form solutions for most systems of recurrences in our studies are solved analytically and/or are computed via the mathematical and analytical software Maple. Complete detailed formulations of all recurrence systems, derivations/proofs of their solutions, and verifying computer programs for Hilbert and z-order curve families are available from the authors.

2   Our Analytical and Combinatorial Approach

For an H_k^2- and Z_k^2-structural grid spaces, we obtain the exact formulas for the mean number of clusters over all rectangular $2^q \times 2^q$ subspaces by computing the edge cuts in and between its subgrids that are decomposed recursively. The idea behind this derivation is to count the total number of edges that are cut by the sides of all possible identically shaped $2^q \times 2^q$ subspaces—by noting that the entry and exit grid points of a cluster connect to grid points outside of this subspace (two cuts by side(s) of this subspace) and every cluster has two cuts by the subspace. A cut on an edge by a side of a subspace is called an "edge cut". We give an overview of the derivation for both curve families as follows.

1. Compute the number of edges that are cut by the sides of all possible $2^q \times 2^q$ subspaces, which are exactly inside of one of the four quadrants, and the number of edges cut by the sides of subspaces across different quadrants, respectively. The number of cuts on edges is twice the number of clusters over all identically shaped subspaces.

2. Categorize the edge cuts caused by subspaces across quadrants into: edge cuts within $2^q \times 2^q$ corner boundaries of the quadrants and within side boundaries (of 2^q rows/columns) of the quadrants. (Note that the edges that are cut by subspaces across quadrants only in the boundary regions (sides and corners) of the quadrants.) By decomposing these corner and side boundaries, we derive recurrences for each of them:

(a) For edge cuts within one of the four corner boundaries: edge cuts within upper (left or right) corner and lower (left or right) corner boundaries are inter-recurrence related.
(b) For edge cuts within one of the four side boundaries:
    i. Edge cuts in left boundary (same as right boundary) consists of substructures of left boundary, bottom boundary and two lower-corner boundaries,
    ii. Edge cuts in bottom boundary consists of substructures of two left boundaries and two upper-corner boundaries, and
    iii. Edge cuts in the top boundary consists of substructures of two top boundaries and two upper-corner boundaries. These inter-recurrent relations are based upon the construction of canonical H_k^2 and Z_k^2 (see Fig. 1(a) and (b)).
3. After obtaining the numbers of edge cuts in boundaries that are derived from Step 2, we solve the recurrence for the total number of edge cuts, then divide it by 2 to get the total number of clusters. To obtain the mean number of clusters, the total number of clusters needs to be divided by the total number of subspaces of size $2^q \times 2^q$, which is $(2^k - 2^q + 1)^2$.

3   Clustering Statistics of 2-Dimensional Hilbert Curves

With respect to the canonical orientation of H_k^2 shown in Fig. 1(a), we cover the 2-dimensional k-order grid with 2^k rows $(R_{k,1}, R_{k,2}, \ldots, R_{k,2^k})$, indexed from the bottom, and 2^k columns $(C_{k,1}, C_{k,2}, \ldots, C_{k,2^k})$, indexed from the left. We denote:

1. For a grid point $v \in [2^k]^2$, its x- and y-coordinate by $X(v)$ and $Y(v)$, respectively (that is, v is the intersection grid point of the row $R_{k,X(v)}$ and the column $C_{k,Y(v)}$),
2. For a rectangular query subgrid with its lower-left corner at grid point (x, y) and upper-right corner at grid point (x', y') ($1 \leq x \leq x' \leq 2^k$ and $1 \leq y \leq y' \leq 2^k$) covering $\cup_{\alpha=x}^{x'} R_{k,\alpha} \cap \cup_{\beta=y}^{y'} C_{k,\beta}$ by $G(x, y, x', y')$ ($= \{v \in [2^k]^2 \mid x \leq X(v) \leq x'$ and $y \leq Y(v) \leq y'\}$). The size of the query subgrid $G(x, y, x', y')$ is $(x' - x + 1) \times (y' - y + 1)$.

Note that the $(+\frac{\pi}{2})$-rotation and $(-\frac{\pi}{2})$-rotation correspond to the 90°-clockwise rotation and 90°-counterclockwise rotation from x-axis to y-axis, respectively.

Remark 1. *For most self-similar m-dimensional order-k space-filling curve C_k^m indexing the grid $[2^k]^m$, we can view C_k^m as a C_{k-q}^m-curve interconnecting $2^{2(k-q)}$ C_q^m-subcurves for all $q \in [k]$.*

The remark above motivates our analytical study of clustering performances to be based upon query subgrids of size $2^q \times 2^q$.

For a 2-dimensional order-k Hilbert curve H_k^2, let $\mathcal{S}_{k,2^q}(H_k^2)$ denote the summation of all numbers of clusters over all $2^q \times 2^q$ query subgrids of an H_k^2-structural grid space $[2^k]^2$.

Remark 2. *Within a query subgrid G, the number of clusters is half of the number of edges of underlying space-filling curve that are cut by the sides of G [22].*

Denote by $\bar{n}(C, G)$ the number of edges within a subcurve C that are cut by sides of subgrid G (without counting edge(s) connecting between C and other subcurves). Remark 2 translates the computation of the summation of all numbers of clusters over all identically shaped subgrids G to the computations of $\frac{1}{2}(\sum_{\text{all } G} \bar{n}(C, G) + 2)$ (the contribution of 2 is the number of cuts for connecting edges of C to other curves). For $2^q \times 2^q$ subgrids G, we denote by $E_q(C)$ all numbers of clusters over all identically shaped $2^q \times 2^q$ subgrids G, which is $\sum_{\text{all } G} \bar{n}(C, G) + 2$.

The recursive decomposition of H_k^2 (see Fig. 1(b)) gives that

$$E_q(H_k^2) = 4E_q(H_{k-1}^2) + \varepsilon_{k,q}(H_k^2),$$

where $\varepsilon_{k,q}(H_k^2)$ denotes the summation of all edge cuts over all identically shaped $2^q \times 2^q$ query subgrids, each of which overlaps with more than one quadrant (that is, two or four). These query subgrids are contained in the boundary regions of neighboring quadrants as shown in Fig. 3.

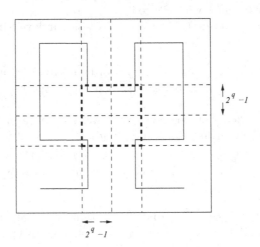

Fig. 3. The boundary regions of neighboring quadrants are organized into nine regions.

Remark 3. *For a 2-dimensional Hilbert curve H_k^2, the connecting edge between $Q_1(H_k^2)$ and $Q_2(H_k^2)$ is on the first column (left-most column), that between $Q_2(H_k^2)$ and $Q_3(H_k^2)$ is on the $2^{k-1}+1$-st row (the lowest row of these two quadrants), and that between $Q_3(H_k^2)$ and $Q_4(H_k^2)$ is on the 2^k-th column (right-most column).*

We denote the connecting edge between two quadrants $Q_i(H_k^2)$ and $Q_j(H_k^2)$ by a pair $(Q_i(H_k^2), Q_j(H_k^2))$. The previous remark tells the locations of the connecting edges. In addition to the cuts on connecting edges, the computation of $\varepsilon_{k,q}(H_k^2)$ is divided into two parts according to the overlaps of subspaces:

For a $2^q \times 2^q$ query subgrid G, G overlaps with:

1. exactly $Q_i(H_k^2)$ and $Q_{i \bmod 4+1}(H_k^2)$,
2. $Q_i(H_k^2)$ for all $i \in \{1,2,3,4\}$.

We develop combinatorial lemmas in the following two subsections to support the computations. We denote by $G(x,y,x',y')$ the query subgrid, in which the lower-left grid point is (x,y) and the upper-right grid point (x',y').

3.1 $\sum \bar{n}(H_k^2, G)$ over Subgrids G Overlapping with Two Quadrants

Consider an arbitrary $2^q \times 2^q$ query subgrid G that exactly overlaps two quadrants $Q_i(H_k^2)$ and $Q_{i \bmod 4+1}(H_k^2)$, where $i \in \{1,2,3,4\}$. The side-length is from 1 to $2^q - 1$ for the side across two quadrants. Since the quadrants are isomorphic to a canonical H_{k-1}^2 via symmetry (reflection and rotation), we consider the following system of summations $\Omega_{k,2^q} = (\Omega_{k,2^q}^L, \Omega_{k,2^q}^R, \Omega_{k,2^q}^B, \Omega_{k,2^q}^T)$ in a general context of a canonical H_k^2:

$$\Omega_{k,2^q}^L = \sum_{x=1}^{2^q-1} \sum_{y=1}^{2^k-2^q+1} \bar{n}(H_k^2, G(1,y,x,y+2^q-1))$$

— for left boundary (see Fig. 4(a)),

$$\Omega_{k,2^q}^R = \sum_{x=2^k-2^q+2}^{2^k} \sum_{y=1}^{2^k-2^q+1} \bar{n}(H_k^2, G(x,y,2^k,y+2^q-1))$$

— for right boundary,

$$\Omega_{k,2^q}^B = \sum_{x=1}^{2^k-2^q+1} \sum_{y=1}^{2^q-1} \bar{n}(H_k^2, G(x,1,x+2^q-1,y))$$

— for bottom boundary,

$$\Omega_{k,2^q}^T = \sum_{x=1}^{2^k-2^q+1} \sum_{y=2^k-2^q+2}^{2^k} \bar{n}(H_k^2, G(x,y,x+2^q-1,2^k))$$

— for top boundary, and

$$\mathcal{N}_{k,2^q}^S = \sum_{x=1}^{2^q-1} \sum_{y=1}^{2^k-2^q+1} 1$$

— for the number of rectangular subgrids in a boundary for $\Omega_{k,2^q}$.

We will establish a system of recurrences (in k) for $\Omega_{k,2^q}$ (see Lemma 4 below). The system of recurrence involves another system of summations as prerequisites, as demonstrated in the following example. Consider a recursive

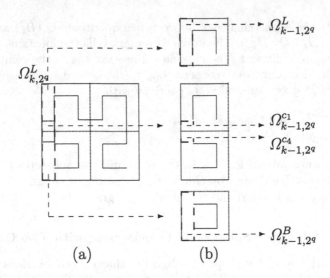

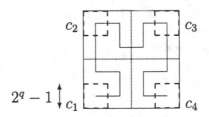

Fig. 4. (a) $\Omega_{k,2^q}^L$ for a canonical H_k^2; (b) its recursive decomposition.

decomposition of $\Omega_{k,2^q}^L$, illustrated in Fig. 4(b), into four parts: (1) $\Omega_{k-1,2^q}^B$, (2) $\Omega_{k-1,2^q}^{c4}$, $\Omega_{k-1,2^q}^{c1}$, (3) $\Omega_{k-1,2^q}^L$, and (4) the number of cuts on connecting edges. The part $\Omega_{k-1,2^q}^{c1}$ ($\Omega_{k-1,2^q}^{c4}$) computes $\sum \bar{n}(H_k^2, G)$ over all "incomplete" rectangular subgrids G (with one side-length at most $2^q - 1$) overlapping both $Q_1(H_k^2)$ and $Q_2(H_k^2)$. Each of the three parts $\Omega_{k-1,2^q}^B$, $\Omega_{k-1,2^q}^{c1}$ ($\Omega_{k-1,2^q}^{c4}$), and $\Omega_{k-1,2^q}^L$ is defined with respect to a canonical H_{k-1}^2. Note that $\Omega_{k,2^q}^L = \Omega_{k,2^q}^R$ because of the left-right symmetry property of H_k^2.

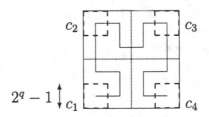

Fig. 5. The four $(2^q - 1) \times (2^q - 1)$ corners of a canonical H_k^2.

The recursive decompositions of all four parts in $\Omega_{k,2^q}^L$, $\Omega_{k,2^q}^R$, $\Omega_{k,2^q}^B$, and $\Omega_{k,2^q}^T$ lead us to consider the following prerequisite system of summations $\Omega_{k,2^q}^c = (\Omega_{k,2^q}^{c1}, \Omega_{k,2^q}^{c2}, \Omega_{k,2^q}^{c3}, \Omega_{k,2^q}^{c4})$ in a more general context of a canonical H_k^2 (see Fig. 5):

$$\Omega_{k,2^q}^{c1} = \sum_{x=1}^{2^q-1} \sum_{y=1}^{2^q-1} \bar{n}(H_k^2, G(1,1,x,y))$$

— for lower-left corner,

$$\Omega^{c_2}_{k,2^q} = \sum_{x=1}^{2^q-1} \sum_{y=2^k-2^q+2}^{2^k} \bar{n}(H^2_k, G(1, y, x, 2^k))$$

— for upper-left corner,

$$\Omega^{c_3}_{k,2^q} = \sum_{x=2^k-2^q+2}^{2^k} \sum_{y=2^k-2^q+2}^{2^k} \bar{n}(H^2_k, G(x, y, 2^k, 2^k))$$

— for upper-right corner,

$$\Omega^{c_4}_{k,2^q} = \sum_{x=2^k-2^q+2}^{2^k} \sum_{y=1}^{2^q-1} \bar{n}(H^2_k, G(x, 1, 2^k, y))$$

— for lower-right corner,

$$\mathcal{N}^c_{k,2^q} = \sum_{x=1}^{2^q-1} \sum_{y=1}^{2^q-1} 1$$

— for the number of incomplete rectangular subgrids in a corner.

These four summations involve rectangular subgrids contained in $(2^q - 1) \times (2^q - 1)$ corners. Note that $\Omega^{c_1}_{k,2^q} = \Omega^{c_4}_{k,2^q}$ and $\Omega^{c_2}_{k,2^q} = \Omega^{c_3}_{k,2^q}$ because of the left-right symmetry property of H^2_k. As suggested by Remark 1, we zoom in on the $2^q \times 2^q$ H^2_q-structural corners, and consider the following system of summations $\overline{\Omega}^c_{q,2^q} = (\overline{\Omega}^{c_1}_{q,2^q}, \overline{\Omega}^{c_2}_{q,2^q})$:

$$\overline{\Omega}^{c_1}_{q,2^q} = \sum_{x=1}^{2^q} \sum_{y=1}^{2^q} \bar{n}(H^2_q, G(1, 1, x, y))$$

— for lower(-left) corner,

$$(= \sum_{x=1}^{2^q} \sum_{y=1}^{2^q} \bar{n}(H^2_q, G(x, 1, 2^q, y)))$$

— for lower(-right) corner,

$$\overline{\Omega}^{c_2}_{q,2^q} = \sum_{x=1}^{2^q} \sum_{y=1}^{2^q} \bar{n}(H^2_q, G(1, y, x, 2^q))$$

— for upper(-left) corner,

$$(= \sum_{x=1}^{2^q} \sum_{y=1}^{2^q} \bar{n}(H^2_q, G(x, y, 2^q, 2^q)))$$

— for upper(-right) corner,

$$\overline{\mathcal{N}}^c_{q,2^q} = \sum_{x=1}^{2^q} \sum_{y=1}^{2^q} 1$$

— for the number of rectangular subgrids in a $2^q \times 2^q$ corner.

Thus far, we learn that the system of recurrences for $\Omega_{k,2^q}$ can be defined and solved via the prerequisite system $\Omega^c_{k,2^q}$, which is related to the system $\overline{\Omega}^c_{q,2^q}$ (see Lemma 3 below). The system $\overline{\Omega}^c_{q,2^q}$, which involves subgrids (with both side-lengths at most 2^q) of a canonical H^2_q, represents the basis of the recursive decompositions (in k to q) of $\Omega_{k,2^q}$ and $\Omega^c_{k,2^q}$. Similar to the reduction of $\Omega_{k,2^q}$ to $\Omega^c_{k,2^q}$, we develop a system of recurrences (in q) for $\overline{\Omega}^c_{q,2^q}$ via a prerequisite system.

The recursive decompositions of $\overline{\Omega}^{c_1}_{q,2^q}$ and $\overline{\Omega}^{c_2}_{q,2^q}$ lead us to consider a prerequisite system of summations $\overline{\Pi}_q = (\overline{\Pi}^v_q, \overline{\Pi}^h_q)$ in a general context of a canonical H^2_q:

$$\overline{\Pi}^v_q = \sum_{y=1}^{2^q} \bar{n}(H^2_q, G(1,1,2^q,y)) - \text{number of vertical edges (edges cut by}$$

$$\text{top (horizontal) sides of } G \text{ that covers lower part of } H^2_q)$$

$$\overline{\Pi}^h_q = \sum_{x=1}^{2^q} \bar{n}(H^2_q, G(1,1,x,2^q)) - \text{number of horizontal edges (edges cut by}$$

$$\text{right (vertical) sides of } G \text{ that covers left part of } H^2_q)$$

We develop and solve a system of recurrences for $\overline{\Pi}_q$ and reverse the sequence of reductions to obtain the closed-form solutions for $\Omega_{k,2^q}$, which are summarized in the following four lemmas.

Lemma 1. *For a canonical H^2_q,*

$$\overline{\Pi}^v_q = \begin{cases} 2\overline{\Pi}^v_{q-1} + 2\overline{\Pi}^h_{q-1} + 2 & \text{if } q > 1 \\ 2 & \text{if } q = 1 \end{cases}$$

$$\overline{\Pi}^h_q = \begin{cases} 2\overline{\Pi}^v_{q-1} + 2\overline{\Pi}^h_{q-1} + 1 & \text{if } q > 1 \\ 1 & \text{if } q = 1 \end{cases}$$

The closed-form solutions for $\overline{\Pi}_q$ are employed to establish a system of recurrences for $\overline{\Omega}^c_{q,2^q}$.

Lemma 2. *For a canonical H^2_q,*

$$\overline{\Omega}^{c_1}_{q,2^q} = \begin{cases} 3\overline{\Omega}^{c_1}_{q-1,2^{q-1}} + \overline{\Omega}^{c_2}_{q-1,2^{q-1}} + 3 \cdot 2^{q-1} \cdot \overline{\Pi}^v_{q-1} + 2^{q-1} \cdot \overline{\Pi}^h_{q-1} \\ \quad +3 \cdot 2^{q-1} + 1 & \text{if } q > 1, \\ 4 & \text{if } q = 1; \end{cases}$$

$$\overline{\Omega}^{c_2}_{q,2^q} = \begin{cases} \overline{\Omega}^{c_1}_{q-1,2^{q-1}} + 3\overline{\Omega}^{c_2}_{q-1,2^{q-1}} + 2^{q-1} \cdot \overline{\Pi}^v_{q-1} + 3 \cdot 2^{q-1} \cdot \overline{\Pi}^h_{q-1} \\ \quad +3 \cdot 2^{q-1} + 2 & \text{if } q > 1, \\ 5 & \text{if } q = 1. \end{cases}$$

The closed-form solutions for $\overline{\Omega}^c_{q,2^q}$ and $\overline{\Pi}_q$ are employed to obtain exact formulas for $\Omega^c_{k,2^q}$.

Lemma 3. *For a canonical H^2_k structured as an H^2_{k-q}-curve interconnecting $2^{2(k-q)}$ H^2_q-subcurves,*

$$\Omega^{c_1}_{k,2^q} = \overline{\Omega}^{c_1}_{q,2^q} - \overline{\Pi}^h_q - \overline{\Pi}^v_q,$$
$$\Omega^{c_2}_{k,2^q} = \overline{\Omega}^{c_2}_{q,2^q} - \overline{\Pi}^h_q - \overline{\Pi}^v_q.$$

The exact formulas for $\Omega^c_{k,2^q}$ are employed to establish a system of recurrences for $\Omega_{k,2^q}$.

Lemma 4. *For a canonical H^2_k structured as an H^2_{k-q}-curve interconnecting $2^{2(k-q)}$ H^2_q-subcurves,*

$$\Omega^L_{k,2^q} = \begin{cases} \Omega^B_{k-1,2^q} + \Omega^L_{k-1,2^q} + 2\Omega^{c_1}_{k-1,2^q} + 2(2^q - 1) & \text{if } k > q, \\ \overline{\Pi}^h_q & \text{if } k = q; \end{cases}$$

$$\Omega^B_{k,2^q} = \begin{cases} 2\Omega^L_{k-1,2^q} + 2\Omega^{c_2}_{k-1,2^q} & \text{if } k > q, \\ \overline{\Pi}^v_q & \text{if } k = q; \end{cases}$$

$$\Omega^T_{k,2^q} = \begin{cases} 2\Omega^T_{k-1,2^q} + 2\Omega^{c_2}_{k-1,2^q} & \text{if } k > q, \\ \overline{\Pi}^v_q & \text{if } k = q. \end{cases}$$

We obtain the closed-form solutions for $\Omega_{k,2^q}$ analytically and by using the mathematical software Maple.

3.2   Query Subgrids Overlapping with All Quadrants

For a $2^q \times 2^q$ query subgrid G that overlaps four quadrants around the center of H^2_k, when zooming in on the incomplete rectangular subgrid $G \cap G_1$ (with both side-lengths at most $2^q - 1$), where G_1 denotes the subspace of $Q_1(H^2_k)$, we reduce $\sum_{\text{all } G \cap G_1} \bar{n}(H^2_k, G \cap G_1)$ to $\Omega^{c_3}_{k-1,2^q} (= \Omega^{c_2}_{k-1,2^q})$ after $(-\frac{\pi}{2})$-rotating and left-right reflecting $Q_1(H^2_k)$ into a canonical H^2_{k-1}. Similar consideration leads to reductions of $\sum_{\text{all } G \cap G'} \bar{n}(H^2_k, G \cap G')$ to $\Omega^{c_4}_{k-1,2^q} (= \Omega^{c_1}_{k-1,2^q})$, $\Omega^{c_1}_{k-1,2^q}$ and $\Omega^{c_2}_{k-1,2^q}$ when $G \cap G'$ denotes the subspace for G overlapping $Q_2(H^2_k)$, $Q_3(H^2_k)$, or $Q_4(H^2_k)$, respectively.

Thus, the summation of numbers of edge cuts for all $2^q \times 2^q$ query subgrids G that overlap all four quadrants is

$$2\Omega^{c_2}_{k-1,2^q} + 2\Omega^{c_1}_{k-1,2^q}.$$

3.3   The Big Picture: Computing $E_q(H^2_k)$

The results in the previous three subsections yield $\varepsilon_{k,q}(H^2_k)$. Hence, we have the following lemma for $E_q(H^2_k)$.

Lemma 5. *For a canonical* H_k^2, *the recurrence for total number of cuts on edges by all* $2^q \times 2^q$ *subgrids* G:

$$
E_q(H_k^2) = \begin{cases}
\begin{aligned}
& 4E_q(H_{k-1}^2) + (\Omega_{k-1,2^q}^L + \Omega_{k-1,2^q}^B + (2^q - 1)) \\
& + (\Omega_{k-1,2^q}^L + \Omega_{k-1,2^q}^L) \\
& + (\Omega_{k-1,2^q}^L + \Omega_{k-1,2^q}^B + (2^q - 1)) \\
& + (\Omega_{k-1,2^q}^T + \Omega_{k-1,2^q}^T) \\
& + (2\Omega_{k-1,2^q}^{c_1} + 2\Omega_{k-1,2^q}^{c_2})
\end{aligned} & \text{if } k > q, \\[2em]
2 & \text{if } k = q.
\end{cases}
$$

Therefore, the exact formula for $E_q(H_k^2)$ is:

$$
E_q(H_k^2) = 2^{2k+q+1} - 2^{k+2q+2} + 2^{k+q+1} + 2^{k-q+1} + 2^{3q+1} - 2^{2q+1}.
$$

The summation of all numbers of clusters over all identically shaped $2^q \times 2^q$ query subgrids of an H_k^2-structural grid space $[2^k]^2$ is

$$
\mathcal{S}_{k,2^q}(H_k^2) = \frac{E_q(H_k^2)}{2}.
$$

The mean number of cluster within a subspace of size $2^q \times 2^q$ for H_k^2 is

$$
\frac{\mathcal{S}_{k,2^q}(H_k^2)}{(2^k - 2^q + 1)^2} = \frac{E_q(H_k^2)}{2(2^k - 2^q + 1)^2}.
$$

Consequently, we obtain the exact formula for the mean number of cluster within a subspace $2^q \times 2^q$ for H_k^2 below.

Theorem 1. *The mean number of cluster over all identical subspaces* $2^q \times 2^q$ *for* H_k^2 *is*

$$
\frac{2^{2k+q+1} - 2^{k+2q+2} + 2^{k+q+1} + 2^{k-q+1} + 2^{3q+1} - 2^{2q+1}}{2(2^k - 2^q + 1)^2}.
$$

4   Clustering Statistics of 2-Dimensional z-Order Curves

With respect to the canonical orientation of Z_k^2 shown in Fig. 1(d), we apply an analogous analytical approach to navigate through the recursive combinatorial structures of the boundary-edge and -vertex regions of Z_k^2, and prove the following clustering statistics of Z_k^2:

1. The exact formula for the summation of all numbers of edge cuts over all identically shaped $2^q \times 2^q$ subgrids, $E_q(Z_k^2)$, is:

$$
\begin{aligned}
E_q(Z_k^2) = {} & 2^{2k+q+2} - 2^{2k+2} + 3 \cdot 2^{2k-q} - 2^{2k-2q} - 2^{k+2q+3} + 3 \cdot 2^{k+q+2} \\
& - 2^{k+3} + 2^{k-q+2} + 2^{3q+2} - 2^{2q+3} + 2^{q+2},
\end{aligned}
$$

2. The summation of all numbers of clusters over all identically shaped $2^q \times 2^q$ query subgrids of an Z_k^2-structural grid space $[2^k]^2$ is

$$\mathcal{S}_{k,2^q}(Z_k^2) = \frac{E_q(Z_k^2)}{2},$$

and

3. The mean number of cluster within a subspace $2^q \times 2^q$ is

$$\frac{\mathcal{S}_{k,2^q}(Z_k^2)}{(2^k - 2^q + 1)^2} = \frac{E_q(Z_k^2)}{2(2^k - 2^q + 1)^2}.$$

The clustering statistics above yield the following exact formula for the mean number of cluster within a subspace $2^q \times 2^q$ for Z_k^2.

Theorem 2. *The mean number of cluster over all identical subspaces $2^q \times 2^q$ for Z_k^2 is*

$$(2^{2k+q+2} - 2^{2k+2} + 3 \cdot 2^{2k-q} + 2^{2k-2q} - 2^{k+2q+3} + 3 \cdot 2^{k+q+2} + 2^{k+3}$$
$$+ 2^{k-q+2} + 2^{3q+2} + 2^{2q+3} + 2^{q+2})/(2(2^k - 2^q + 1)^2).$$

5 Concluding Remarks

For a space-filling curve C_k indexing the grid space $[2^k]^2$, denote by $\mathcal{S}_{k,q}(C_k)$ the mean number of clusters over all $2^q \times 2^q$ subgrids of the C_k-structural grid space.

The exact formulas for $E_q(H_k^2)$ and $E_q(Z_k^2)$ give the exact formulas for $\mathcal{S}_{k,2^q}(H_k^2)$ and $\mathcal{S}_{k,2^q}(Z_k^2)$. We simplify the exact results asymptotically as follows. For sufficiently large k and q with $k \gg q$ (typical scenario for range queries),

$$\frac{\mathcal{S}_{k,2^q}(Z_k^2)}{\mathcal{S}_{k,2^q}(H_k^2)} = \frac{E_q(Z_k^2)}{E_q(H_k^2)} \approx 2$$

With respect to the $\mathcal{S}_{k,q}$-statistics, the Hilbert curve family clearly performs better than the z-order curve family over the considered ranges for k and q.

The analytical study of the clustering performances of 2-dimensional order-k Hilbert and z-order curve families are based upon the clustering statistics $\mathcal{S}_{k,2^q}$— mean number of clusters over all $2^q \times 2^q$ identically shaped subgrids, respectively. By taking advantage of self-similar properties of Hilbert and z-order curve, we derive their exact formulas for $\mathcal{S}_{k,2^q}$. The exact results allow us to compare their relative performances with respect to this measure. For sufficiently large k and q with $k \gg q$, Hilbert curve family performs significantly better than z-order curve family with respect to $\mathcal{S}_{k,2^q}$.

References

1. Alber, J., Niedermeier, R.: On multi-dimensional curves with Hilbert property. Theory Comput. Syst. **33**(4), 295–312 (2000). https://doi.org/10.1007/s002240010003
2. Asano, T., Ranjan, D., Roos, T., Welzl, E., Widmayer, P.: Space-filling curves and their use in the design of geometric data structures. Theoret. Comput. Sci. **181**(1), 3–15 (1997)
3. Bader, M.: Space-Filling Curves - An Introduction with Applications in Scientific Computing. Texts in Computational Science and Engineering, vol. 9. Springer, Heidelberg (2013). https://doi.org/10.1007/978-3-642-31046-1
4. Ban, X., Goswami, M., Zeng, W., Gu, X., Gao, J.: Topology dependent space filling curves for sensor networks and applications. In: Proceedings of the IEEE INFOCOM 2013, Turin, Italy, 14–19 April 2013, pp. 2166–2174. IEEE (2013)
5. Böhm, C., Berchtold, S., Keim, D.A.: Searching in high-dimensional spaces – index structures for improving the performance of multimedia databases. ACM Comput. Surv. **33**(3), 322–373 (2001)
6. Chen, H.-L., Chang, Y.-I.: Neighbor-finding based on space-filling curves. Inf. Syst. **30**(3), 205–226 (2005)
7. Dai, H.K., Su, H.C.: Approximation and analytical studies of inter-clustering performances of space-filling curves. In: Proceedings of the International Conference on Discrete Random Walks (Discrete Mathematics and Theoretical Computer Science, Volume AC (2003)), pp. 53–68, September 2003
8. Dai, H.K., Su, H.C.: On the locality properties of space-filling curves. In: Ibaraki, T., Katoh, N., Ono, H. (eds.) ISAAC 2003. LNCS, vol. 2906, pp. 385–394. Springer, Heidelberg (2003). https://doi.org/10.1007/978-3-540-24587-2_40
9. Dai, H.K., Su, H.C.: Norm-based locality measures of two-dimensional Hilbert curves. In: Dondi, R., Fertin, G., Mauri, G. (eds.) AAIM 2016. LNCS, vol. 9778, pp. 14–25. Springer, Cham (2016). https://doi.org/10.1007/978-3-319-41168-2_2
10. Dai, H.K., Su, H.C.: On norm-based locality measures of 2-dimensional discrete Hilbert curves. In: Dang, T.K., Küng, J., Takizawa, M., Chung, T.M. (eds.) FDSE 2020. LNCS, vol. 12466, pp. 169–184. Springer, Cham (2020). https://doi.org/10.1007/978-3-030-63924-2_10
11. Dai, H.K., Whang, K.-Y., Su, H.C.: Locality of corner transformation for multidimensional spatial access methods. Electron. Notes Theor. Comput. Sci. **212**, 133–148 (2008). Proceedings of the 2008 International Conference on Foundations of Informatics, Computing and Software
12. Gaede, V., Günther, O.: Multidimensional access methods. ACM Comput. Surv. **30**(2), 170–231 (1998)
13. Gotsman, C., Lindenbaum, M.: On the metric properties of discrete space-filling curves. IEEE Trans. Image Process. **5**(5), 794–797 (1996)
14. He, Z., Owen, A.B.: Extensible grids: uniform sampling on a space filling curve. J. Roy. Stat. Soc. Ser. B **78**(4), 917–931 (2016)
15. Jagadish, H.V.: Analysis of the Hilbert curve for representing two-dimensional space. Inf. Process. Lett. **62**(1), 17–22 (1997)
16. Kaddoura, M., Ou, C.-W., Ranka, S.: Partitioning unstructured computational graphs for non-uniform and adaptive environments. IEEE Parallel Distrib. Technol. **3**(3), 63–69 (1995)
17. Kinney, N., Hickman, M., Anandakrishnan, R., Garner, H.R.: Crossing complexity of space-filling curves reveals entanglement of S-phase DNA. Public Libr. Sci. One **15**(8), e0238322 (2020)

18. Lawder, J.K.: The application of space-filling curves to the storage and retrieval of multi-dimensional data. Ph.D. thesis, Birkbeck College, University of London, December 1999
19. Lee, M.-J., Whang, K.-Y., Han, W.-S., Song, I.-Y.: Adaptive row major order: a new space filling curve for efficient spatial join processing in the transform space. J. Syst. Softw. **78**(3), 257–269 (2005)
20. Lempel, A., Ziv, J.: Compression of two-dimensional images. In: Apostolico, A., Galil, Z. (eds.) Combinatorial Algorithms on Words. NATO ASI Series (Series F: Computer and Systems Sciences), vol. 12, pp. 141–156. Springer, Heidelberg (1984). https://doi.org/10.1007/978-3-642-82456-2_10
21. Mitchison, G., Durbin, R.: Optimal numberings of an $N \times N$ array. SIAM J. Algebraic Discrete Methods **7**(4), 571–582 (1986)
22. Moon, B., Jagadish, H.V., Faloutsos, C., Saltz, J.H.: Analysis of the clustering properties of the Hilbert space-filling curve. IEEE Trans. Knowl. Data Eng. **13**(1), 124–141 (2001)
23. Pérez, A., Kamata, S., Kawaguchi, E.: Peano scanning of arbitrary size images. In: Proceedings of the International Conference on Pattern Recognition, pp. 565–568. IEEE Computer Society (1992)
24. Platzman, L.K., Bartholdi, J.J., III.: Spacefilling curves and the planar travelling salesman problem. J. ACM **36**(4), 719–737 (1989)
25. Sagan, H.: Space-Filling Curves. Springer, New York (1994). https://doi.org/10.1007/978-1-4612-0871-6
26. Song, J.-W., Whang, K.-Y., Lee, Y.-K., Lee, M.-J., Han, W.-S., Park, B.-K.: The clustering property of corner transformation for spatial database applications. Inf. Soft. Technol. **44**(7), 419–429 (2002)
27. Song, J.-W., Whang, K.-Y., Lee, Y.-K., Lee, M.-J., Kim, S.-W.: Spatial join processing using corner transformation. IEEE Trans. Knowl. Data Eng. **11**(4), 688–695 (1999)
28. Wang, C., Jiang, H., Dong, Y.: Connectivity-based space filling curve construction algorithms in high genus 3D surface WSNs. ACM Trans. Sens. Netw. **12**(3), 22:1-22:29 (2016)
29. Xu, P., Tirthapura, S.: On the optimality of clustering through a space filling curve. In: PODS 2012 Proceedings of the 31st Symposium on Principles of Database Systems, pp. 215–224, May 2012

Attendance Monitoring Using Adjustable Power UHF RFID and Web-Based Real-Time Automated Information System

Quoc-Hung Chiem[1,2], Kha-Tu Huynh[1,2], Minh-Thien Nguyen[1,2], Minh-Duy Tran[1,2], Xuan-Phuc Phan Nguyen[1,2], and Tu-Nga Ly[2,1(✉)]

[1] International University, Ho Chi Minh City, Vietnam
{hktu,nmthien}@hcmiu.edu.vn
[2] Vietnam National University, Ho Chi Minh City, Vietnam
ltnga@hcmiu.edu.vn

Abstract. The attendance system has a significant impact on university and students' academic achievement. The design of a university automatic attendance system using Ultra-High Frequency RFID technology with four circularly polarized antennas with coverage distances ranging from 1 to 3 m is presented in this study. When integrated with web technology, RFID technology assures rapid speed, high accuracy, and user-friendliness, upgradability, and data protection in the digital age. The proposed system was evaluated with 142 samples in two classrooms measuring 5 m × 5 m and 5 m × 8 m. The team offers a technique to automatically calibrate card reader power based on when students enter or leave the class and the operating distance measured using experiment equipment; in contrast, the card reader will reduce its capacity while ensuring attendance. This mechanism saves power while also extending the life of the RFID reader. The conventional circular polarization antenna design fulfills the requirement, according to the axis ratio value and radiation patterns measurements. The proposed software allows the lecturer to directly examine the student's status by displaying the student's name in real-time in case of card fraud. Furthermore, our system supports a feature that allows teachers to activate manual attendance when students forget or lose their cards. Finally, it automatically provides the student's learning status for each subject weekly. It also provides the instructor with the overall proportion of students engaging in the subjects so that the teacher can get the complete picture. With the initial assessment, our system plays a positive role in reminding students of the number of lessons via email so that they are more aware of the situation and more inclined to visit class, as well as automatically digitizing and visualizing the sessions' attendance for the teacher in terms of statistics.

Keywords: RFID · Long-range RFID · Attendance monitoring · Web-based system · Automated information system

1 Introduction

There is no question that class attendance and subject expertise are significantly correlated. The degree of interaction between students and lecturers has a good influence on

T. K. Dang et al. (Eds.): FDSE 2021, LNCS 13076, pp. 392–407, 2021.
https://doi.org/10.1007/978-3-030-91387-8_25

students, who are better able to grasp the material after participating in class and will be able to utilize this information in their future employment. Technology has revolutionized the globe and transformed the way of education exceedingly. The attendance system has also gained traction in the last few years. Such an automated system reduces the resource and time expense for turnout management. There are multiple methods in terms of developing attendance systems. While biometric authentication processes (fingerprint, face, voice recognition) are some of the lowest in speed and accuracy due to their requirements in subject's distance and direction [1], RFID equipment is proved to be superior in these characteristics. RFID in attendance systems [2] guarantees quick speed, high accuracy, and user-friendliness despite expensive installation and operating costs. Furthermore, it eliminates illegal access, data loss, and data manipulation due to combining RFID technology with web technology, recently low-cost, upgradeable, and highly configurable. Presently, RFID for identification in academic research [2–6] has almost been achieved with short-distance readers. An automated attendance system based on radio frequencies with long-range distances for universities has yet to be developed.

Our paper presents the development of a safe, automated real-time attendance tracking system that utilizes long-range Ultra-high Frequency (UHF) RFID technology. Unlike short-range Low/High-Frequency RFID, barcode, QR code, UHF RFID can use its output parameters to develop localization functionality for tracking student traffic [7, 8]. Moreover, by considering UHF RFID setting factors, approaches can be established to improve energy efficiency.

The main contributions of our paper are:

1. Employed RFID reader and tags' technical parameters and measurements, with the initial technical constants delivered by RFID technology provider, and the measuring methods using high-tech laboratory equipment as [9, 10], data is established to ensure the feasibility of these products towards the attendance system.
2. Implementation of adjustable power of RFID reader to reduce electric consumption as [11].
3. Requirement analysis for the integrated web-based system.
4. Information and statistics are gathered to determine the outcome of our built system.

All related works are presented in Sect. 2, where the authors will review previous researches to lay the foundation for this system. In Sect. 3, executions from hardware attribute measurement to software development are demonstrated. The results of the experiments using the established system are produced in Sect. 4. Finally, the authors give a conclusion, including completed work, and suggest future improvements.

2 Literature Review

A research team [2] has built a real-time RFID identification system of employees, including an RC522 card reader, a Web-based interface stored on a server, and a database. However, the card reading distance is about 5cm. Similar to this work, the author [3] provides the formula for calculating average reading time and automatically calculating student attendance percentage based on collected and stored data. In addition, the author [5] has proposed two methods to process the facial images, such as the Viola-Jones method used to detect face objects in the image and the Local Binary Patterns Histogram algorithm for facial recognition. Here, the author has experimented with 11 samples by RFID scanning cards, but the distance was short, about from 1 cm to 4 cm.

In terms of web technology for the time attendance system, software that aids in storing and analyzing data from RFID systems is critical. Proposed RFID-based attendance systems [4] have been reviewed and evaluated for system functionality and primary conclusions for future research directions. Meanwhile, the authors [12] have proposed the toolkit for automating services related to academic counseling, course registration (planning of course sections, course availability, and handling waiting lists), schedule exams (detect conflicts), track course attendance, help students find suitable internships and track student progress in their study. These technologies have been obtained as programming languages and libraries in web development but provide weak performance for real-time and optimized data traffic features. Besides that, more advanced tools and frameworks are applied, such as Nodejs and MongoDB, to increase performance.

For the office checkup task in a surveillance monitoring system, the author [6] has provided a solution based on a mix of face recognition and RFID tags. In particular, the system is linked to a SQL Server database, which offers better synchronization than traditional surveillance management systems. This system scanned tags with a short distance RC522 card reader and verified 200–300 images with 93.5–95.3% accuracy. International and local research programs have yet to establish an automatic attendance system based on radio frequencies with long-range distances. In comparison with short-range RFID, the UHF RFID ensures faster speed for reducing the checking time and customizable power to reduce energy consumption.

3 Proposed System

We propose a fully developed system that utilizing the Impinj R2000 reader, C# Winform [13] for desktop middleware, Node.js [14], and Express.js for backend, MongoDB [15], Mongoose for database, and React.js [16], Ant Design [17] for frontend application.

3.1 System Structure

Figure 1 presents the overall structure of the system. Each classroom requires a reader placed in a suitable position to capture students' tags going in and out of the classroom (the recommended position is determined in the session of simulation result), and a computer installed with attendance middleware to receive tag data read from the reader. Both reader and desktop must be connected to a LAN (local area network) of the university. The middleware can at least fetch current course data of the classroom from and send attendance checking requests to the backend application and display the attendance results to students. The Nodejs server is hosted on the Heroku cloud. It connects and controls the MongoDB cloud storage data and then sends report emails to both students and lecturers. The frontend application is written with ReactJs, decorated by Ant Design. The webpage serves the admin who manages the course and student data and lecturers who manage their courses' attendance.

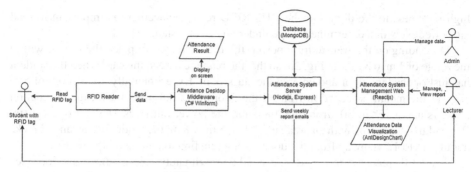

Fig. 1. System structure

3.2 Hardware Placement

The transceiver distance of the RFID reader from the centre of the hardware and with a full 360° circle with the centre as the reader's antenna is measured. The best position for each measurement angle will be recorded to calculate the average value that the reader can identify student attendance cards quickly and accurately. This is considered because students can approach the card reader in many different directions with various received signal strengths, thereby choosing the most suitable power for the reader.

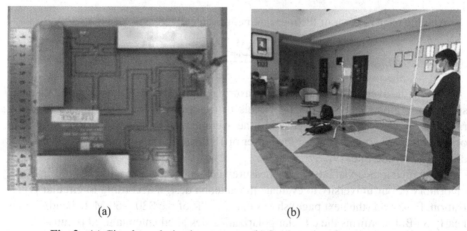

(a)   (b)

Fig. 2. (a) Circular polarization antenna, (b) System setup for measurement.

In addition, a long-range RFID card reader with a circular polarization antenna arrangement, including four antenna bars arranged on four square sides is proposed. Each antenna bar has dimensions of 7.2 cm × 2.3 cm (see Fig. 2a). These four antenna bars are divided into two parts at the antenna input: the left hand two circular polarization (LHCP: Left Hand Circular Polarization) antenna bars and the right hand two antenna bars (RHCP: Right Hand Circular Polarization). Each classroom requires a UHF RFID reader, as shown in Fig. 2b, to be mounted on top of the door or mounted on a 1.65 m high rack next to the door. For the UHF RFID reader, placement is important to read tags effectively in a determined range under any circumstances [18].

Depending on the transmitting power, the team chooses to place the mount within the range of 1 m to 3 m (see Fig. 2b) so that the reader can read the card when the student has just opened the door and entered the classroom. Experimentally, for a room of 5 m × 5 m, the mount is placed in front of the door about 0.8 m to 1.2 m, the inclination angle is about 15° to 30° from the door, and the power output is 18 dBm or 24 dBm. Particularly for rooms with an area larger than 5 m × 8 m, the reader has arranged racks about 1 m to 1.5 m away from the door, with an inclination angle of 15° to 30°.

The RFID reader is mounted on a stand approximately 1.65 m above the floor (see Fig. 2b) and connected to a laptop using Hopeland's RFID Manager software interface (see Fig. 3a, the next page). Setting up an interval of 15° from 0° to 360° is considered here. For each determined point, the tester starts near the centre then slowly moves away until the demo application GUI shows that the Speed(T/S) at the bottom right corner is reduced to 0 (see Fig. 3a, the next page). If the speed reaches exactly 0, they move closer to or further away from the reader to identify if the speed can be increased or not. Then, if the reading speed continues to be 0 for longer than 5 s, they mark the distance down. This procedure is repeated five times for each point to find their average [19], see Fig. 3b. The formula for the average maximum read distance is as follows

$$\underline{x} = \frac{\sum_{i=1}^{5} x_i}{5} \tag{1}$$

And the relative error of the measurement is calculated as:

$$\delta = \frac{\sum_{i=1}^{5} \frac{x_i - \underline{x}}{\underline{x}}}{5} \tag{2}$$

Taking the average value for each measuring angle, and then the operating distance spectrum is plotted as [9] with two levels of transmit power, 18 dBm and 24 dBm, respectively, shown in Fig. 3c. It shows the average read ranges while facing the front of the reader are about 1.8 m for the power of 18 dBm (red line) and about 2.9 m for 24 dBm power (blue line) [9].

By applying the axial ratio (AR) measurement technique [10] with the well-invested laboratory of our university to evaluate the antenna design according to circular polarization. Figure 3d (the next page) shows that the AR of the 820–960 MHz bandwidth is nearly 3 dB. It confirms the circular polarization designed antenna used is suitable.

To evaluate half power of radiation patterns shown in Fig. 4, the radiation of LHCP (blue line) at −3dB with frequency 860 MHz and 900 MHz is about 104 and 116°, respectively. This helps us to set up the UHF RFID reader with the main beamforming of LHCP (blue line) and other is RHCP (red line).

3.3 Power Adjustment

There are two methods [11] to reduce the energy consumption of a long-range RFID reader: using firmware options and reducing RF power. The author recommends using both methods for minimum consumption.

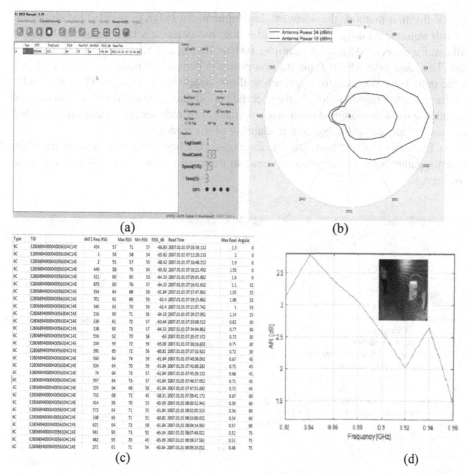

Type	TID	ANT1 Read	RSSI	Max RSSI	Min RSSI	RSSI_dB	Read Time	Max Read	Angular
6C	E28068940000400563D4C14E	454	57	71	57	-66.83	2007.01.01 07:03:39.112	1.3	0
6C	E28068940000400563D4C14E	3	58	58	54	-65.92	2007.01.01 07:11:29.132	2	0
6C	E28068940000400563D4C14E	2	55	57	55	-68.42	2007.01.01 07:10:48.212	1.9	0
6C	E28068940000400563D4C14E	449	58	75	54	-65.92	2007.01.01 07:10:21.492	1.55	0
6C	E28068940000400563D4C14E	321	60	65	53	-64.33	2007.01.01 07:05:01.882	1.8	0
6C	E28068940000400563D4C14E	875	60	76	57	-64.33	2007.01.01 07:16:42.632	1.1	15
6C	E28068940000400563D4C141	354	64	68	58	-61.84	2007.01.01 07:17:47.892	1.05	15
6C	E28068940000400563D4C14E	701	63	69	59	-62.4	2007.01.01 07:19:15.862	1.06	15
6C	E28068940000400563D4C14E	340	63	70	59	-62.4	2007.01.01 07:21:07.742	1	15
6C	E28068940000400563D4C14E	116	60	71	56	-64.33	2007.01.01 07:29:27.092	1.14	15
6C	E28068940000400563D4C14E	330	61	72	57	-63.64	2007.01.01 07:33:08.522	0.82	30
6C	E28068940000400563D4C14E	138	60	73	57	-64.33	2007.01.01 07:34:04.802	0.77	30
6C	E28068940000400563D4C14E	556	62	70	58	-63	2007.01.01 07:35:37.372	0.73	30
6C	E28068940000400563D4C14E	104	59	72	59	-65.09	2007.01.01 07:36:16.602	0.75	30
6C	E28068940000400563D4C14E	382	65	72	56	-60.81	2007.01.01 07:37:32.622	0.72	30
6C	E28068940000400563D4C14E	560	64	74	59	-61.84	2007.01.01 07:40:36.062	0.67	45
6C	E28068940000400563D4C14E	524	64	70	59	-61.84	2007.01.01 07:42:09.282	0.75	45
6C	E28068940000400563D4C14E	74	64	73	57	-61.84	2007.01.01 07:45:29.152	0.68	45
6C	E28068940000400563D4C14E	597	64	73	57	-61.84	2007.01.01 07:46:57.052	0.71	45
6C	E28068940000400563D4C14E	193	64	69	58	-61.84	2007.01.01 07:51:01.082	0.72	45
6C	E28068940000400563D4C14E	732	68	72	45	-58.31	2007.01.01 07:59:41.172	0.67	60
6C	E28068940000400563D4C14E	414	59	70	53	-65.09	2007.01.01 08:00:52.942	0.59	60
6C	E28068940000400563D4C14E	572	64	71	55	-61.84	2007.01.01 08:02:05.523	0.56	60
6C	E28068940000400563D4C14E	338	65	71	51	-60.81	2007.01.01 08:03:00.652	0.54	60
6C	E28068940000400563D4C14E	621	64	72	58	-61.84	2007.01.01 08:04:14.902	0.57	60
6C	E28068940000400563D4C14E	961	59	73	52	-65.09	2007.01.01 08:07:40.022	0.52	75
6C	E28068940000400563D4C14E	482	59	70	45	-65.09	2007.01.01 08:08:37.582	0.51	75
6C	E28068940000400563D4C14E	371	61	71	54	-63.64	2007.01.01 08:09:24.012	0.48	75

Fig. 3. (a) RFID Manager UI, (b) Operation distance, (c) Captured data for measurement, and (d) Measured axial ratio.

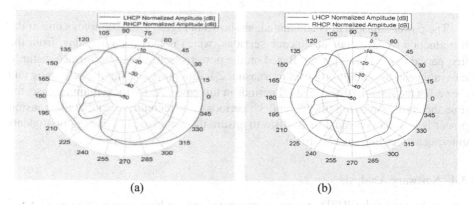

Fig. 4. Measured radiation patterns at (a) 860 MHz, (b) 900 MHz.

For the first method, the software development kit (SDK) provided by Hopeland controls automatic idling. When the reader does not detect any tags for about 20 ms, it will rest for a specified time, for example, 100 ms, then goes back to continue reading the card. The longer the RF off time, the less power is consumed; however, the readability of the reader is reduced. This approach will not affect the read range. The card reader also has a tag filtering function. In the specific period, the same card content can only be uploaded once. Compared to the above approach, this reduces power consumption but does not reduce the read range and readability of the reader.

As for the second method, usually, for a class, the beginning and the end are when students gather a lot in the entrance area, so the reader capacity needs to have wide coverage.

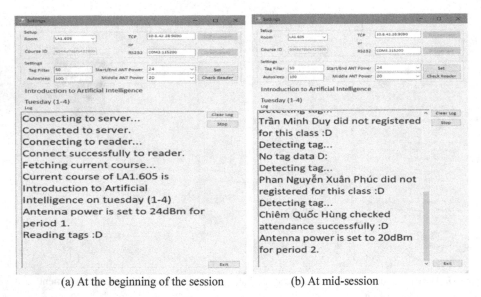

(a) At the beginning of the session   (b) At mid-session

Fig. 5. Automated power adjustment via desktop middleware

The power adjustment is automated, as shown in Fig. 5. This example is one of the Introduction to Artificial Intelligence sessions, which is taught on Tuesday, from the first period to the fourth period. Based on the power spectrum and performed distance measurement, the author proposes to transmit 24 dBm power during the beginning and the end of the session (the 1st and 4th periods in this case); As for the remaining case, the time the class is studying (the 2nd and 3rd periods), it is recommended that the transmit power is less than or equal to 20 dBm to ensure the automatic monitoring of students entering and leaving the classroom.

3.4 Software Analysis

The ambiance of the RFID system, if presented with a high volume of metallic and liquid components, can cause a destabilized electromagnetic condition [20]. Therefore, regulation is required to ensure successful attendance monitoring. For students to check attendance, see Fig. 6. To reduce reading sensitivity, they are obligated to carry their RFID student card separated from metal objects (vacuum flask, phone, …), liquid, and other RFID cards. The student can walk past the reader and wait for their result to appear on the screen. The server requires the course's ID and the student tag's ID from the middleware to determine if the student is registered for the course or not. If the student registered, the server stores the attendance data in the database and responses success to the middleware. The result is then displayed on the web application authenticated by the classroom teacher.

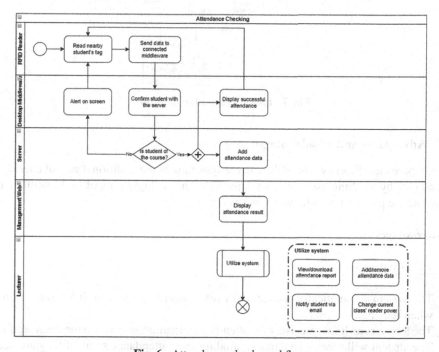

Fig. 6. Attendance check workflow.

For a MongoDB database, a collection contains documents [15], similar to a table containing rows of data in a SQL database, see Fig. 7. Each row of data requires an elicit primary key in SQL database, as for MongoDB, each document is generated with its ObjectId, thus requires no primary or foreign key. Both users, admin, and lecturer have their table, which stores the name, email, and encrypted password. Lecturer also has information about courses that they are teaching. Each student has an RFID tag ID. 200 tags are used during testing, and the tags' IDs are stored in the database to process students' attendance more quickly. A course, or a class, contains the time and location of the class to help determine where and when its sessions will be held. Each attendance record stores ObjectIds of the course and student, and the check times store as times that the reader reads the student's tag. Check-in, check-out, leave for a break can be determined via the CheckTimes attribute.

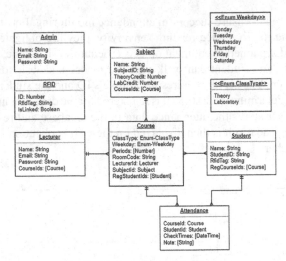

Fig. 7. Entity-relationship diagram

3.5 Advantages and Disadvantages

To compare the efficiency cost of the current system with the traditional way of checking attendance by teaching assistants, we have listed a few highlights of both methods to illustrate the pros and cons between these types.

- **Advantages**

For RFID system:

1. The system automatically does all the work concerning the data. It helps save time when running.
2. The lecturer can know precisely the attendance circumstance within the class section.
3. The student will receive an email reminding their attendance status if they are near the allowed day-offs.
4. The lecturer will receive an email to notify the attendance status of the class weekly.
5. The university can save the high cost when applying for the entire university.

For manual checking by a Teaching Assistant (TA):

1. High authenticity
2. Relatively low pay per class

- **Disadvantages:**

For RFID system:

1. Medium authenticity because teachers need to double-check.

2. High cost if just applying for few classes.

For manual checking by a teaching assistant (TA):

1. The TAs must come to class regularly to collect the attendance information.
2. The TAs must summarize all the data again manually. This consumes time and effort.

Although there are still some obstacles if we only apply it on a small scale, the current RFID-based attendance system still supports lecturers' attendance and assessing student progress. As a result, we can save time and effort for teaching assistants and lecturers in taking attendance and synthesizing attendance data. Besides, we have many ways to reduce the current system's cost by scaling up the operation, performing preliminary manual assembly but still keeping the main components to maintain the operational efficiency of the present motion.

4   Simulation Results

Based on equipment installation, system structure, and attendance process, the authors experimented with two laboratory rooms (LA1.605 and LA1.302) at our university with a list of classes from 11 to 53 students. The room size of LA1.605 and LA2.302 are 5 m × 5 m and 5 m × 8 m, respectively, arranged with an RFID reader on a rack (see Fig. 2b) according to the system layout by distance.

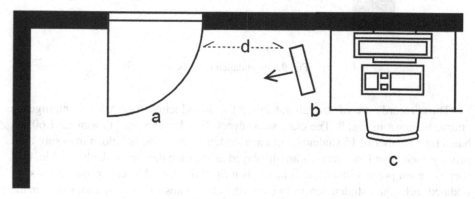

Fig. 8. System layout in LA1.605.

Figure 8 shows our system layout used in the room LA1.605, with label a is the door (width of 60 cm), label b is the RFID reader, which is placed at a d distance (80 cm) from the door and its front side (indicated by the arrow line) is fixed at 15° tilted from the wall, and label c is the computer used for the developed desktop middleware. The reader is tested with a different layout, for example, no tilt or 45° tilt, d distance of 0 cm to 120 cm. By having a tilted degree, the reader can perform well for both check-in and check-out times, if there is no tilt, check-out performance is not as effective as check-in, and having a reasonable distance from the door, the reader cannot read tags that are out of the boundary, outside the room, near the door. The layout is mainly based on the authors' experience as it depends on the structure of the room, the wall and the door's thickness and material, other components inside the room, and the number of students and RFID tags.

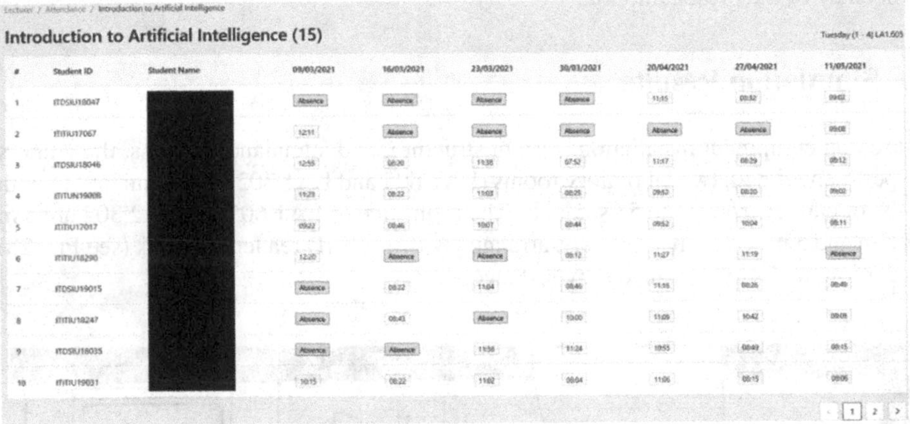

Fig. 9. Attendance records.

These records are of the laboratory of the Introduction to Artificial Intelligence course is shown in Fig. 9. The class starts every Tuesday at 8 am in room LA1.605. It had a total number of 15 students. For each student, their name is hidden to ensure their privacy, their attendance records are displayed as the time they arrived, checked in, or if they were not present, the records are shown as "Absence." Moreover, the records are updated each time a student leaves or enters the class. These timestamps are stored in the CheckTimes attribute. The last time a student interacts with the online reader is stored in the UpdatedAt feature, the attendance record. Later, these data can help the lecturer determine how long a student stays in the class and if their records are satisfying.

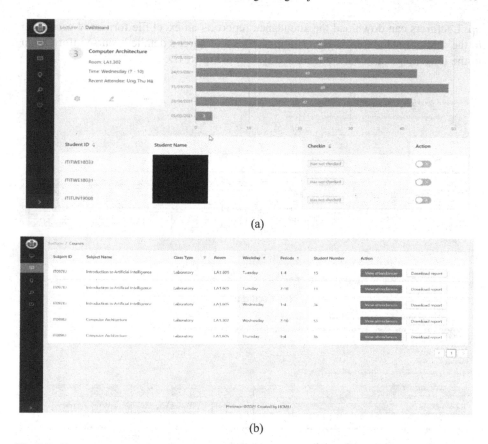

(a)

(b)

Fig. 10. Current course and course view for lecturer in real-time. (a). Current course information, (b). Course view for lecturer

Each time a class starts, the lecturer can log in and view the current course they are teaching in real-time. The view consists of an existing course card, a bar graph demonstrating previous sessions' total attendee counts, and a table displaying students who register the course and their check-in status ("Has not checked" if they have not arrived, the time of check-in if they arrive). Figure 10a helps lecturers visualize the status of students participating in the course according to bar graph parameters. All operations to display attendance records are performed on the web application in real-time, view the current number of students and the names of students attending, as well as a bar graph for comparison, compare the number of students who "attend" previous sessions and allow the instructor to check if the student has the correct card. This is practical for avoiding card fraud. When a student enters the class, the student's first and last name is immediately updated on the dashboard. For example, in Fig. 10a, student Ung Thu Ha has just entered the classroom. In addition, the software is designed for teachers to update attendance manually if a student forgets or loses their card. This is done by changing the "Action" button state. At the end of the day, instructors can view their course attendance reports, discerning students who are late or on time or do not show up. Lecturers can download the attendance report as an excel file formatted according to the school's standards. Figure 10b illustrates a lecturer's detailed course information in the long-range automatic attendance system using RFID.

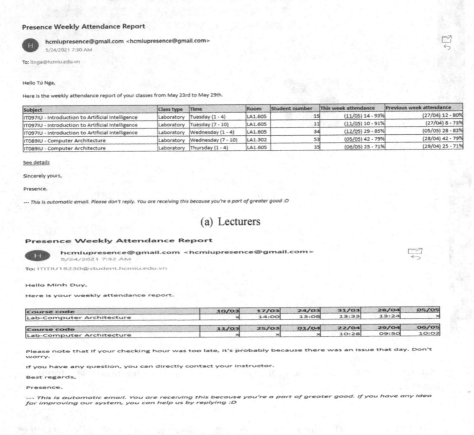

Presence Weekly Attendance Report

hcmiupresence@gmail.com <hcmiupresence@gmail.com>
5/24/2021 7:30 AM

To: ltnga@hcmiu.edu.vn

Hello Tú Nga,

Here is the weekly attendance report of your classes from May 23rd to May 29th.

Subject	Class type	Time	Room	Student number	This week attendance	Previous week attendance
IT097IU - Introduction to Artificial Intelligence	Laboratory	Tuesday (1 - 4)	LA1.605	15	(11/05) 14 - 93%	(27/04) 12 - 80%
IT097IU - Introduction to Artificial Intelligence	Laboratory	Tuesday (7 - 10)	LA1.605	11	(11/05) 10 - 91%	(27/04) 8 - 73%
IT097IU - Introduction to Artificial Intelligence	Laboratory	Wednesday (1 - 4)	LA1.605	34	(12/05) 29 - 85%	(05/05) 28 - 82%
IT089IU - Computer Architecture	Laboratory	Wednesday (7 - 10)	LA1.302	53	(05/05) 42 - 79%	(28/04) 42 - 79%
IT089IU - Computer Architecture	Laboratory	Thursday (1 - 4)	LA1.605	35	(06/05) 25 - 71%	(29/04) 25 - 71%

See details

Sincerely yours,

Presence.

--- This is automatic email. Please don't reply. You are receiving this because you're a part of greater good :D

(a) Lecturers

Presence Weekly Attendance Report

hcmiupresence@gmail.com <hcmiupresence@gmail.com>
5/24/2021 7:32 AM

To: ITITIU18230@student.hcmiu.edu.vn

Hello Minh Duy,

Here is your weekly attendance report.

Course code	10/03	17/03	24/03	31/03	28/04	05/05
Lab-Computer Architecture	x	14:00	13:08	13:33	13:24	x

Course code	11/03	25/03	01/04	22/04	29/04	06/05
Lab-Computer Architecture	x	x	x	10:28	09:50	10:02

Please note that if your checking hour was too late, it's probably because there was an issue that day. Don't worry.

If you have any question, you can directly contact your instructor.

Best regards,

Presence.

--- This is automatic email. You are receiving this because you're a part of greater good. If you have any idea for improving our system, you can help us by replying :D

(b) Students

Fig. 11. Weekly report email

Moreover, weekly reports personalized for each student and lecturer can help tremendously have quick updates on their statuses shown in Fig. 11. Lecturers can learn about their classes' attendance behavior and adjust their sessions based on these reports. Students can view their attendance progress and receive warnings if they miss too many sessions and are about to be prohibited from taking finals.

In addition, we can have a short analysis of the circumstance of both experiment lab classes. More specifically, checking attendance every day using the RFID system will improve the percentage of students attending the course, as shown in Fig. 12 (the next page).

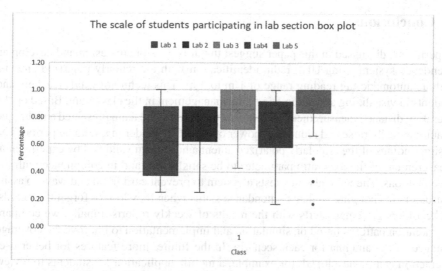

Fig. 12. The scale of students participating in all five labs.

To sum up, Table 1 shows a short brief about our dataset. We will know more clearly about how the students attend each class by attending each lab session.

Table 1. The information of datasets.

	Introduction to artificial intelligence lab	Computer architecture lab
Number of sections	8	6
Total students	60	82
The average proportion of the student get involved in class	77,89%	81,23%
The number of students who did not attend any lab section	0	0
The minimum percentage value that the students attend the class	25%	17%
The maximum percentage value that the students attend the class	100%	100%
The percentage of students who attend classes regularly	100%	100%
The first quartile (Q1)	0,625	0,67
The third quartile (Q3)	1	1
The outlier	None	None

5 Conclusions

Experiments discussed in this paper suggest that it is possible to design and develop an attendance system using UHF radio identification with a circularly polarized antenna with an automatic tag reading range of 1 m to 3 m. This technology aids teachers and students in visualizing and automating time management in the classroom. Based on the operating distance measurement spectrum of the system, the team presented a technique to automatically raise and reduce the power of the RFID reader in a suitable period. The design findings of the circularly polarized antenna have been validated by experimental measurement of the axis ratio parameter to be satisfactory and in compliance with the specifications. The team also suggests a system to prevent card fraud and ways to assist instructors in changing students' attendance status when they lose or forget their cards. Stakeholders get email alerts with the results of weekly reports. Finally, we compare the radiation patterns based on simulation and implementation to improve the coverage distance of the antenna for each scenario. In the future, more features for better user interaction can be developed, for example, a mobile application for students to review their attendance and get notified of upcoming classes. To increase the system's performance, by developing data streaming pipelines, issues of significant data traffic, different data formats can be overcome; an algorithm can be developed to automate the power adjustment based on data of check-in times from students.

References

1. Kotevski, Z., Blazheska-Tabakovska, N., Bocevska, A., Dimovski, T.: On the technologies and systems for student attendance tracking. Int. J. Inf. Technol. Comput. Sci. **10**(10), 44–52 (2018)
2. Koppikar, U., Hiremath, S., Shiralkar, A., Rajoor, A., Baligar, V.P.: IoT based smart attendance monitoring system using RFID. In: 2019 1st International Conference on Advances in Information Technology (ICAIT), pp.193–197. IEEE (July 2019)
3. Adeniran, T., Sanni, Y., Faruk, N., Olawoyin, L.A.: Design and implementation of an automated attendance monitoring system for a Nigerian university using RFID. Afr. J. Comput. ICT **12**, 72–89 (2019)
4. Rjeib, H.D., Ali, N.S., Al Farawn, A., Al-Sadawi, B., Alsharqi, H.: Attendance and information system using RFID and web-based application for academic sector. Int. J. Adv. Comput. Sci. Appl. **9**(1) (2018)
5. Basthomi, F.R., et al.: Implementation of RFID attendance system with face detection using validation viola-jones and local binary pattern histogram method. In: 2019 International Symposium on Electronics and Smart Devices (ISESD), pp.1–6. IEEE (October 2019)
6. Hoang, V.D., Dang, V.D., Nguyen, T.T., Tran, D.P.: A solution based on combination of RFID tags and facial recognition for monitoring systems. In: 2018 5th NAFOSTED Conference on Information and Computer Science (NICS), pp.384–387. IEEE (November 2018)
7. Kharrat, I., Duroc, Y., Vera, G.A., Awad, M., Tedjini, S., Aguili, T.: Customized RSSI method for passive UHF RFID localization system. J. Telecommun. **10**(2) (2011)
8. Yusof, R.J.R., Qazi, A., Inayat, I.: Student real-time visualization system in classroom using RFID based on UTAUT model. Int. J. Inf. Learn. Technol. (2017)
9. Lu, J.H., Chang, B.S.: Planar compact square-ring tag antenna with circular polarization for UHF RFID applications. IEEE Trans. Antennas Propag. **65**(2), 432–441 (2016)

10. Yuan, J., Wu, S., Chen, Z., Xu, Z.: A compact low-profile ring antenna with dual circular polarization and unidirectional radiation for use in RFID readers. IEEE Access **7**, 128948–128955 (2019)
11. Tertium, C.: How to save power consumption in an RFID reader? Homepage. https://iotlab.tertiumcloud.com/2020/09/11/how-to-save-power-consumption-in-an-rfid-reader-2/. Accessed 28 Aug 2021
12. Hallal, H.H., Aloul, F., Alawnah, S., Kolli, P., Alnabulsi, A.: Improving student experience using automated toolset of academic services. In: Proceedings of the 2020 6th International Conference on Computer and Technology Applications, pp. 97–101 (April 2020)
13. Microsoft. C# Winform Document Homepage. https://docs.microsoft.com/en-us/visualstudio/ide/create-csharp-winform-visual-studio?view=vs-2019. Accessed 28 Aug 2021
14. Nodejs Document Homepage. https://nodejs.org/en/docs/. Accessed 28 Aug 2021
15. MongoDB Document Homepage. https://docs.mongodb.com/guides/. Accessed 28 Aug 2021
16. Reactjs Document Homepage. https://reactjs.org/. Accessed 28 Aug 2021
17. Ant Design Document Homepage. https://ant.design/docs/react/introduce. Accessed 28 Aug 2021
18. Gan, O.P.: Placement of passive UHF RFID tags and readers using graph models. In: 2019 24th IEEE International Conference on Emerging Technologies and Factory Automation (ETFA), pp. 640–645. IEEE (September 2019)
19. Carolina, T.U.o.N. Measurements and Error Analysis Homepage. https://www.webassign.net/question_assets/unccolphysmech11/measurements/manual.html. Accessed 28 Aug 2021
20. Expósito, I., Cuiñas, I.: Exploring the limitations on RFID technology in traceability systems at beverage factories. Int. J. Antennas Propag. (2013)

Integrating Deep Learning Architecture into Matrix Factorization for Student Performance Prediction

Thanh-Nhan Huynh-Ly[1,2,3(✉)], Huy-Thap Le[1], and Nguyen Thai-Nghe[4]

[1] Department of Information Technology, Lac Hong University,
Bien Hoa, Dong Nai Province, Vietnam
[2] Department of Information Technology, An Giang University,
Long Xuyen, An Giang Province, Vietnam
hltnhan@agu.edu.vn
[3] Vietnam National University, Ho Chi Minh City, Vietnam
[4] College of Information and Communication Technology,
Can Tho University, Can Tho City, Vietnam
ntnghe@cit.ctu.edu.vn

Abstract. In universities using the academic credit system, choosing elective courses is a crucial task that significantly affects student performance. Because of poor performances, numerous students have been receiving formal warnings and expulsions from universities. Certainly, a good study plan from course recommendation methods plays an important role in obtaining a good study performance. In addition, early warnings that release on challenging courses enable students to prepare better for such courses. Predicting student learning performance is a vital factor in the courses recommendation system and is an essential task of an academic advisor. Many research methods solved this problem with diverse approaches such as association rules, deep learning, and recommender systems (RS). It recently built the courses recommendation system, which is used for personalized recommendation, especially the matrix factorization (MF) technique; But, the prediction accuracy of the MF still need to be improved. So, many studies try to integrate more information (e.g., social networks, course relationships) into the model. Besides, deep learning addresses the student performance prediction, which currently is state of the art, but it usually is general rules (not a personalized prediction). Indeed, deep learning and matrix factorization have advantages and disadvantages, so they need to compound together to get better. This paper proposes an approach to predict student performance that utilizes the deep learning architecture to carry out the MF method to enhance prediction accuracy, called deep matrix factorization. Experimental results of the proposed approach are positive when we perform on the published educational dataset.

Keywords: Educational data mining · Deep learning · Matrix factorization · Courses recommendation · Student performance prediction

© Springer Nature Switzerland AG 2021
T. K. Dang et al. (Eds.): FDSE 2021, LNCS 13076, pp. 408–423, 2021.
https://doi.org/10.1007/978-3-030-91387-8_26

1 Introduction

The original meaning of designing and developing such systems was the vision that Artificial Intelligence (AI) could give a promising solution to the limitations educational professionals face. Challenges in education include optimizing the faculty to student ratio, classifying students to improve individual student performance, and predicting student learning accurately to give appropriate suggestions for personalized students. It is possible to apply the advantages of current computer science development as classification methods, machine learning, decision-making recommendation system. Thus, the study [1] gave a systematic review and assessed the impact of AI on education. It is a qualitative research study, leveraging literature review as a research design and method was used.

AI researchers were keenly seeking a meaningful venue for their enthusiasm to spread the power of AI in many traditional fields when AI was blossoming. Computer scientists, cognitive scientists, educational professionals viewed the newborn Intelligent Tutoring Systems (ITS) to fulfil their various goals. In the paper [2], the authors reviewed the historical survey of ITS development. ITS uses AI techniques and support quality learning for individuals with little or no human assistance. As a result, ITS research is a multidisciplinary effort. It requires seamless collaborations of various disciplines, such as education, cognitive science, learning science, and computer science. In which (i) Artificial Intelligence (Computer Science) addresses how to reason about intelligence and thus learning, (ii) Psychology (Cognitive Science) tackles how people think and learn, and (iii) Education focuses on supporting teaching/learning.

Although the ITSs have diverse structures, the principal structure of an ITS contains four components such as Student-Model, Tutoring-Model, Domain-Model, and User-Interface. Although student modelling exists as one of four major components in the classic architecture of ITSs, it is a vital component in any ITSs [3]. It observes student behaviours in the tutor and creates a quantitative representation of student properties of interest necessary to customize instruction, respond effectively, engage students' interest, and promote learning. To ensure the student model's positive feedback feature, predicting student performance (PSP) is first researched.

Many types of studies may be using the learner's behaviour or learner grade [4]. Using learner behaviour is an implicit method in which researchers can predict student performance by observing the student's learning activities through the application system. Nevertheless, using the grade or mark of students is an explicit and straightforward method because all schools have a student grading system. Therefore, this method is widely used and in this article too.

The principal concern was that the effectiveness of advising improves with small students to teacher ratios. New expected performance prediction techniques are necessary to learning planning and predicting the risk of failing or dropping a class to solve the student retention problem. In [5], Personalized multi regression and matrix factorization approaches based on recommender systems, initially developed for e-commerce applications, accurately forecast students' grades in future courses as well as on in-class assessments. Accordingly, [6] gave the personalized prediction much more effective than the general rule prediction for the whole group of students. Their study was conducted with 772 students registered in e-commerce and e-commerce technologies modules at higher educational institutions. The study aimed to predict student's overall performance at the end of the semester using video learning analytics and data mining techniques. So that, the recommender system for personalized advising is better than the traditional data mining for suggesting general prediction. Many researchers address the student performance prediction using the recommender system, but these approaches still need improvement.

In the applications of artificial intelligence in general and machine learning in particular, it is essential to have a large enough dataset to mine well. Fortunately, there is the educational data mining competition [7], and registration for the competition and the dataset was entirely free, in line with the goals of promoting educational data mining. Even though the competition has already been finished, interested researchers can still get the dataset with permission. So, many studies in educational data mining used this dataset (including this study) because of its usefulness.

Recently, Deep Learning (DL) has outperformed well-known Machine Learning (ML) techniques in many domains, e.g., cybersecurity, natural language processing, bioinformatics, robotics and control, and medical information processing, among many others. [8] propose a more holistic approach to provide a more appropriate starting point to develop a complete understanding of DL. They outline the importance of DL and present the architecture of DL techniques and networks, and we can apply the deep learning architecture to other methods.

It is possible to apply these architectures to other methods. [9] has applied deep learning architecture into matrix factorization for improving the prediction accuracy in the entertainment field. Each method will be suitable for a specific problem and specific data samples. The research had a positive result in entertainment so that it can be good perform in education.

2 Related Works

Many research addresses predicting student performance by several data mining methods. However, each applied method has both advantages and disadvantages. Therefore, in [10], The survey synthesizes the intelligent models and paradigms applied in education to predict student performance. The survey identifies several key challenges and provides recommendations for future research in educational data mining. They proposed many traditional data mining algorithms to start-of-the-art methods such as statistical models, neural networks, Tree-based models, Bayesian-based models, Support Vector Machines, Instance-based models, and others.

The author of the paper [11] listed and compared implementing three different decision tree algorithms. They showed that J48 is the best decision tree algorithm used as a prediction and classification road map of students' actions. Additionally, decision tree graphs were affected by the number of input attributes and the end class attributes. Another work used k-NN and decision tree classification methods to predict employee performance on the internal dataset. Many comparative results are also interested in and studied by researchers. For example, [12] compared two methods, Decision Tree and Bayesian Network algorithms, to predict student academic performance.

The intelligent course recommendation system uses association rules to recommend courses to the student by common rules; however, this system is not personalized for each student. Moreover, Huu-Quang Nguyen et al. [13] have used the sequential rules algorithm applied to predicting student performance to give suggestions for students to choose elective courses. Another study proposed a system for academic advising using case-based reasoning (CBR) that recommends the student the most suitable major in his case after comparing the historical case with the student case [14].

In [15], the authors focus on designing a recommender system that recommends a set of learning objects to multiple students. Moreover, to deal with multi-decision group recommendations, they model the recommendation process as a non-cooperative game to achieve Nash equilibrium and demonstrate the effectiveness of their proposed model with a case study experiment. Furthermore, they built the system to help university students choose elective courses using a hybrid multi-criteria recommendation system with genetic optimization [16].

Rivera A.C. et al. [17] had a systematic mapping study about education recommender systems (RS). Thus, they have statistics several methods to address the problem of predicting student performance by using RS. In the paper [18], the author's proposed methods can build course recommendation systems, such as user/student k-nearest neighbours (student-kNN), item/course-kNN, standard MF, and biased MF. These methods are analyzed and validated using an actual data set before selecting the appropriate methods. They presented the framework for building the course recommendation system. However, this study focuses on the application systems and uses baseline methods.

Several works considered integrating social networks into RS, e.g., [19] have shown that the prediction accuracy can be improved by utilizing users' social networks in many ways. They have compared methods to integrate social networks into the MF. Recently, there has been a rapidly growing amount of online social networks like Facebook, Twitter. Many researchers have increasingly considered approaches for a recommendation based on social networks because they believe they affect each other. Several experiments confirmed that the social network provides independent sources of information, which can be exploited to improve the quality of recommendations. Indeed, [20] discovered that the relationship between classroom members is integrated into the training model, making the prediction better accurate. However, the algorithm is restricted only to used for data sets with user relationships. Additionally, the paper [21] proposed an approach to gather the relationships of the courses (e.g., knowledge/skills) and use those relationships to integrate into the Matrix Factorization to solve the PSP problem in the ITS.

Likewise, in the paper [22], the authors proposed to exploit multiple relationships by using multi-relational factorization models (MRMF) to improve accuracy for the PSP problems in Student-Model. However, these methods have not taken advantage of social relationships that can be integrated. In [23], the authors proposed an approach that aims to provide a solution to student performance predicting problems in ITS by combining Multiple Linear Regression (modelling Emotional Impact) and a Weighted MultiRelational Matric Factorization model advantage both students cognitive and emotional faculties. Their method considers the relationships that exist between students, tasks and skills, and their emotions.

In recent years, it has been common to transfer knowledge from one domain to another has gained much consideration among scientists. Tsiakmaki M. et al. [24] used transfer learning. A machine learning approach (deep neural networks) aims to exploit the knowledge retrieved from one problem to improve the predictive performance of a learning model. Likewise, in the study [25], the authors proposed deep learning models (Long Short Term Memory and Convolutional Neural Networks) to predict the student performance prediction problem in educational data mining. They used some techniques for data pre-processing (e.g., Quantile Transforms, MinMax Scaler) before fetching them into deep learning models and robust machine learning such as Linear Regression to do prediction tasks.

This study will be introducing applying deep learning architecture into matrix factorization technology, improving predicting student performance. First, we present an overview of problem definitions that can predict student's marks. Then, we introduce baseline methods such as matrix factorization and biased-matrix factorization. Next, the approach of integrating the deep learning architecture is conducted. Last, the result and the comparison of this study were presented.

3   Proposed Method

3.1   Problem Definition

In the studies [18, 20, 21], the author presented the mapping the predicting student performance problem to recommendation prediction task. In recommender systems, there are three main terms, which are user, item, and rating. The recommendation task predicts the user's rating for all un-rated items and recommends top-N highest predicted scores. Similarly, the PSP problem contains three essential objects: student, course, and performance (correct/incorrect). The task predicts the course's results that the students have not learned or solved in this setting. As presented in Fig. 1, there is a similar mapping between the PSP and RS. Where student, course, and grading would become user, item, and rating, respectively.

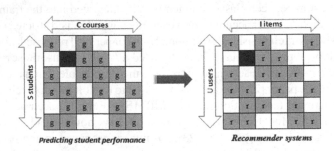

Fig. 1.   The similar mapping between PSP and RS

Student scoring management systems seem to be available to all universities, but they have not exploited them effectively. The problem is that we have to use them to predict student performance by using computer science methods. Although different datasets may have diverse structures, the principal structure contains three main fields (student-id, course-id, performance).

This study used the ASSISTments dataset, a web-based math tutoring system, first created in 2004 as joint research conducted by Worcester Polytechnic Institute and Carnegie Mellon University. Its name, ASSISTments, came from the idea of combining assisting the student with the automated assessment of the student's proficiency at a fine-grained level [7]. Thousands of middle and high school students use ASSISTments for their daily learning, homework, and preparing the MCAS (Massachusetts Comprehensive Assessment System) tests. 2010–2011, more than 20,000 students and 500 teachers used the system, which was considered part of their regular math classes in and out of Massachusetts. The snapshot of the ASSISTments dataset is displayed in Fig. 2. Some fields are necessary for mining, such as "User_id", "Problem_id", and "Correct".

	A	B	C	D	E	F	G
1	user_id	student_class_id	assignment_id	assistment_id	problem_id	skill_id	correct
2	77759	12138	245748	12914	12914	231	1
3	77759	12138	245748	15320	15320	231	1
4	77759	12138	245748	14529	14529	231	1
5	77912	12138	245698	1159	1159	100	0
6	77912	12138	245698	1647	1647	93	1
7	77912	12138	245698	2705	2705	100	1
8	77912	12138	245698	2186	2186	35	1
9	77912	12138	245698	1653	1653	35	1

Fig. 2. A snapshot of the data sample

Figure 3 shows an example of how we can factorize the students and problems (the performance is correct/1 or incorrect/0). From this point to the rest of the paper, we call course, problem, task interchangeably.

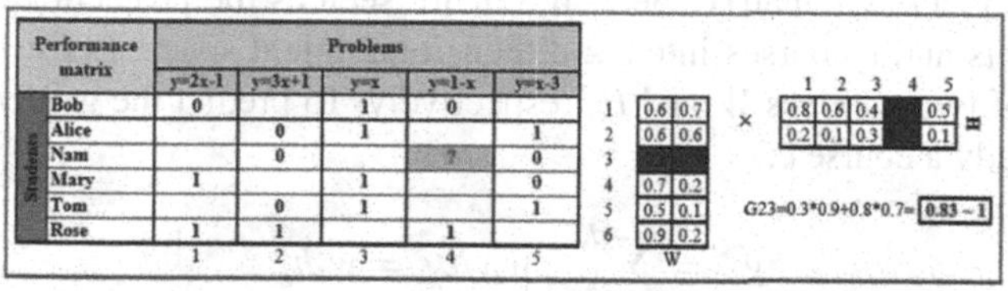

Fig. 3. An example of factorizing on students and problems

For improving the accuracy of prediction, many researchers integrated some information from independent sources. These studies have a better result.

3.2 Baseline Methods

Recommender systems are typically used by a list of recommendations using collaborative filtering (CF), content-based filtering, or a hybrid approach. Collaborative filtering methods are classified as memory-based and model-based collaborative filtering. A well-known example of memory-based processes is user-based algorithms and item-based algorithms. Model-based approaches are the latent factor models, especially the matrix factorization method.

Matrix Factorization Method

Matrix Factorization is an effective method for latent factor models. The matrix factorization is a flexible model in dealing with various datasets, applications, and fields. Approximating a matrix $X \in R^{|S| \times |C|}$ by a product of two smaller matrices, W and H, is the main idea of matrix factorization. Figure 4 below describes a model that factorized the matrix and gives a predicted grading for the student to learn a course.

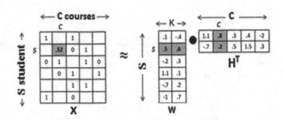

Fig. 4. The prediction process of matrix factorization

Based on [18], This method can be formalized as follows. The predicting student performance is dealing with s students and c courses, whose grades are collected in a matrix $X = (g_{s,c}) \in G_{s \times c}$ where $g_{s,c}$ is the grading that the student s learnt the course c (typically, float values between 0 and 1), or $g_{s,c} = \varnothing$ if the student s has not learnt the course c. We look for a factorization of G of the form $G \approx W \cdot H^T$. Where W is a $s \times k$ matrix, and H is a $k \times c$ matrix. Here, W, H are seen as the projection/co-projection of the s students and c courses into a k-dimensional latent space. Let w_{sk} and h_{ck} are the elements of two matrices W and H, respectively. To predict the grade/mark g for a student s to study a course c:

$$\widehat{g_{sc}} = \sum_{k=1}^{K} w_{sk} h_{ck} = w_s h_c^T \qquad (1)$$

Root Mean Square Error (RMSE) is a criterion to find optimal values for the parameters W and H. It is determined:

$$RMSE = \sqrt{\frac{1}{|D^{test}|_{s,c,g \in D^{test}}} \sum (g_{si} - \widehat{g_{si}})^2} \qquad (2)$$

In the MF technique [18], training the model is to find the optimal parameters W and H. These matrices are initialized with some random values (from the normal distribution). Besides, the error function is added as a term for preventing over-fitting. The error function is determined:

$$o^{MF} = \sum_{(s,c,g) \in D^{train}} \left(g_{sc} - \sum_{k=1}^{K} w_{sk} h_{ck} \right)^2 + \lambda \left(\|W\|_F^2 + \|H\|_F^2 \right) \qquad (3)$$

Where $\| \cdot \|_F^2$ is a Frobenius[1] norm, λ is a regularization weight. The error function O^{MF} can be derived to w_s and h_c resulting in the following updated rules for learning the model parameters. The w_{sk} and h_{ck} are updated by the equations below (where $e_{sc} = g_{sc} - \hat{g}_{sc}$, and w'_{sk} is the updated value of w_{sk}, and h'_{ck} is the updated value of h_{ck}) The values of w'_{sk} and h'_{ck} are carried out respectively.

$$w'_{sk} = w_{sk} + \beta(2e_{sc}h_{ck} - \lambda w_{sk}) \qquad (4)$$

$$h'_{ck} = h_{ck} + \beta(2e_{sc}w_{sk} - \lambda h_{ck}) \qquad (5)$$

Where β is the learning rate. We update the values of W and H iteratively until the error converges to its minimum $(O_{n-1}^{MF} - O_n^{MF} < \varepsilon)$ or reaching a predefined number of iterations. Finally, the performance of student s on courses c is now determined by Eq. (6) and Fig. 4:

$$\hat{g}_{sc} = \sum_{k=1}^{K} w_{sk}h_{ck} = w_s h_c^T \qquad (6)$$

Biased Matrix Factorization Method

We have presented the standard matrix factorization to encode the student/course latent factors. Now, we introduce how to use the biased matrix factorization (BMF) to deal with the problem of "user effect" ("user bias") and "item effect" ("item bias") [18]. The user and item biases are on the educational setting, respectively, the student and course biases/effects. The student effect (student bias) models how good/clever/bad a student is (i.e., how likely is the student to perform a course correctly), and the course effect (course bias) models how difficult/easy the course is (i.e., how likely is the course to be performed correctly). With these biases, the prediction function for student s on course c is presented by

$$\hat{g}_{sc} = \mu + b_s + b_c + \sum_{k=1}^{K} w_{sk}h_{ck} \qquad (7)$$

$$\mu = \frac{\sum (s, c, g) \in D^{train}}{|D^{train}|} \qquad (8)$$

$$b_s = \frac{\sum_{(s',c,g) \in D^{train}|s'=s} (g - \mu)}{|\{(s', c, g) \in D^{train}|s' = s\}|} \qquad (9)$$

$$b_c = \frac{\sum_{(s,c',g) \in D^{train}|c'=c} (g - \mu)}{|\{(s, c', g) \in D^{train}|c' = c\}|} \qquad (10)$$

Moreover, the error function is also changed by adding these two biases to the regularization:

$$o^{BMF} = \sum_{(s,c,g) \in D^{train}} \left(g_{sc} - \mu - b_s - b_c - \sum_{k=1}^{K} w_{sk}h_{ck}\right)^2 + \lambda\left(\|W\|_F^2 + \|H\|_F^2 + b_s^2 + b_c^2\right) \qquad (11)$$

---

[1] https://en.wikipedia.org/wiki/Matrix_norm#Frobenius_norm.

3.3 Deep Learning Matrix Factorization Method

A current trend in the field of predicting student performance is to improve the quality of the predictions utilizing different techniques of MF, such as Biased-MF [18], Social-MF [20], and CourseRelationship-MF [21]. However, all these techniques rely on the same approach: they pose an optimization problem through an error function by integrating more information that measures the divergence of the model, and the model with lower errors is better.

The main idea of integrating deep learning architecture into the MF technique is recursing the matrix factorization, repeatedly approximate the output matrix until it meets the best result. Each step in the process is a baseline method (matrix factorization); the complete model summarises the stack of the models in stages.

We are trying to break with the MF paradigm, and based on the paper [9], we present an integrating model that uses the DL principles to refine the model's output through successive training. It is called the Deep Learning Matrix Factorization (DLMF). Figure 5 illustrates the operation of DLMF. As we can observe, the model is initialized with the standard input of a CF-based RS: a matrix X that contains the student's grading/score/mark to the course/problem/exercise.

As in the classical MF method, this matrix X will also be called $X = G^0$ which is the beginning of the process. Approximating a matrix $X \in G^{|S| \times |C|}$ by a product of two smaller matrices, W^0 and H^0, is a form $\widehat{G}^0 = W^0 \cdot H^0$. The matrix $\widehat{G}^0$ provides all the predicted gradings, and they were stored in the stack at the first step. At this step, the recursive is begins. A new matrix $G^1 = X - \widehat{G}^0$ is built by computing the attained errors between the original gradings matrix X and the predicted gradings stored in $\widehat{G}^0$. A factorization again approximates this new matrix G^1 into two new small rank matrices $\widehat{G}^1 = W^1 \cdot H^1$, which produces the errors at the second step $G^2 = G^1 - \widehat{G}^1$. This process is repeated many times by generating and factorizing successive error matrices $G^1, ..., G^T$. Presumably, this sequence of error matrices converges to zero, so we get preciser predictions as we add new layers to the model.

Similar to the standard MF method was presented at the baseline methods section above. Two small k-rank matrices are trained such that the product $W^0 \cdot H^0$ is a good approximation of the rating matrix $G^0 = X$, that is, in the usual Euclidean distance. The term $W^0 \in G^{0|s| \times |k^0|}$ is a matrix where each row s is a vector w_s (rendering the student s) and has k^0 latent factors. Similarly, the term $H^0 \in G^{0|c| \times |k^0|}$ is a matrix where each row c is a vector h_c (rendering the course c) and has k^0 latent factors. The approximation can be expressed as follows:

$$G^0 \approx \widehat{G}^0 = W^0 \cdot H^0 \tag{12}$$

To implement the deep learning model, we subtract the approximation performed by $\widehat{G}^0$ to the original matrix X, to obtain a new sparse matrix G^1 that contains the prediction error at the first iteration:

$$G^1 = X - \widehat{G}^0 = G^0 - W^0 \cdot H^0 \tag{13}$$

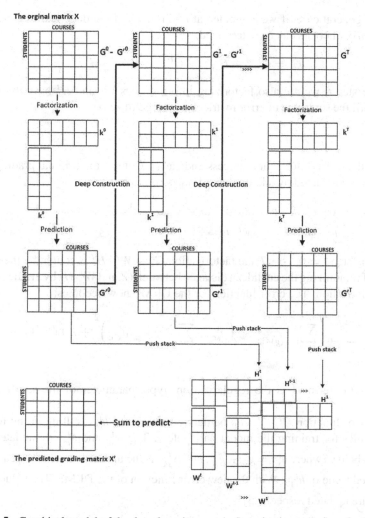

Fig. 5. Graphical model of the deep learning matrix factorization technique for PSP

Note that positive values in the matrix G^1 mean that the prediction is low and need to be increased. Similarly, the negative values in the matrix G^1 mean that the prediction is high and need to be decreased. Indeed, this adjustment is the main idea of applying the deep learning approach. To do this, we need to perform a new factorization to the error matrix G^1 in such a way that

$$G^1 \approx \widehat{G}^1 = W^1 \cdot H^1 \tag{14}$$

The approximation process in each step is performed similarly. However, two matrices W^1 and H^1 have orders $s \times k^1$ and $k^1 \times c$ for a definite number of latent factors k^1. Note that we should take $k^1 \neq k^0$ to get various resolutions in the factorization.

In the general case, if we computed at $t-1$ steps of the deep learning procedure, the tth matrix of errors can be determined:

$$G^t = G^{t-1} - \widehat{G}^{t-1} = G^{t-1} - W^{t-1} \cdot H^{t-1} \qquad (15)$$

At the step t, we are also factorizing into matrices W^t and H^t have the k^t latent factors until the sequence of error matrices converge to zero.

$$G^t \approx \widehat{G}^t = W^t \cdot H^t \qquad (16)$$

Once the deep factorization process ends after T steps, the original grading matrix X can be reconstructed by adding the estimates of the errors as

$$X \approx \widehat{X} = \widehat{G}^0 + G^1 = \widehat{G}^0 + \widehat{G}^1 + G^2 = \cdots = \widehat{G}^0 + \widehat{G}^1 + \widehat{G}^2 + \cdots + \widehat{G}^T = W^0 \cdot$$
$$H^0 + W^1 \cdot H^1 + \cdots + W^T \cdot H^T = \sum_{t=0}^{T} W^t \cdot H^t \qquad (17)$$

For any step $t = 0, \cdots, T$ the factorization $G^t \approx W^t \cdot H^t$ is sought by the standard method of minimizing the euclidean distance between G^t and $W^t \cdot H^t$ by gradient descent with regularization. The error function for the DLMF now becomes:

$$O^{DLMF} = \sum_{t=0}^{T} \left(\sum_{(s,c,g) \in D^{t\,train}} \left(g_{sc}^t - \sum_{k^t=1}^{K} w_{sk^t}^t h_{k^t c}^t \right)^2 + \lambda^t \left(\|W^t\|_F^2 + \|H^t\|_F^2 \right) \right) \qquad (18)$$

Where the term λ^t is the regularization hyper-parameter of the step t to avoid overfitting.

The error function O^{DLMF} can be derived to w_s^t and h_c^t resulting in the following updated rules for training the model parameters. The w_{sk}^t and h_{ck}^t are updated by the equations below (where $e_{sc}^t = g_{sc}^t - \widehat{g}_{sc}^t$, and $w_{sk}^{t\,\prime}$ is the updated value of w_{sk}^t, and $h_{ck}^{t\,\prime}$ is the updated value of h_{ck}^t). With the new error function of the DLMF, The values of $w_{sk}^{t\,\prime}$ and $h_{ck}^{t\,\prime}$ are updated respectively

$$h_{ck}^{t\,\prime} = h_{ck}^t + \beta^t \left(2 e_{sc}^t w_{sk}^t - \lambda h_{ck}^t \right) \qquad (19)$$

$$w_{sk}^{t\,\prime} = w_{sk}^t + \beta^t \left(2 e_{sc} h_{ck}^t - \lambda w_{sk}^t \right) \qquad (20)$$

Where β^t is the learning rate hyper-parameter of step t to control the learning speed.

In this way, after finishing the nested factorization, all the predicted ratings are collected in the matrix $\widehat{X} = (\widehat{g}_{sc})$, where the predicted the grading of the student s to the course c is given by

$$\widehat{g}_{sc} = \sum_{t=0}^{T} \sum_{k=1}^{K} w_{sk}^t h_{kc}^t = \sum_{t}^{T} w_s^t * \left(h_c^t \right)^T \qquad (21)$$

Note that this method consists of successive repetitions of an MF process using the results of the previous MF as input. All the hyper-parameters are stored in the stack, so we may easily use a recursive approach and recursive implementation algorithm.

Proposed Algorithm

The algorithm receives inputs as the original matrix X and the model hyper-parameters. Similarly, the output of the algorithm will be a stack containing the pairs $\langle W, H \rangle$ that fully represent the predictability of the deep learning process.

Note that these hyper-parameters were stacked so that each of the factorizations performed uses different hyper-parameters. The hyper-parameters of the first factorization will be pushed at the top of the stack, and the hyper-parameters of the second factorization in the next one. And so on until the parameters of the last factorization will be pushed at the bottom of the stack. This allows us to define the stopping criteria of the algorithm as the depth of stack, which is usually around four layers.

Details of the proposed method that integrates the deep learning architecture into Matrix Factorization are presented in the function below "Deep-Learning-Matrix-Factorization – DLMF". This DLMF is recursively factorizing student and course using the stacks of stochastic gradient descent with k latent factors, β learning rate, λ regularization weight, stopping condition, and the depth. For example, we pop the hyper-parameters to carry out the block of MF statements in each depth. In each MF statement block in lines 1–13, we perform as standard MF method. Then we recursive call and push the complete training models to the stack in lines 18–19.

```
Function Deep-Learning-Matrix-Factorization
Input: D^train, K, betas, lamdas, stopping condition, depth
Output: W, H
1.  Let s ∈ S be a student, c ∈ C a course, g ∈ G a grade
2.    Let  W[|S|][K], H[|C|][K]  be latent factors of
students, courses
3.    W ← N(0,σ²) and H ← N(0,σ²)
4.    k, t, β, λ ← pop (K, T, Betas, Lamdas)
5.    while (the stopping condition is NOT met) do
6.         for each (s,c,g_sc) from D^train
7.             ĝ_sc ← Σ_k^K(W[s][k] * H[c][k])
8.             e_sc = g_sc − ĝ_sc
9.             for  k = 1..K do
10.                W[s][k] = W[s][k] + β * (2e_si * H[c][k] − λ * W[s][k])
11.                H[c][k] ← H[c][k] + β * (2e_sc * W[s][k] − λ * H[c][k])
12.             end for
13.         end for
13.    end while
14.    if (is-empty(K)) return new stack (⟨W,H⟩)
15.    else
16.        Ĝ = G − W · H
17.        Params-Stack = Deep-Learning-Matrix-
Factoriztion(Ĝ, K, T, Betas, Lamdas)
18.        Return push(⟨W,H⟩, Params-Stack)
19.    end else.
20.    end function.
```

4 Result

4.1 Dataset

The ASSISTments[2] dataset is published by the ASSISTments Platform. It is a web-based tutoring system that assists students in learning mathematics and gives teachers an assessment of their students' progress. It allows teachers to write individually, and each ASSISTments is composed of questions and associated hints, solutions, web-based videos. After preprocessing, this dataset contains 8519 students (users), 35978 tasks (items), and 1011079 gradings (ratings).

4.2 Evaluation

In this work, predicting student marks is the task of rating prediction (explicit feedback), so we use a popular measure in RS is Root Mean Squared Error (RMSE), for model evaluation. We have used the hold-out approach (2/3 of data is used for training, and 1/3 of data is used for testing) for experimenting with the models.

The accuracy of the prediction depends on the parameters that feed to the algorithm. If the parameters were unsuitable, the prediction accuracy would not be good even though the algorithm is correct. Thus, finding the best parameter is significant.

The hyper-parameters search, a searching parameter method, is applied to search all the parameters of the approached models [18]. The hyper-parameters search has two stages based on grid search: raw search (for the long segments) and smooth search (for the short segments). First, the raw search stage is carried out to find the best hyper-parameters in the long data segments. Then, we perform a smooth search to find the nearby best hyper-parameters. For example, using RMSE as a criterion, the hyper-parameter search results for the models on the ASSISTments dataset are presented in Table 1.

Table 1. Hyper-parameters on ASSISTments dataset

Methods	Hyper parameter
MF	$\beta = 0.03$, #iter $= 50$, K $= 4$, $\lambda = 0.05$
BMF	$\beta = 0.0015$, #iter $= 50$, K $= 2$, $\lambda = 0.1$
DLMF	Deep (T) $= 4$ numFactors (K) $= [3, 6, 3, 3]$; numIters (Iter) $= [50, 50, 50, 50]$; learningRate (β) $= [0.01, 0.1, 0.1, 0.1]$; regularization (λ) $= [0.01, 0.1, 0.1, 0.01]$)

After having the best hyper-parameters, we use them for training and testing each respective model. However, the training time is slower than without deep learning. How much deeper training, the time delay that many times.

---

[2] https://sites.google.com/site/assistmentsdata/home.

4.3   Experimental Result

We have compared integrating deep learning architecture into matrix factorization (DLMF) to predict student performance in the ITS with other methods, such as standard MF, BMF, and DLMF. Fortunately, many open-source libraries implemented these algorithms, such as LibRec (librec.net), MyMediaLite (mymedialite.net), Collaborative Filtering For Java (CF4J), that we can inherit from them.

We conducted three experiments; the experimental results are displayed in Fig. 6. Comparing with others, the RMSE of the proposed approach (DLMF) is the smallest one (0.419) on the dataset. The smallest error demonstrates the best model.

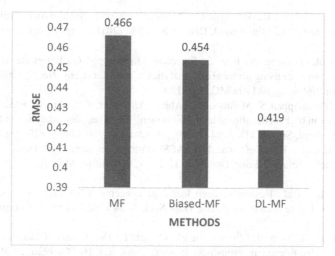

Fig. 6. Experiment results on the ASSISTment dataset

5   Conclusion

This paper has introduced an approach that utilizes the deep learning architecture to carry out the MF method to enhance the accuracy of student performance prediction. We can take advantage of the deep learning principles to refine the model's output (predicted matrix) through successive training for building the prediction model with this approach. Thus, the prediction results can be improved significantly. Conducting experiments on the published competition datasets shows that the proposed process works well.

Applying deep learning architecture to matrix factorization make the training time will be slow. However, it is easy for us to implement a parallel algorithm to solve this problem. This work uses deep learning architecture for the standard matrix factorization without using other complex integrating techniques. Future research is how to find the meta-data for integrating to get highly effective and fast algorithms.

References

1. Chen, L., Chen, P., Lin, Z.: Artificial intelligence in education: a review. IEEE Access **8**, 75264–75278 (2020). https://doi.org/10.1109/ACCESS.2020.2988510
2. Guo, L., Wang, D., Gu, F., Li, Y., Wang, Y., Zhou, R.: Evolution and trends in intelligent tutoring systems research: a multidisciplinary and scientometric view. Asia Pac. Educ. Rev. **22**(3), 441–461 (2021). https://doi.org/10.1007/s12564-021-09697-7
3. Khodeir, N.: Student modeling using educational data mining techniques. In: ACCS/PEIT 2019 - 2019 6th International Conference on Advanced Control Circuits and Systems (ACCS) & 2019 5th International Conference on New Paradigms in Electronics & Information Technology, pp. 7–14 (2019). https://doi.org/10.1109/ACCS-PEIT48329.2019.9062874
4. Bogarín, A., Cerezo, R., Romero, C.: A survey on educational process mining. Wiley Interdiscip. Rev. Data Min. Knowl. Discov. **8**, e1230 (2018). https://doi.org/10.1002/widm.1230
5. Elbadrawy, A., Polyzou, A., Ren, Z., Sweeney, M., Karypis, G., Rangwala, H.: Predicting student performance using personalized analytics. Computer (Long. Beach. Calif). **49**, 61–69 (2016). https://doi.org/10.1109/MC.2016.119
6. Hasan, R., Palaniappan, S., Mahmood, S., Abbas, A., Sarker, K., Sattar, M.: Predicting student performance in higher educational institutions using video learning analytics and data mining techniques. Appl. Sci. **10**(11), 3894 (2020). https://doi.org/10.3390/app10113894
7. Patikorn, T., Baker, R.S., Heffernan, N.T.: ASSISTments longitudinal data mining competition special issue: a preface. J. Educ. Data Min. **12**, i–xi (2020). https://doi.org/10.5281/ZENODO.4008048
8. Alzubaidi, L., et al.: Review of deep learning: concepts, CNN architectures, challenges, applications, future directions. J. Big Data **8**(1), 1–74 (2021). https://doi.org/10.1186/s40537-021-00444-8
9. Lara-Cabrera, R., González-Prieto, Ángel., Ortega, F.: Deep matrix factorization approach for collaborative filtering recommender systems. Appl. Sci. **10**(14), 4926 (2020). https://doi.org/10.3390/app10144926
10. Namoun, A., Alshanqiti, A.: Predicting student performance using data mining and learning analytics techniques: a systematic literature review. Appl. Sci. **11**(1), 237 (2020). https://doi.org/10.3390/app11010237
11. Ayyappan, G.: Ensemble classifications for student academics performance data set. Indian J. Comput. Sci. Eng. **10**, 31–34 (2019). https://doi.org/10.21817/INDJCSE/2019/V10I1/191001009
12. Ghorbani, R., Ghousi, R.: Comparing different resampling methods in predicting students' performance using machine learning techniques. IEEE Access **8**, 67899–67911 (2020). https://doi.org/10.1109/ACCESS.2020.2986809
13. Nguyen, H.Q., Pham, T.T., Vo, V., Vo, B., Quan, T.T.: The predictive modeling for learning student results based on sequential rules. Int. J. Innov. Comput. Inf. Control. **14**, 2129–2140 (2018). https://doi.org/10.24507/IJICIC.14.06.2129
14. Hasnawi, M., Kurniati, N., Mansyur, S.H., Irawati, Hasanuddin, T.: Combination of case based reasoning with nearest neighbor and decision tree for early warning system of student achievement. In: Proceedings of 2nd East Indonesia Conference on Computer and Information Technology Internet Things Ind. EIConCIT 2018, pp. 78–81 (2018). https://doi.org/10.1109/EICONCIT.2018.8878512
15. Zia, A., Usman, M.: Elective learning objects group recommendation using non-cooperative game theory. In: Proceedings of 2018 International Conference on Frontiers of Information Technology, FIT 2018, pp. 194–199 (2019). https://doi.org/10.1109/FIT.2018.00041

16. Esteban, A., Zafra, A., Romero, C.: Helping university students to choose elective courses by using a hybrid multi-criteria recommendation system with genetic optimization. Knowl.-Based Syst. **194**, 105385 (2020). https://doi.org/10.1016/J.KNOSYS.2019.105385

17. Rivera, A.C., Tapia-Leon, M., Lujan-Mora, S.: Recommendation systems in education: a systematic mapping study. In: Rocha, Á., Guarda, T. (eds.) ICITS 2018. AISC, vol. 721, pp. 937–947. Springer, Cham (2018). https://doi.org/10.1007/978-3-319-73450-7_89

18. Thanh-Nhan, H.L., Nguyen, H.H., Thai-Nghe, N.: Methods for building course recommendation systems. Proc. - 2016 8th International Conference on Frontiers of Information Technology, KSE 2016, pp. 163–168 (2016). https://doi.org/10.1109/KSE.2016.7758047

19. Chen, R., et al.: A novel social recommendation method fusing user's social status and homophily based on matrix factorization techniques. IEEE Access **7**, 18783–18798 (2019). https://doi.org/10.1109/ACCESS.2019.2893024

20. Thanh-Nhan, H.L., Huy-Thap, L., Thai-Nghe, N.: Toward integrating social networks into intelligent tutoring systems. In: Proceedings of 2017 9th International Conference on Knowledge and Systems Engineering, KSE 2017, 2017-January, pp. 112–117 (2017). https://doi.org/10.1109/KSE.2017.8119444

21. Huynh-Ly, T.N., Le, H.T., Nguyen, T.N.: Integrating courses' relationship into predicting student performance. Int. J. Adv. Trends Comput. Sci. Eng. **9**, 6375–6383 (2020). https://doi.org/10.30534/IJATCSE/2020/320942020

22. Thai-Nghe, N., Schmidt-Thieme, L.: Multi-relational factorization models for student modeling in intelligent tutoring systems. In: Proceedings - 2015 IEEE International Conference on Knowledge and Systems Engineering, KSE 2015, pp. 61–66. Institute of Electrical and Electronics Engineers Inc. (2015). https://doi.org/10.1109/KSE.2015.9

23. Assielou, K., Théodore, C., Tra, B., Lambert, T., Daniel, K.: Emotional impact for predicting student performance in intelligent tutoring systems (ITS). Int. J. Adv. Comput. Sci. Appl. **11**(7), 219–225 (2020). https://doi.org/10.14569/IJACSA.2020.0110728

24. Tsiakmaki, M., Kostopoulos, G., Kotsiantis, S., Ragos, O.: Transfer learning from deep neural networks for predicting student performance. Appl. Sci. **10**(6), 2145 (2020). https://doi.org/10.3390/app10062145

25. Dien, T., Hoai, S., Thanh-Hai, N., Thai-Nghe, N.: Deep learning with data transformation and factor analysis for student performance prediction. Int. J. Adv. Comput. Sci. Appl. **11**(8), 711–721 (2020). https://doi.org/10.14569/IJACSA.2020.0110886

Author Index

Printed in the United States
by Baker & Taylor Publisher Services